Springer Proceedings in Physics 424

Indexed by Scopus

The series Springer Proceedings in Physics, founded in 1984, is devoted to timely reports of state-of-the-art developments in physics and related sciences. Typically based on material presented at conferences, workshops and similar scientific meetings, volumes published in this series will constitute a comprehensive up to date source of reference on a field or subfield of relevance in contemporary physics. Proposals must include the following:

- Name, place and date of the scientific meeting
- A link to the committees (local organization, international advisors etc.)
- Scientific description of the meeting
- List of invited/plenary speakers
- An estimate of the planned proceedings book parameters (number of pages/articles, requested number of bulk copies, submission deadline).

Please contact:
For Americas and Europe: Dr. Zachary Evenson; zachary.evenson@springer.com
For Asia, Australia and New Zealand: Dr. Loyola DSilva; loyola.dsilva@springer.com

S. K. Ghosh · Anirban Bera ·
Ayan Kumar Bandyopadhyay
Editors

Proceedings of the National Conference on Emerging Trends in Vacuum Electronic Devices and Applications

VEDA2024, 18–20 Nov, Pilani, India

Springer

Editors
S. K. Ghosh
Central Electronics Engineering Research Institute
Pilani Town, Rajasthan, India

Anirban Bera
Central Electronics Engineering Research Institute
Pilani Town, Rajasthan, India

Ayan Kumar Bandyopadhyay
Central Electronics Engineering Research Institute
Pilani Town, Rajasthan, India

ISSN 0930-8989 ISSN 1867-4941 (electronic)
Springer Proceedings in Physics
ISBN 978-981-96-5173-3 ISBN 978-981-96-5174-0 (eBook)
https://doi.org/10.1007/978-981-96-5174-0

This Springer imprint is published by the registered company Springer Nature Singapore Pte Ltd.
The registered company address is: 152 Beach Road, #21-01/04 Gateway East, Singapore 189721, Singapore

Preface

Council of Scientific & Industrial Research (CSIR), Central Electronics Engineering Research Institute (CEERI), Pilani, Rajasthan, India, has a rich legacy in the R&D of vacuum electronics devices, since the inception of the institute. Vacuum Electronics, although a topical area of research, has immense applications towards strategic, commercial, and societal areas; and also ushers newer avenues of research towards solving futuristic problems. CSIR-CEERI is one of the few institutes around the globe which is engaged in R&D of microwave and millimeter wave tubes for instance: Travelling Wave Tubes, Klystrons, Magnetrons, Gyrotrons and Gyro Devices, Plasma Devices, Cathodes, High Power Microwaves, THz Vacuum Electronics Devices and Systems, etc. For years, CSIR-CEERI has harnessed numerous technologies required for successful and indigenous realizations of technologies for vacuum electronics devices.

The National Conference on Emerging Trends on Vacuum Electron Devices and Applications (VEDA – 2024), held during 18th November 2024 to 20th November 2024 at CSIR-CEERI marked a grand success with numerous scientific experts, academicians, industrial partners, and scholars attending the event and exchanging their ideas and work in the niche field of vacuum electron devices. Numerous technical sessions were organized having almost 90 technical paper presentations of varying topics ranging from vacuum electron devices like magnetrons, gyrotrons, klystrons, travelling wave tubes, high power passive components, antenna and materials research, high power microwave devices, microwave systems and their applications. Scientific experts, research scholars, academicians and students shared their research work and state of the art design work in these technical papers and fruitful discussions were performed in the fraternity of VEDs and this niche domain.

We are proud to present a diverse array of technical topics, with each abstract offering insights into the latest trends, methodologies, and applications. This technical journal abstract book serves as a comprehensive collection of cutting-edge research and advancements across various scientific and engineering vacuum electron devices and their developments. The articles contained within have undergone rigorous peer review by experts in their respective fields, ensuring the highest standards of quality and credibility. Each contribution reflects the dedication of researchers to advancing knowledge and pushing the boundaries of innovation.

The scope of the articles in this book spans a wide array of technical domains in VEDs, from theoretical advancements to practical applications. Whether focusing on new methodologies, innovative technologies, or emerging trends, the content reflects the dynamic nature of modern scientific inquiry. It highlights the important work being done in laboratories, research institutes, and universities around the world, and serves as a testament to the passion, dedication, and intellectual curiosity that drive scientific progress. The primary objective of this book is to provide a platform for the dissemination of groundbreaking research in vacuum electron device fraternity and to foster collaboration among professionals, researchers, and practitioners. The peer review process plays a pivotal role in validating the relevance and integrity of the content, providing readers with trustworthy and impactful research. It serves not only as a mechanism to validate the quality of the research but also as a means to maintain the credibility and relevance of the information presented. Each article has been carefully evaluated by experts in the respective fields to guarantee that the findings are both accurate and valuable to the broader scientific community.

It is our hope that this book serves as a valuable resource for researchers, students, and professionals, providing them with a deeper understanding of the technical advancements in the area of VEDs and further inspires exploration and fosters collaborative efforts among professionals and scholars, contributing to the continued progress of VEDs development. We trust that these articles will inspire further research, collaboration, and innovation, contributing to the development of new ideas and solutions that will shape the future.

In conclusion, the technical journal abstract book you hold in your hands represents the culmination of rigorous academic effort and the spirit of scientific discovery. It reflects the vibrant and ever-evolving nature of technical research, and we are proud to present it as a resource for those committed to pushing the boundaries of knowledge and technological progress.

Pilani, India

S. K. Ghosh
Anirban Bera
Ayan Kumar Bandyopadhyay
Proceedings of the VEDA-2024

Contents

Simulation of THz BWO with Sheet Beam Using SUNRAY-1D Code Simulation

Vishnu Srivastava(✉)

CSIR-Central Electronics Engineering Research Institute, Pilani, Rajasthan 333031, India
vsceeri@gmail.com

Abstract. SUNRAY-1D large signal model developed earlier for a TWT with a cylindrical electron beam was modified for one-dimensional nonlinear analysis of beam-wave interaction of a planar THz BWO with a rectangular sheet beam. The cold RF characteristics of the SWS along with the input dc beam voltage, beam current and beam size are used to determine the RF performance of a BWO with the setting of proper boundary conditions. The RF field is set very low (0.001 μW) at the collector end and the initial complex RF voltage at each integration plane is set with the RF wave group velocity negative but the phase velocity positive at the RF frequency corresponding to the slow space-charge wave. The RF wave interacts with the forward-moving electron beam and provides beam velocity modulation and beam bunching (density modulation) which grows near the output. The induced RF field at each integration plane due to the modulated electron beam provides the backward wave growing from the output (collector end) to the gun end. A fairly good agreement of backward wave growth from the output to the input as well as BWO tuning range by varying the input dc beam voltage as simulated by SUNRAY-1D code are found with the published results simulated by the CST-3D simulation code.

Keywords: Backward wave oscillator · Slow wave structure · Beam wave interaction analysis · One dimensional large signal model

1 Introduction

THz frequencies from 0.1 THz to 10 THz of the e. m. spectrum are being explored for many new emerging applications related to space exploration, medical imaging and bio sensing, microscopy, spectroscopy, wideband communication, high-resolution radar, astronomy, etc. For above applications, compact, coherent, high power, high efficiency, wide tuning THz sources are being investigated [1–4]. Among vacuum electronic devices, backward-wave oscillator (BWO) particularly O-type, as shown in Fig. 1, is the most preferred device because of its simplicity and easy wideband tuning.

BWO is basically like a TWT amplifier but the electron beam in THz BWO interacts with either the fundamental (n = −1) or next higher (n = −2) backward space harmonic of the circuit field whose group velocity (Vg) is negative but phase velocity (Vph) is positive, as shown in Fig. 2. The growing RF field output is at the gun end but beam bunching takes near the output (collector) end. BWO is therefore a low-efficiency device,

S. K. Ghosh et al. (Eds.): VEDA 2024, SPPHY 424, pp. 1–7, 2025.
https://doi.org/10.1007/978-981-96-5174-0_1

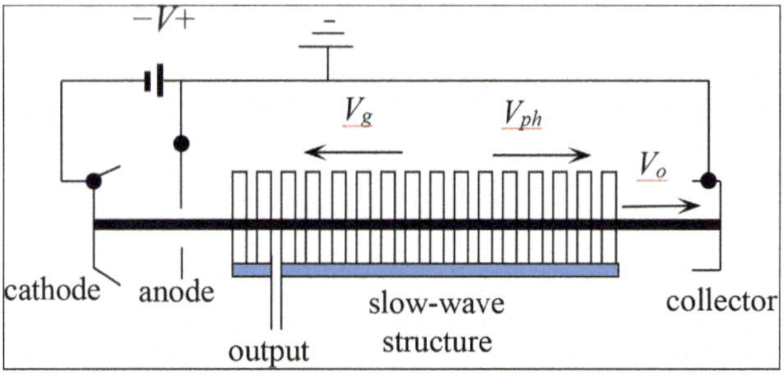

Fig. 1. Schematic of a BWO with RF output near gun end.

and efforts are needed to enhance the interaction efficiency over wideband by selecting RF slow-wave structure of high interaction impedance and wide bandwidth.

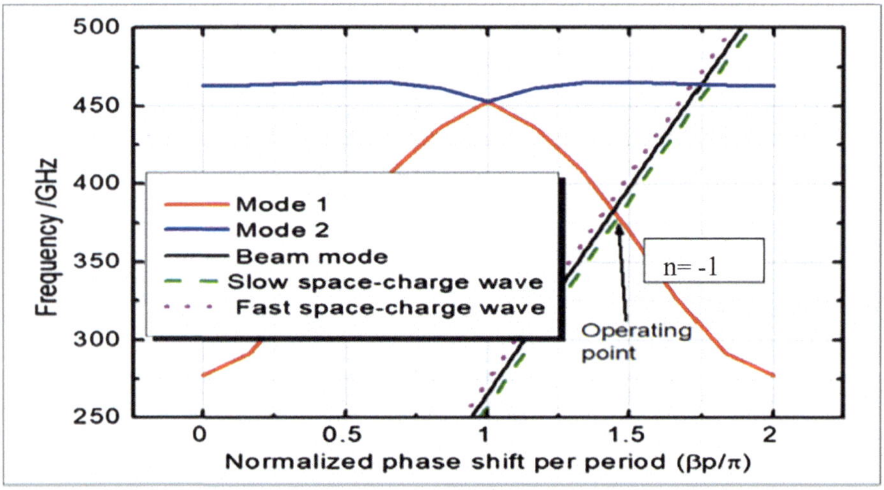

Fig. 2. Dispersion diagram of mode 1 and mode 2 for RF-SWS; beam velocity with slow and fast space charge waves synchronous with n = −1 space harmonic.

Many designs of RF-SWS for a THz BWO along with the rectangular sheet electron beam have been explored in the recent past for achieving high power, high efficiency, and a wide tuning-range [5–8]. The sheet beam has several advantages over a cylindrical beam of the same beam current because of the relatively large cross-sectional area. It has a relatively much lower current density and much larger interaction area. For increasing the RF output power and interaction efficiency of a BWO, effort is also made by optimum feedback coupling some of the generated RF signal from the gun end to the output (collector end) at the proper phase. Another example of increasing the efficiency of a BWO is by using an inclined electron beam at an optimum angle [7]. Higher RF output power is also achieved by extracting the backward growing energy from many positions along the tube to the common waveguide [8] by adopting the degenerate synchronization concept. Under this concept, the distributed gain and power extraction along the SWS are used. TWT amplifier with optimized feedback from the output (collector end) to the input (gun end) is also used as a high power oscillator. In most of the published designs of BWOs, 3D particle-in-cell simulation code like CST-3D, MAFIA-3D, and ANSYS, is used, which is very time consuming and expensive. Indigenous analytical design tools [9] are developed and successfully used to evaluate designs of various published BWOs. The design of RF-SWS for a published 1.0 THz 110 mW BWO [10] was analyzed using indigenous design tools. It is a vane loaded rectangular wave structure with sheet beam as shown in Fig. 3. One-dimensional large signal code is developed to perform beam-wave interaction analysis of a THz BWO with sheet beam for optimizing the RF performance of a BWO.

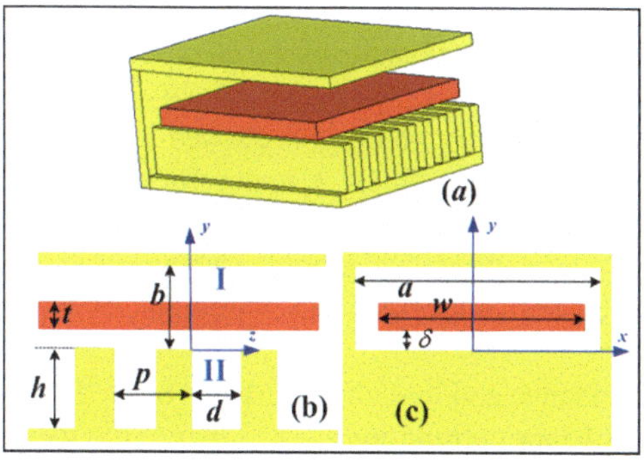

Fig. 3. Single-grating rectangular waveguide SWS with rectangular sheet beam

2 One-Dimensional Large Signal Analysis for BWO

SUNRAY-1D large signal model [11, 12] developed earlier for a TWT with a cylindrical electron beam was modified for one-dimensional nonlinear beam-wave interaction analysis of a planar THz BWO with a rectangular sheet beam. The BWO oscillation frequency is decided by the slow space charge wave frequency that is synchronous with the backward space harmonic ($n = -1$). The frequency (ω') of the slow space charge wave velocity is decided by RF frequency (ω), the plasma frequency (ω_p) and the plasma frequency reduction factor (R) [13, 14]. It is defined as:

$$\omega_q = R.\omega_p \tag{1a}$$

$$\omega' = \omega - \omega_q \tag{1b}$$

The plasma frequency (ω_p) is decided by the space charge density and therefore the increase in beam current at fixed beam voltage will reduce the BWO oscillation frequency. Similarly, the factor (R) is decided by the beam parameters and the beam tunnel size. All these effects are to be considered.

The rectangular sheet beam of thickness (t) and width (w), as shown in Fig. 3, is represented by equivalent cylindrical beam of same cross-sectional area as per Eq. (2). The rectangular beam tunnel is represented by the cylindrical tunnel of radius (r_a) equals to the equivalent beam radius (r_b) plus the beam to the vane spacing (δ), as per Eq. (3):

$$\pi r_b^2 = t \times w \tag{2}$$

$$r_a = r_b + \delta \tag{3}$$

This approach is made to keep the space charge field of equivalent cylindrical beam almost same as that for the sheet beam. The RF circuit field on equivalent cylindrical beam in cylindrical tunnel is also kept same by using the average interaction impedance of the actual RF-SWS on the sheet beam.

For one-dimensional large signal analysis, the electron beam in one RF beam wavelength ($\lambda e = u_0/f$, where u_0 is input dc beam velocity and f is RF frequency) was represented by large particles ($N_D > 48$) and tracked in the RF-SWS in small-steps of forward integration ($dz = \lambda/N_Z$, where λ is one RF wavelength and $N_Z > 32$) by the Runge-Kutta method of forward integration in presence of the axial components of the RF circuit field and the ac space charge field. The density modulated electron beam defined by the RF beam current induces the backward induced voltage at each integration plane.

The dispersion, impedance and loss characteristics of the SWS along with the operating beam voltage and beam current are used to determine the RF performance of a BWO with the setting of proper boundary conditions. The RF circuit field is set very low (0.001 μW) at the collector end and the initial complex backward RF voltage component (VB_k) at each integration plane is set with the RF wave group velocity negative but the phase velocity positive, as per Eq. (4) from the last integration plane ($k = n_2$) at the collector end to the input plane ($k = n_1$) at the gun end.

$$VB_{k-1} = VB_k e^{-(\alpha + j\beta dz)} \tag{4}$$

where, α is the attenuation constant and β is the axial propagation constant of the SWS.

The backward RF wave interacts with the forward-moving electron beam and provides beam velocity modulation and beam bunching (density modulation) which grows near the output. The induced RF field at each integration plane due to the modulated electron beam (as per Eq. 5) provides the backward wave growing from the output (collector end) to the gun end, as per Eq. (6).

$$DVB_K = (Z/2)I_{rf}[(1 - e^{\gamma 0.dz}/e^{\gamma \varepsilon.dz})/(1 + \gamma 0/\gamma \varepsilon)] \tag{5}$$

$$VB_k = VB_{k+1} e^{-(\alpha + j\beta dz)} + DVB_k \tag{6}$$

where, VB_k and VB_{k+1} are complex backward voltage components at planes k and k + 1; DVB_k is complex induced backward voltage component at plane k due to the RF beam current (I_{rf}) flowing over small distance step dz from integration plane k to k + 1; γ_0 and γ_e are respectively the complex propagation constants for the RF field and the

RF beam; and Z is average interaction impedance for synchronous (−1) backward wave frequency.

The backward wave growth from the output to the input can therefore be simulated using the modified SUNRAY-1D code. The SUNRAY-1D code also considers the circuit loss as well as reflections at both the ends. The effect of the induced forward voltage components on the growing backward voltage can also be analyzed. Therefore the model can be used to analyze band edge oscillation and the regenerative oscillation in TWTs due to reflections at the terminations. A fairly good agreement of the backward wave growth from the output to the input as well as BWO tuning range by varying beam voltage as simulated by SUNRAY-1D code are found with the published results simulated by CST-3D code.

3 Results and Discussion

The RF performance of the 1 THz BWO [10] was simulated using the modified large-signal analysis code. Fig. 4 shows the backward wave growth for the 1 THz BWO as simulated using the indigenously developed SUNRAY-1D code and comparison with the published CST simulated result [10]. The interaction length is taken 6.6 mm with 132 periods for simulating the RF performance of the 1 THz BWO, using values of the propagation constant and the circuit impedance as given in [10] at 1.03 THz. The growing RF output power was achieved around 130 mW at frequency 1.03 THz for beam voltage 12.534 kV and beam current 8 mA. The SUNRAY-1D simulated result (star marked) in Fig. 4 is fairly comparable with the simulated result (solid curve) by the CST-3D code [10].

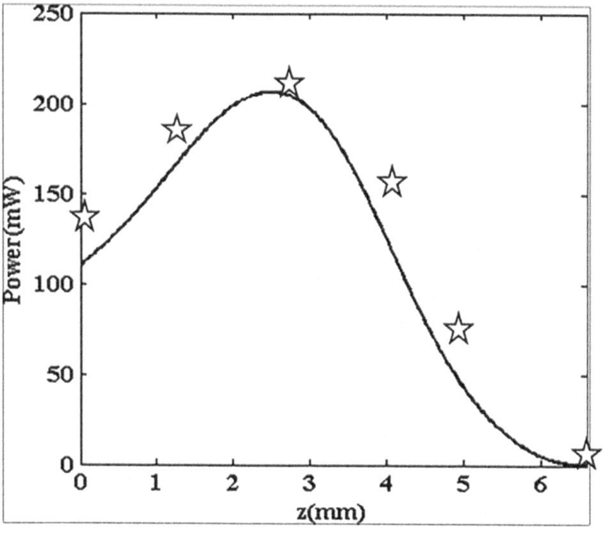

Fig. 4. BWO power profile from (z = 6.6 mm) to input (z = 0) at frequency 1.03 THz; solid line as published [10], star marked points as determined by SUNRAY-1D code.

The beam voltage decides the oscillation frequency for which the beam velocity corresponding to the slow space charge wave is synchronous with the phase velocity of (n= −1) backward space harmonic. The SUNRAY-1D large-signal analysis code was used to simulate the BWO output power by varying the beam voltage at the synchronous frequency. The SUNRAY-1D simulated BWO output powers over the tuning range 0.96 THz to 1.10 THz varying beam voltage from 10 kV to 16.5 kV at beam current 8 mA, are found comparable with the simulated output powers by the CST-3D code, as shown in Table 1.

Table 1. BWO power and frequency varying beam voltage at beam current 8 mA

Voltage (kV) / BWO Freq. (THz)	SUNRAY (power)	Published CST (power) [10]
10.0 / 0.96	0.12 W	0.11 W
11.5 / 0.98	0.14 W	0.11 W
13.0 / 1.01	0.16 W	0.12 W
14.0 / 1.04	0.15 W	0.14 W
15.0 / 1.07	0.14 W	0.13 W
16.5 / 1.10	0.13 W	0.12 W

The biggest advantage of the SUNRAY-1D code is it takes only tens of seconds to simulate RF performance of a BWO for one beam voltage and one frequency, which is at least 1000 times faster than the CST-3D simulation time. Therefore, the developed SUNRAY-1D code for the large-signal analysis of a BWO can interactively be used for analysing the design as well as achieving the initial design of a THz BWO.

4 Conclusions

Compact, coherent, high-power pulsed as well as CW THz sources are being investigated for spectroscopy and many other applications. Among different THz sources, BWO is most preferred as a coherent, high-power, wide-tunable source with a compact and simple design. In-house developed one-dimensional SUNRAY-1D large-signal code developed earlier for a TWT with cylindrical electron beam is modified for beam wave interaction analysis of a planar THz BWO operating with the rectangular sheet beam. Equivalent cylindrical approach is followed with the setting of proper boundary conditions to analyze the performance of a THz BWO. The simulated RF output power as well as tuning range by varying input beam voltage by the modified SUNRAY-1D code for a 1 THz BWO are found to be comparable with the published results that have been achieved using the CST-3D code. The present one-dimensional large-signal analysis will be extended for the 3-dimeninal analysis of a THz BWO with sheet beam for better accuracy. Also, the present large signal analysis using single sheet beam will be extended for multi-beam analysis, needed for a high power THz BWO.

Acknowledgment. The author is thankful to his parent organizations, CSIR-CEERI, Pilani, and CSIR, New Delhi for their support during his service period (from Jan.1976 till Nov.2018). He also expresses his thanks to the VEDA-2024 organising committee for needful support.

References

1. Srivastava, V.: THz vacuum microelectronic devices. J. Phys. **114**, 1–10 (2008)
2. Paoloni, C., Gamzina, D., Luhmann, N.: THz backward wave oscillators for plasma diagnostic in nuclear fusion. IEEE Trans. Plasma Sci. **44**, 369–376 (2016)
3. Bao, R., Wang, H., Li, Y., Liu, C.: Low peak-to-average ratio 850GHz BWO for THz communication. Phys. Plasmas **30**, 1–8 (2023)
4. Srivastava, V.: Design of a micro-fabricated planar slow wave structure for a 0.22 THz TWT for communication, imaging & remote sensing. IETE Tech. Rev. **36**, 501–509 (2019)
5. Latif, J., Wang, Z., Feng, J., Gong, Y.: Study and experimental validation of a 0.34THz double-corrugated waveguide interaction structure for a BWO with sheet beam. J. Infrared Millim. Terahertz Waves, 1–12 (2024)
6. Zhang, F., Zhang, Z., Gao, D., Wang, X.: A 3D frequency-domain nonlinear theory of BWO with inclined electron beam. IEEE Trans. Plasma Sci. **48**, 3040–3046 (2020)
7. Mealy, T., Abbdelshafy, A.F., Capolino, F.: Backward wave oscillator with distributed power extraction based on exceptional point of degeneracy. In: IEEE-IVEC, Korea, 1–2 (2019)
8. Seidfaraji, H., Elfrgani, A., Christodouleu, C., Schamiloglu, E.: A multi-beam metamaterial backward wave oscillator. Phys. Plasmas **26**, 1–7 (2019)
9. Srivastava, V.: Determination of design parameters of SDV-SWS for a planar high power THz BWO. Asian J. Phys. **22**, 425–433 (2023)
10. Xie, W., Wang, Z., Luo, J., Zhao, D.: Three dimensional nonlinear analysis of a single grating rectangular waveguide Cerenkov maser. Phys. Plasmas **22,** 042307-1 to 7 (2015)
11. Srivastava, V.: SUNRAY-1D and SUNRAY-2.5D codes for large signal analysis of TWT. In: Proceedings of the 14th IEEE-IVEC, Paris (2013)
12. Srivastava, V.: Nonlinear analysis of beam-wave interaction in a planar THz travelling-wave tube amplifier. J. Electromagn. Waves Appl. **32**, 190–203 (2018)
13. Datta, S.K., Kumar, L.: Plasma frequency reduction factor. Def. Sci. J. **58**, 768–770 (2008)
14. Tian, H., Guo, H., Wang, S., et al.: Plasma frequency reduction factor of sheet electron beam in rectangular waveguide. In: PIERS Symposium, China, pp. 763–767 (2021)

Comparative Analysis of Port Excitation in Microwave Oven

Bharti Singhal, Ekta Singh, and Vikram Kumar(✉)

Department of Electronic Science, Delhi University, South Campus, India
{bhartisinghal,vikram}@south.du.ac.in

Abstract. This paper aims to find the most effective port position for a microwave oven operated in TE_{10} mode at 2.45 GHz with 1.1 kW of input power. We examine the electromagnetic field distribution inside the cavity using waveguides made with ports at different locations, focused on dimensions a (wider) and b (narrower) of rectangular waveguides and their behaviour at a quarter wavelength (λ/4) in the x, y, and z directions is studied. The λ/4 transformer is employed to adjust the impedance for maximum power transfer and to avoid reflections. Among the configurations designed, there is one with dimension a along the y-axis, dimension b along the z-axis, and λ/4 (quarter wavelength) along the x-axis, placed distantly from the microwave cavity, that perform best with S_{11} −21.01 dB, power output of 1100.5 W, and power efficiency of 99.2%. This study can help improve microwave oven design in relation to electric field distribution and exposure uniformity both at domestic and industrial applications.

Keywords: Microwave oven cavity · Port excitation · Impedance matching

1 Introduction

This demands research and comparison of the port position of the microwave oven to ensure that the appliance is operated in the best possible configuration [1]. The port position of the microwave oven guides the microwaves into the cooking cavity thus it is fundamentally appealing to the strategy to ensure evenly food exposure. Uneven food exposure always implies the presence of hotspots, partly undercooked food, and in some areas overcooking, which reduces quality and food safety in most cases. Understanding and optimizing port positions enables manufacturers to design microwave ovens that will cook uniformly for customers, thereby reducing cooking time while increasing user satisfaction and energy efficiency [2]. By understanding and optimizing port positions, manufacturers can design microwave ovens that deliver more uniform cooking results, enhancing user satisfaction and energy efficiency while reducing cooking times. Waveguides carry the microwaves from the magnetron to the cooking chamber. They are usually metal tubes of rectangular cross section [3]. Magnetrons get heated during their operation and also experience varying load value in cavity. Therefore, high reflection loss is preferred at input port and could be enhanced using different port positions.

S. K. Ghosh et al. (Eds.): VEDA 2024, SPPHY 424, pp. 8–13, 2025.
https://doi.org/10.1007/978-981-96-5174-0_2

Other applications of microwave ovens are used in industrial fields, e.g. pasteurization of vegetables, drying of paper or textiles, thermal treatment of pharmaceutical products and vulcanization of rubber and elastomers [3]. In few experiment the microwave oven is turned on after a short delay the plasma sparks, to study low cost plasma source [4].

It is required to consider port positioning during the design phase to balance modelling, and functionality, and manufacturability. Efficient port design can lead to more compact and effective microwave oven designs. Also, proper port positioning can help prevent localized overheating that might cause fires or damage to the microwave oven.

2 Design and Methodology

This section describes the systematic process developed in designing the best port excitation waveguide for enhanced microwave oven performance. In designing a microwave oven using CST Studio, we will need two rectangular waveguides: one for the oven cavity and the other for the magnetron defining the port as shown in Fig. 1.

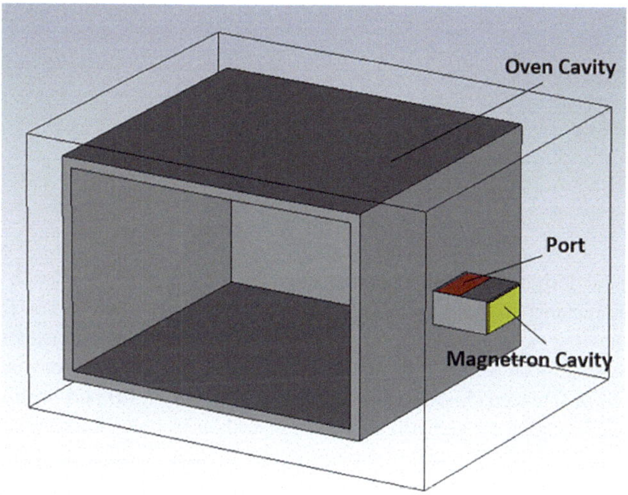

Fig. 1. Design of microwave oven using CST studio.

2.1 Design and Parameters

The dimensions of the cavity of the Microwave oven are 31.1 cm (width) × 20.2 cm (height) × 30.1 cm (depth) with 1.1 KW power. The dimension of the waveguide is calculated based on the operating frequency and mode. The oven operates at the TE_{10} mode at a frequency of 2.45 GHz. With the use of following formula (1), dimensions a and b of the waveguide can be calculated [3], where c is the speed of light (approximately

3×10^8 m/s), fc is the operating frequency (2.45 GHz), and m and n are mode numbers, while a is wider and b is narrower dimension of rectangular waveguide port [5].

$$fc = \frac{c}{2}\sqrt{\left(\frac{m}{a}\right)^2 + \left(\frac{n}{a}\right)^2} \tag{1}$$

We get the value of a = 60.24 mm and in TE_{10} mode a > b also b = 0.5*a. So, dimensions of the rectangular port are a = 60.24 mm and b = 30.12 mm.

Figure 2 shows CST Studio design with different port positions where, a is the wider dimension, b is the narrower dimension of the rectangular waveguide for TE_{10} mode.

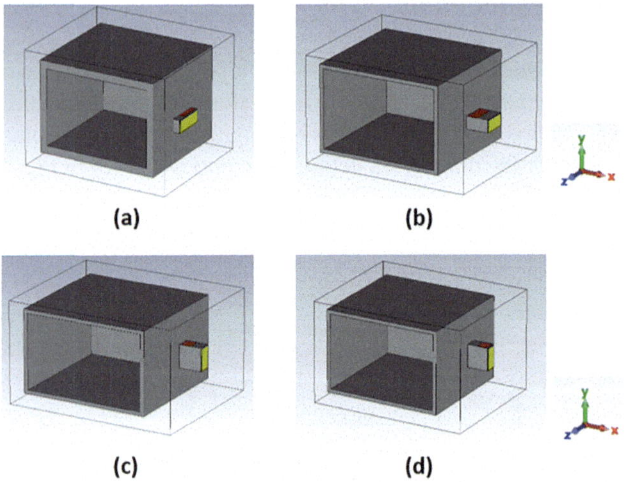

Fig. 2. CST Studio design with different port positions (a is the wider dimension, b is the narrower dimension of the rectangular waveguide for TE_{10} mode): **(a)** port position a on the z-axis, b on y-axis, and λ/4(quarter wavelength) on z-axis, **(b)** port position a on z-axis, b on y-axis, and λ/4(quarter wavelength) on x-axis, **(c)** port position a on y-axis, b on z-axis, and λ/4(quarter wavelength) on x-axis, placed closed to the microwave oven cavity, **(d)** port position a on y-axis, b on z-axis, and λ/4(quarter wavelength) on x-axis, placed far from the microwave oven cavity.

3 Result Results and Discussion

In microwave oven design, the port position plays a critical role in determining the appliance's performance and efficiency. The strategic placement of the port significantly influences the uniformity of heating, power efficiency, and the potential for reflection and interference. Incorrect port placement can result in uneven heating, potentially causing damage to food or materials. Additionally, the port's location can impact the efficiency of power transfer, affecting energy consumption. Moreover, the port's position can influence the amount of microwave energy reflected back into the cavity, which can impact the overall performance and safety of the microwave oven [6].

Table 1. Performance comparison of different Positions of Port configuration

Parameters	Position of port excitation of waveguide			
	a = z axis ***b = y axis*** λ/4 length = z axis	***a = z axis*** ***b = y axis*** λ/4 length = x axis	***a = y axis*** ***b = z axis*** λ/4 length = x axis (close to cavity)	***a = y axis*** ***b = z axis*** λ/4 length = x axis (far from cavity)
S_{11}	−16.72 dB	−12.85 dB	−19.39 dB	−21.01 dB
Power outgoing	783.09 W	768.80 W	1099.48 W	1100.5 W
Power efficiency (%)	97.8	94.8	98.85	99.2

Shown above Table 1 are the analysis results of four various port position designs of a microwave oven operating at 2.45 GHz with 1.1 kW input power in TE10 mode. Considering Fig. 2 design (a) shows moderate performance with S11 of −16.72dB, indicating a fair level of reflection. The power output is 783.09 W, with a power efficiency of 97.8%. This configuration provides fairly reasonable efficiency but fails to achieve optimal power output or minimal reflection. Design (b) performs the worst of the four configurations with S11 −12.85 dB, which means it allows higher reflections and less efficient power transfer. The power outgoing is 768.80 W, and the power efficiency is 94.8%, the lowest attained by any design. This configuration fails to maintain efficient power transfer as well as to reach proper heating. Design (c) shows better performance as with S11 at −19.39 dB, thus it shows less reflection and good impedance matching. It is showing a power output of 1099.48 W and power efficiency as 98.85%. The positioning of the port near the microwave cavity is showing a better transfer of power and field distribution. Design (d) achieves the best performance, with S11 −21.01 dB, indicating excellent output is 1100.5 W, and the power efficiency is 99.2%. This configuration is not only to maximize the power efficiency but also provides for the most uniform electric field distribution inside the cavity shown in Fig. 3, thereby presenting an optimal design for the placement of microwave oven ports.

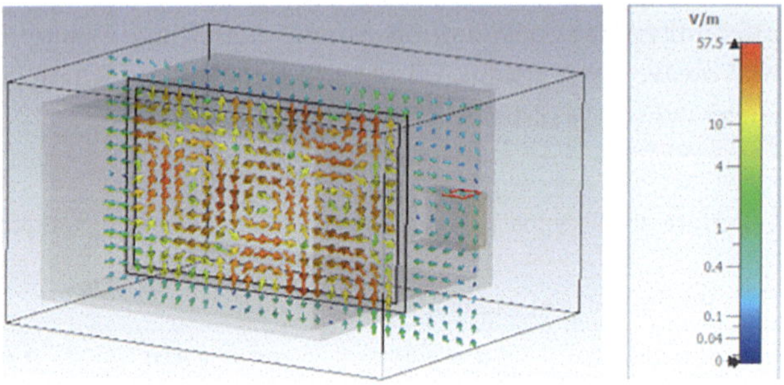

Fig. 3. Electric field distribution for port position a on the y-axis, b on the z-axis, and λ/4(quarter wavelength) on the x-axis, placed far from the microwave oven cavity (a is the wider dimension, b is the narrower dimension of the rectangular waveguide for TE_{10} mode).

4 Conclusion

Among the four port position designs evaluated, Design 2(d) (a = y-axis, b = z-axis, λ/4 = x-axis, placed far from the microwave cavity) shows the best performance. This configuration got S_{11} −21.01 dB, a power output of 1100.5 W, and a power efficiency of 99.2%. These results indicate effective impedance matching, minimal reflections, and best power transfer efficiency. Additionally, design 2(d) provided the most uniform electric field distribution inside the cavity, making it the optimal choice for microwave oven port positioning. In contrast, design 2(b) (a = z-axis, b = y-axis, λ/4 = x-axis) exhibited the poorest performance, with S11 −12.85 dB, power output of 768.80 W, and power efficiency of 94.8%. This configuration showed higher reflections and less efficient power transfer. These findings show the importance of port positioning in optimizing microwave oven performance. By selecting the optimal port configuration, manufacturers can enhance energy efficiency, ensure uniform exposure, and improve overall appliance performance. This research provides valuable insights for designing more efficient microwave ovens, benefiting both domestic and industrial applications.

Acknowledgment. This work was supported in part by the Anusandhan National Research Foundation (ANRF) (Earlier Known as SERB—Science and Engineering Research Board), Department of Science and Technology (ANRF-DST), Government of India under Startup Research Grant SRG/2023/001663.

References

1. Cooper, N.S.: 4—Microwave ovens. In: Erle, U., Pesheck, P., Lorence, M. (eds.) Development of Packaging and Products for Use in Microwave Ovens. Woodhead Publishing in Materials. Woodhead Publishing, 2nd edn, pp. 177–199 (2020). https://doi.org/10.1016/B978-0-08-102713-4.00004-9
2. Osepchuk, J.M.: A review of microwave oven safety. J. Microw. Power **13**(1), 13–26 (1978). https://doi.org/10.1080/16070658.1978.11689069
3. Vollmer, M.: Physics of the microwave oven. Phys. Educ. **39**(1), 74 (2004). https://doi.org/10.1088/0031-9120/39/1/006

4. Barnes, B.K., et al.: Plasma generation by household microwave oven for surface modification and other emerging applications. Am. J. Phys. **89**(4), 372–382 (2021). https://doi.org/10.1119/10.0002706
5. Sadiku, M.: Elements of Electromagnetics. Oxford University Press. Accessed 16 Sep. 2024. https://global.oup.com/ushe/product/elements-of-electromagnetics-9780190698614
6. Horikoshi, S., Schiffmann, R.F., Fukushima, J., Serpone, N.: Microwave Chemical and Materials Processing: A Tutorial. Springer (2017)

Design of Ultra Wide Band Antenna

R. Venkateswaran[1(✉)], D. Kannadassan[2], and S. K. Datta[1]

[1] Microwave Tube Research and Development Centre (MTRDC), Ministry of Defense, DRDO, Bengaluru, India
rvenkat.mtrdc@gov.in

[2] Vellore Institute of Technology (VIT), Vellore, India

Abstract. In this paper, a low dispersion ultra-wideband antenna is designed. Propagation velocities of different frequencies are same. Different frequencies of the applied pulse reach the antenna aperture simultaneously. Hence, dispersion will not be present in this type of antenna when it is used for applications that require wide frequency spectrum form S band to X band. In this paper a 13 dBi gain valentine antenna is designed.

Keywords: Dispersion · Gain · Half Power Beam Width (HPBW)

1 Introduction

Antennas that radiate microwave energy with low distortion and low dispersion find applications in civilian and military systems from S band to X band frequencies [1]. Dispersion is an important parameter when it is used for ultra-wide frequency spectrum. In this paper, a compact sized wideband valentine antenna is designed. This antenna is one of the types of a traveling wave antenna and dispersion is also very low over a wide frequency band. It supports a uniform propagation velocity over a large spectrum and hence dispersion is low over a wide frequency band from S band up to X band. Different frequencies of the applied pulse reach the antenna aperture simultaneously.

2 Valentine Antenna

TEM horn antenna is an antenna which can be used as a wideband antenna without much pulse distortion. But for transient signals, the spread is more. At low frequencies, the size is big [2]. Valentine antenna is a light weight compact traveling wave antenna fabricated with metal strips. Typical configuration of valentine antenna is shown in Fig. 1. The input of the antenna is fed to Teflon block placed in between the metallic strips in the form of a symmetrical double-strip transmission line to avoid electrical breakdown. Radiations

S. K. Ghosh et al. (Eds.): VEDA 2024, SPPHY 424, pp. 14–17, 2025.
https://doi.org/10.1007/978-981-96-5174-0_3

start at the beginning of the exponential profiled metallic strips. After the line section, the strips flare in to an exponential profile and return back according to a circular form. Diameter of the circular part has been increased to reduce the current reflections from the circular profile of the antenna to get more radiation at low frequencies [2].

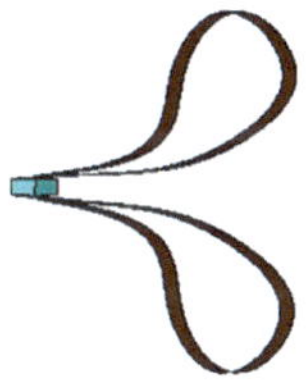

Fig.1. Configuration of valentine antenna

3 Design and Simulation

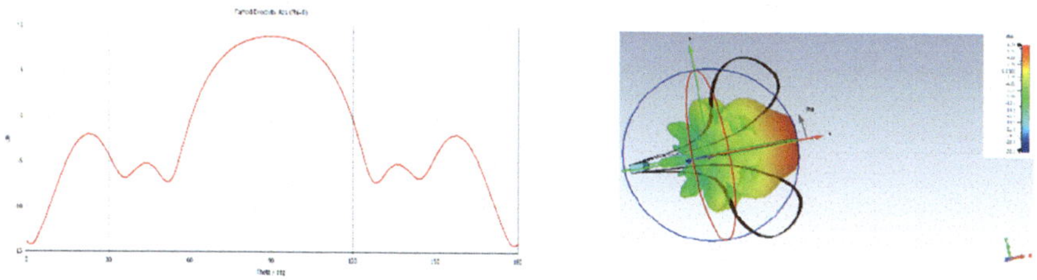

(a) (b)

Fig. 2. The a & b Radiation Pattern at S Band

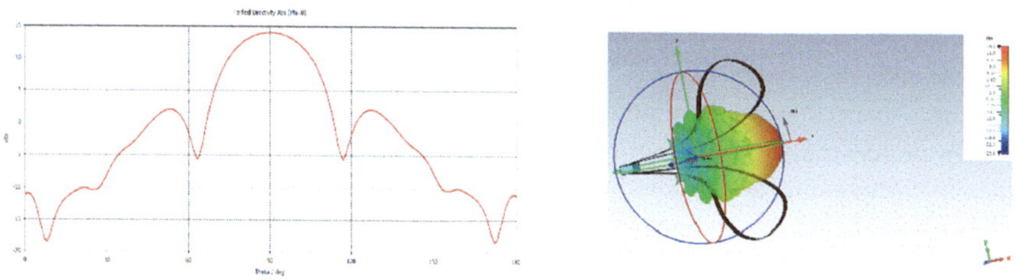

(a) (b)

Fig. 3. The a & b Radiation Pattern at C Band

Valentine antenna was designed for radiating microwaves over a wide frequency range from S band extending up to X band. A compact Valentine antenna with metal strips of length of 285 mm and height of 330 mm is designed with a dielectric slab between two

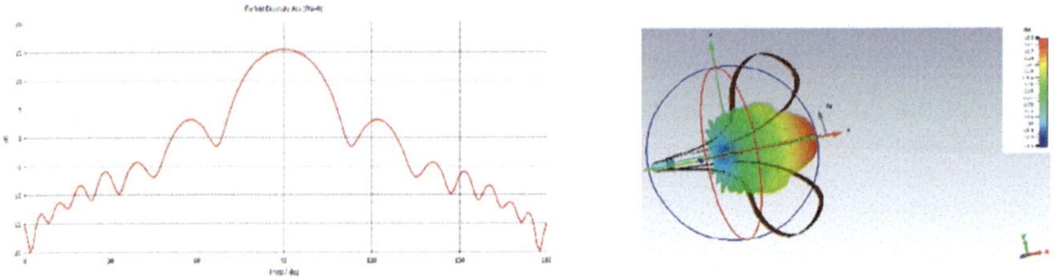

(a) (b)

Fig. 4. The a & b Radiation Pattern at X Band

flaring sections. Simulation is carried out using CST studio software. Radiation pattern obtained from the simulation are given in Figures 2a, b for S band, 3a, b for C band and 4a, b for X band.

4 Results and Discussions

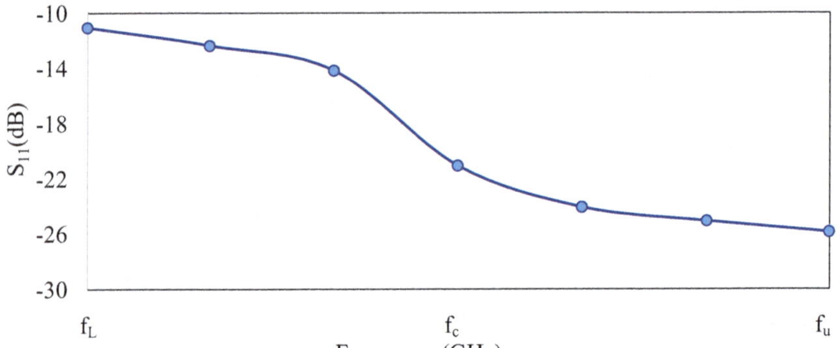

Fig. 5. Reflection coefficient (S_{11}) plot

Gain of 10 dBi at low frequency end of S band and 16 dBi is obtained at high frequency end of X band. Gain of 13 dBi, side lobe level of −13.5 dBi and VSWR of 1.13 are obtained at mid frequency at C band. Diameter of the rounded portion is gradually increasing to minimize low frequency components reflection [2]. The high frequency components are radiated from the beginning narrow portions of the antenna and low frequencies are radiated from the end broader portions [3]. Lesser gain at lower frequencies is due to some of the low frequency components are reflected back. Optimization of diameter of rounding portions will improve reflection of low frequency components and there by gain will enhance (Figs. 5, 6).

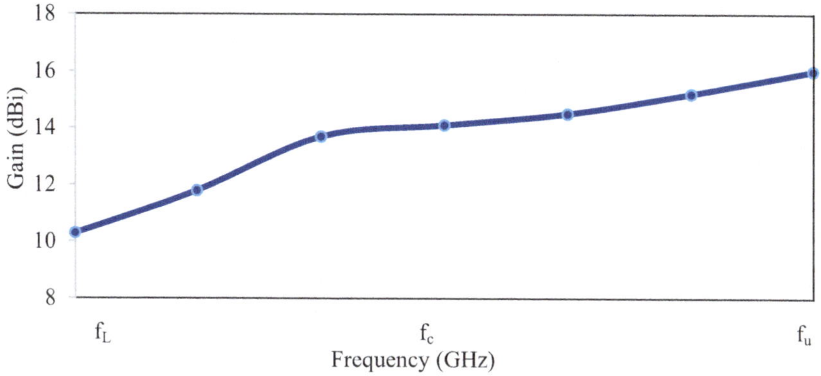

Fig. 6. Gain versus frequency plot

5 Conclusion

An ultra-wide band antenna is designed and simulated from S band to X band frequency range and a gain of 10 to 16 dBi is obtained. Very high gain and breakdown strength can be obtained with a multi element valentine antenna array configuration.

Acknowledgement. Author acknowledges the support extended by M. Vijayakumar, MTRDC. Author is thankful to Centre Head, MTRDC for his valuable comments during this work.

References

1. Benford, J., Sweagle, J., Schamiloglu, E.: High Power Microwaves, 3rd edn. CRC press, Taylor & Francis Group, New York (2016)
2. Cadilhon, B., Cassany, B., Modin, P., Diot, J.C., Bertand, V., Pecasting, L.: Ultra wideband antennas for high pulsed power applications ultra wide band communications:novel trends-antennas and propagation
3. Diot, J.C., et al.: A novel antenna for transient applications in the frequency band 300 MHz–3 GHz: The valentine antenna. IEEE Trans. Antennas Propag. **55**(3), 987–990 (2007)
4. Chevalier, Y., Imbs, Y., Beillard, B., Andrieu, J., Jouvet, M., Jecko, B., Legoff, M., Legros, E.: A new broad band resistive wire antenna for ultra-wide band applications. In: Proceeding of the EUROEM'98, Tel Aviv, Israel, pp. 14–19 (1998)
5. Andrieu, J., Beillard, B., Imbs, Y.: Wide band scissors antenna. France French patent—99 14\940, Nov. 26 (1999)
6. Andrieu, J., Nouvet, S., Bertrand, V., Beillard, B., Jecko, B.: Transient characterization of a novel ultrawide band antenna: The scissors antenna. IEEE Trans. Antennas Propag. **53**(4), 1254–1261 (2005)
7. Mehrrdadian, A., Forooraghi, K.: Design of a UWB combine antenna and an array of miniaturized elements with and without lens. Prog. Electromagn. Res. **39**, 37–38 (2013)

An Optimum Design of C-band Conical Corrugated Horn Antennas for High-Power Microwave Applications

Mrunal A. Marihal(✉), Shesh Kumar, Sourav Biswas, P. Raja Raman Rao, and S. K. Datta

Microwave Tube Research and Development Centre, Bangalore, India
mmrunal2@gmail.com

Abstract. High-power microwave antennas have a great demand in defence technology and conical horn antenna is one of them. This antenna is known for producing symmetric beam radiation pattern, high directive gain. In the proposed work, we designed two conical horn antennas: one for a power of 200 MW over X-Band and the other for a power of 350 MW over C-Band. The electromagnetic simulation of the antennas is carried out using CST software. The results show return loss less than −20 dB, gain more than the required 10 dBi, cross polarization less than −20 dB, voltage standing wave ratio (VSWR) between 1 to 1.2, and side lobe level (SLL) less than −20 dB for both antennas.

Keywords: Circular waveguide · Conical horn antenna · Gain · VSWR

1 Introduction

From decades, the microwave horn antennas are most preparable choice for HIGH-POWER microwave (HPM) sources. At the output of HPM devices, the horn antenna acts as a crucial component in high power generation and radiation [1]. Other than the defence field, these antennas can also be used for geographic information, satellite tracking and communication, wireless networks, radio astronomy and radars [2].

There are four main types of horn antenna namely pyramidal, conical, H-plane sectoral, E-plane sectoral. A horn antenna with the cone shape is known as conical horn antenna [3]. It can produce radiation patterns having a good axial symmetry. The physical dimensions like flare angle, horn length, waveguide diameter (or cone initial diameter) and aperture radius of a conical horn antenna are very important factors to determine the performance of the antenna. These dimensions can be optimized to achieve the required voltage standing wave ratio, gain, sidelobes and cross-polarization level [4, 5].

The modelling and simulation of an antenna can be performed using many softwares like computer simulation technology (CST) Studio Suite, HFSS, MATLAB, and Personal Computer Aided Antenna Design (PCAAD) [2–6].

S. K. Ghosh et al. (Eds.): VEDA 2024, SPPHY 424, pp. 18–24, 2025.
https://doi.org/10.1007/978-981-96-5174-0_4

The conical horn antenna has typically two types of feeding techniques known as a waveguide feeding and a coaxial feeding. Feeding method influences the radius of the initial cone structure and ensures the proper impedance matching between the cone structure of the horn antenna and feed [7]. Our proposed conical horn antennas have the requirement of a waveguide feeding technique.

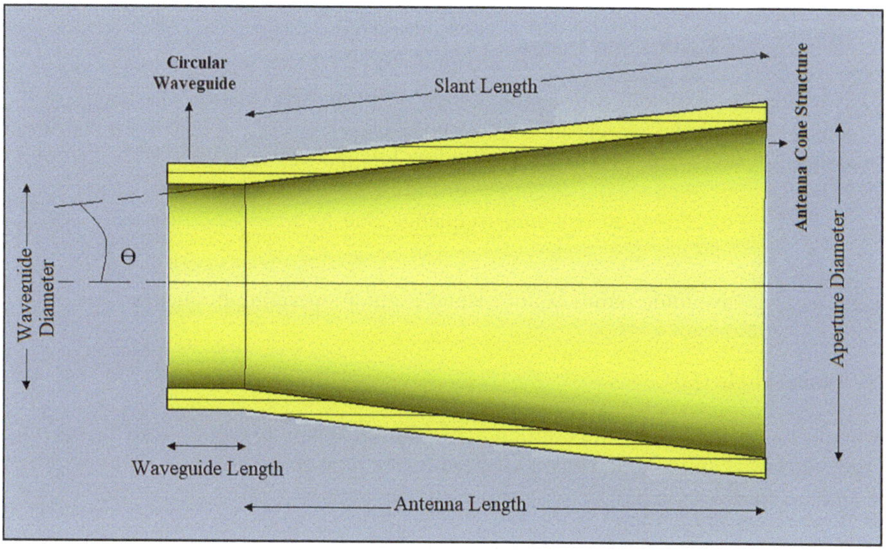

Fig. 1. Two-dimensional model of a conical horn antenna

The paper is organised as follows. Section 2 describes the working principle of conical horn antenna. The designing parameters of conical horn antenna are explained in Sect. 3. Section 4 covers simulation results of the antennas. Section 5 discusses the conclusion of the work.

2 Principle of Operation of Conical Horn Antenna

A conical horn antenna contains a circular waveguide and cone shaped structure as shown in Fig. 1. A circular waveguide is a cylindrical hollow metal tube. This tube is used for electromagnetic (EM) wave propagation. The conical horn is fed or excited by the circular waveguide, whose dominant mode is TE11 mode. When EM waves pass the waveguide into the horn, it produces high output power in TE11 mode. This produces a solid beam which directs EM waves in one direction.

3 Designing Parameters of Conical Horn Antenna

The aim of this work is to design conical horn antennas over X-band and C-band which can provide a gain more than 10 dBi. This gain, bandwidth, radition pattern, antenna efficiency and other performance characteristic of conical horn antenna are determined by the designing key parameters and they are as follows:

3.1 Waveguide Radius and Length

A waveguide has a smooth wall inside, which supports TE11 mode. The length of the waveguide is selected with respect to the guided wavelength (λ_g). The diameter of the waveguide can be calculated using following equation.

$$a_o = \frac{3\lambda_o}{2\pi} \tag{1}$$

where, a_o is a waveguide radius, λ_o is a wavelength at operating frequency

3.2 Flare Angle (Θ)

An angle at which the cone shape structure of the horn antenna expands from the opening of the feed (waveguide) to the antenna aperture. As flare angle increases, beamwidth of the antenna increases and directivity and gain decreases.

3.3 Aperture Diameter (D)

The directive gain and angular beamwidth are depended upon aperture diameter. Decrease in the diameter of the aperture leads to the decrease in the gain and the directivity and increase in the angular beamwidth. The aperture diameter of the conical horn antenna is computed using following formula.

$$G = e_A * \frac{(\pi d)^2}{(\lambda_o)^2} \tag{2}$$

where, G is a gain and e_A is an Aperture Efficiency

3.4 Waveguide Antenna Length (L)

The length of the conical horn antenna can be calculated using Pythagoras theorem with respect to aperture radius and slant length (l) of the antenna. The antenna slant length value is calculated using following equation.

$$d = \sqrt{3\lambda_o l} \tag{3}$$

4 Simulation Results

In this work, we use CST studio suite 2023 to design and simulate the conical horn antennas. First, we discuss the antenna operating over X-Band for the power of 200 MW. The simulation results are shown Fig. 2. We find that the gain is 11.8 dBi which is slightly higher than the desired one. When a signal is transmitted, a part of it gets reflected and this can be measured using a term called return loss. Our target is to achieved a return loss less than −20 dB for a better input-output matching. From the Fig. 2(a), it is observed that −27.61 dB of return loss is achieved at X-band operating frequency (fo). The corresponding VSWR value is 1.08 (Fig. 2(b)).

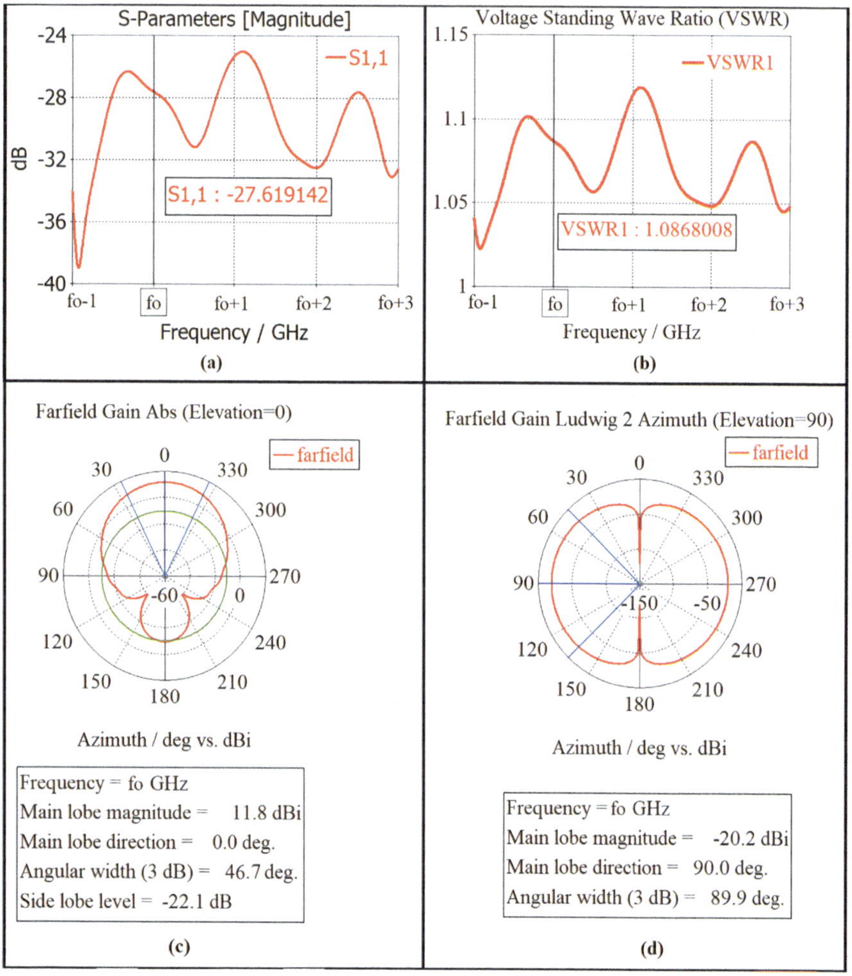

Fig. 2. X-Band antenna's (a) Return loss; (b) VSWR; (c) Gain plot; (d) Cross-polarization plot.

Figure 2(c) shows the gain plot at center frequency, fo. The angular beamwidth result determines the area covered for radiation. Here we can observe that our X-band antenna covers 46.7° of area. In an antenna, side lobes and back lobes are the unwanted signals and measured in dB. Figures 2(c, d) show that a side lobe level (SLL) of −22.1 dB with a cross-polarization of −20.2 dBi is achieved.

The second conical horn antenna in this work is designed and simulated over C-Band for the power of 350 MW. The simulation results of C-band conical horn antenna are represented by Fig. 3. The simulation result shows a gain of 11.8 dBi with an angular beamwidth of 51° and a side lobe level of −22.2 dB as shown in Fig. 3(c).

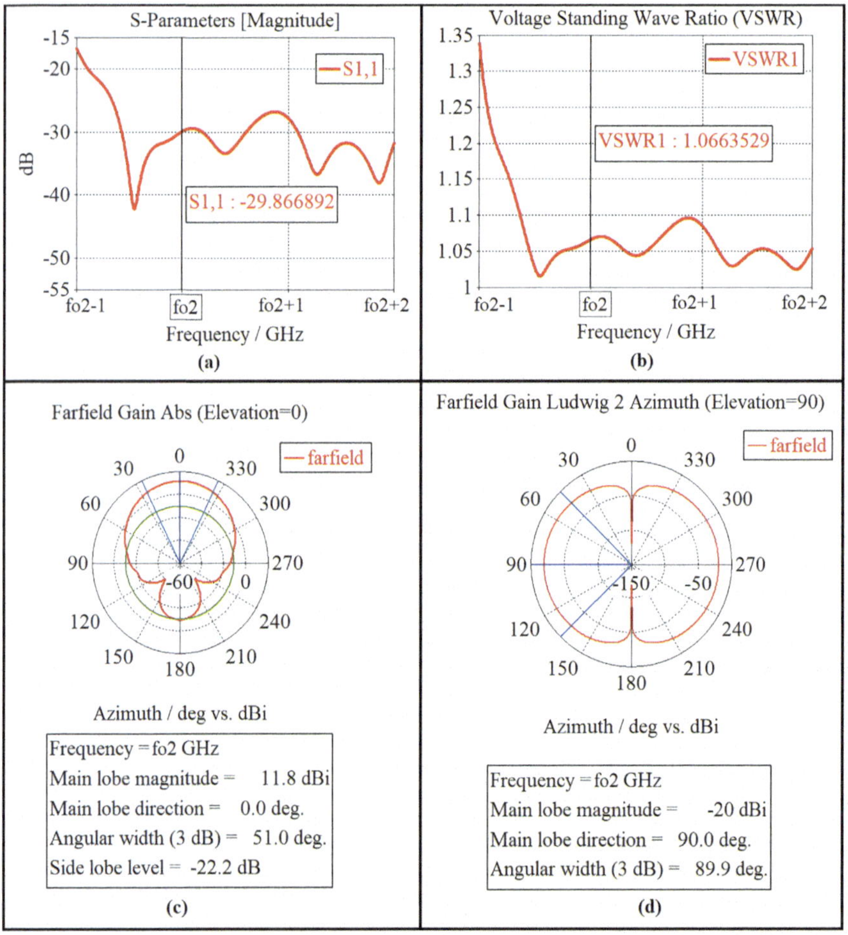

Fig. 3. C-Band antenna's (a) Return loss; (b) VSWR; (c) Gain plot; (d) Cross-polarization plot.

From Fig. 3a, we can observe that the return loss of C-band antenna is −29.86 dB. As shown in the Fig. 3a, b VSWR of 1.06 is achieved. Figure 3d shows the cross-polarization polar plot of the antenna for. Here the cross-polarization value is −20 dBi at operating frequency (fo2). All the obtained simulation results are summarized in Table 1.

Table 1. Simulated results of the conical horn antennas

Sl	Parameters	Simulation results	
1	Frequency band	X-Band	C-Band
2	Bandwidth	88.7%	93%
3	Return loss	−27.61 dB	−29.86 dB
4	VSWR	1.08	1.06
5	Gain	11.8 dBi	11.8 dBi
6	Cross-polarization	−20.2 dBi	−20 dBi
7	Angular width	46.7°	51°
8	Side lobe level	−22.1 dB	−22.2 dB

5 Conclusion

The purpose of this work is to design two different conical horn antennas over two different bandwidths, (over X-band and C-band) to achieve a gain value more than 10 dBi. The designing key parameters play important role in determining the impedance matching, directive gain and radiation pattern of the antennas. As the waveguide feeding technique is used with the smooth inside wall, the antenna produced output in TE11 mode. For antenna simulation, CST software is used and simulation results shows a return loss, cross polarization and SLL less than −20 dB, gain more than required 10 dBi, and VSWR between 1 to 1.2, for both antennas at their operating frequencies.

Acknowledgement. The authors are thankful to Centre Head, Microwave Tube Research and Development Centre (MTRDC), Bengaluru, India for the kind permission to publish this work.

References

1. Granatstein, V.L., Alexeff, I.: High—Power Microwave Sources, Boston, London
2. Singh, J., Dhaliwal, A.S.: Optimization and designing of conical horn antenna. Int. J. Electron. Commun. Technol. **2**(4) (2011)
3. Abousetta, M.M., Ali ALdoukali, H.: Investigation of circular horn antenna for x-band radar and space communications. Al Acad J Basic Appl Sci (AJBAS) **5**(5) (2023)
4. Arun Vijayakumar, B., Sanaa Fathima, B., Thejal, H.: Design and analysis of a linearly polarized S-band conical horn antenna. Int. J. Res. Trends Innov. **6**(3) (2021)
5. Pereira, R.A.M., Costanzo, A., Masotti, D., Carvalho, N.B.: 3D-printed smooth-walled conical horn antennas: manufacturing and quasioptical analysis. In: URSI GASS 2023, Sapporo, Japan (2023)

6. Balanis, C.A.: Antennas Theory: Analysis and Design, 3rd edn. Wiley, New York (2005)
7. Jadhav, R.P., Dongre, V.J., Heddallikar, A.: Design of X-band Coniacl horn antenna using coaxial feed and improved design technique for bandwidth enhancement. In: 2017 Third International Conference on Computing, Communication, Control and Automation (ICCUBEA) (2017)

Design of a Slotted Waveguide Antenna Array

Ashok Bansiwal(✉), P. Shalini, V. Nallasamy, and S. K. Datta

Microwave Tube Research and Development Centre, DRDO, Ministry of Defence, Bangalore, India
ashokbansiwal@gmail.com

Abstract. Design of a high power and high gain slotted waveguide antenna array operating in S-band has been presented. A standard rectangular waveguide WR-284 has been chosen for the slotted waveguide antenna and slots has been made on the narrow wall of the waveguide. A slotted waveguide antenna element with 36 pair of slots has been modeled and simulated using CST microwave studio. An array of 32 number of slotted waveguide antenna elements has been simulated and antenna gain in excess of 40dB could be obtained with 0.4° and 2° of horizontal and vertical beamwidth, respectively.

Keywords: High gain antenna · High power antenna · Slotted waveguide array antenna

1 Introduction

High power and high gain antennas are required for military applications, especially for long range monitoring and tracking radar [1]. Slotted waveguide antennas (SWAs) can provide higher gain with lower side lobes over a narrow bandwidth with higher power handling capability. The SWAs are categorized into two types, one is resonant (standing wave) antennas and another one is non-resonant (travelling wave) antennas. In the case of resonant antennas, one of the ends is made short-circuited and they provide low power loss and better efficiency compared to the non-resonant antennas where short circuit is replaced with matched load at the other end to absorb the non-radiated power which results into lower efficiency [2, 3].

Slotted waveguide antenna element has been designed which consists of a standard rectangular waveguide WR-284 operating in TE_{10} mode and a 36 pair of slots made on the narrow wall of the waveguide [4–6]. The RF input is fed at one end of the waveguide and other end is short circuited. The simulation of antenna element has been carried out using CST microwave studio [7]. Dimensions of the antenna element has been optimized and a gain of 26.4 dBi could be achieved. Further, an array of 32 number of slotted waveguide antenna elements has been designed and simulated. The inter element spacing has been optimized for obtaining the high gain. Slotted waveguide antenna array gain of more than 40 dBi could be achieved through simulations.

S. K. Ghosh et al. (Eds.): VEDA 2024, SPPHY 424, pp. 25–30, 2025.
https://doi.org/10.1007/978-981-96-5174-0_5

In this paper, design and simulation of slotted waveguide antenna array have been discussed in Sect. 2. Simulation results have been discussed in Sect. 3 and finally, conclusions have been made in Sect. 4.

2 Design and Simulation

A standard rectangular waveguide WR-284 operating in TE_{10} mode has been used for the design of a slotted waveguide antenna. The modeling of slotted waveguide antenna has been done using CST Microwave Studio. 3D model of slotted waveguide antenna element with 36 pairs of slots is shown in Fig. 1. The slots are carved on the narrow wall of the waveguide to arrange larger number of antenna elements. Spacing between the two slots is kept as 0.5 times of the guided wavelength (λ_g) at the center frequency (f). Further, the distance between the input port to the first slot and last slot to the short circuited end is kept as $0.25\lambda_g$ as shown in Fig. 2. The guided wavelength (λ_g) at the designed frequency has been calculated using relations (1) and (2) and the slot length (L_s) has been calculated using relation (3). The slot parameters of antenna element such as depth, width and angle have been optimized to obtain a higher gain of more than 24 dBi. The optimized slot parameters are given in Table.1.

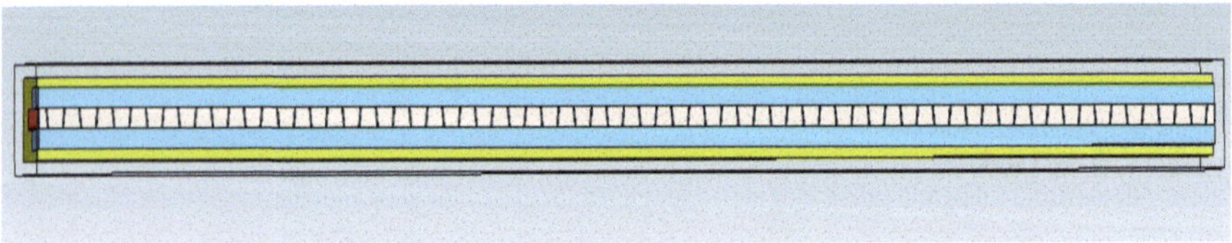

Fig. 1. 3D model of single slotted waveguide antenna element

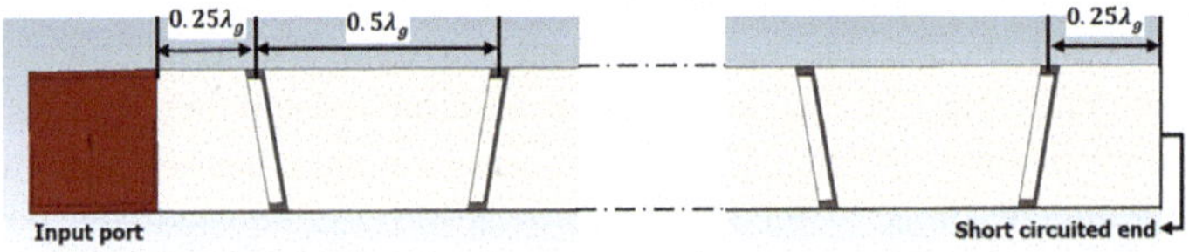

Fig. 2. Schematic of slotted waveguide antenna element with input port and short circuited end

The cutoff frequency f_c of the rectangular waveguide WR-284 operating in TE_{10} is given as [5],

$$f_c = \frac{c}{2a} \tag{1}$$

where 'c' is the speed of light in vacuum and 'a' is the width of the rectangular waveguide. The guided wavelength (λ_g) is given as

$$\lambda_g = \frac{\lambda_0}{\sqrt{1 - (f_c/f)^2}} \tag{2}$$

Table 1. Optimized slot parameters.

Sl. No	Parameter	Value
1	Operating frequency	S-Band
2	Width of rectangular waveguide	72.14 mm
3	Height of rectangular waveguide	34.04 mm
4	Width of slot	5.5 mm
5	Depth of slot	11.8 mm
6	Angle of slot	11°
7	Distance between the slots	0.5 λ_g
8	Distance between first slot to the input port	0.25 λ_g
9	Distance between last slot to the short circuited end	0.25 λ_g

where 'λ_0' is the free space wavelength and "f" is the center frequency. The relation between resonant length and depth of the slot is given by,

$$L_s + 2d \simeq 0.5\lambda_g \quad (3)$$

where, L_s is the length of the slot.

Once the parameters of slotted waveguide antenna element has been optimized, simulation of array of slotted waveguide antenna elements has been carried out to achieve the desired gain of more than 40 dBi.

3 Simulation Results

A slotted waveguide antenna element with 36 pair of slots, as shown in Fig. 1, has been modeled and simulated on CST studio. The dimensions of the antenna element have been optimized for low VSWR and high gain at the operating frequency f_0 in S-band. A fan beam radiation pattern with a gain of 26.4 dBi, maximum side lobe level of −14.5 dB, beamwidth at azimuth and elevation angle is 0.4° and 68°, respectively and VSWR of 1.5:1 have been obtained from the farfield simulation results using Microwave CST Studio [7]. The farfield radiation pattern in azimuth and elevation plane are shown in Figs.3 and 4, respectively.

The analysis has been extended for an array of slotted waveguide antenna with 32 elements to achieve high gain. The slotted waveguide antenna array has been modeled on CST studio and shown in Fig. 5. The spacing between the slotted waveguide antenna elements has been optimized to obtain the highest gain from the array and arrived as 38 mm. The highest gain of 41.6 dB with side lobe level of −16.9 dB in azimuth plane and side lobe level of −14.4 dB in elevation plane with beamwidth in azimuth and

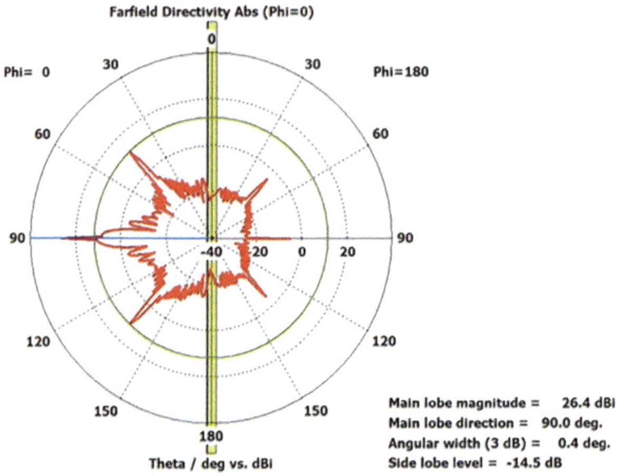

Fig. 3. Farfield results of slotted waveguide antenna element in azimuth plane.

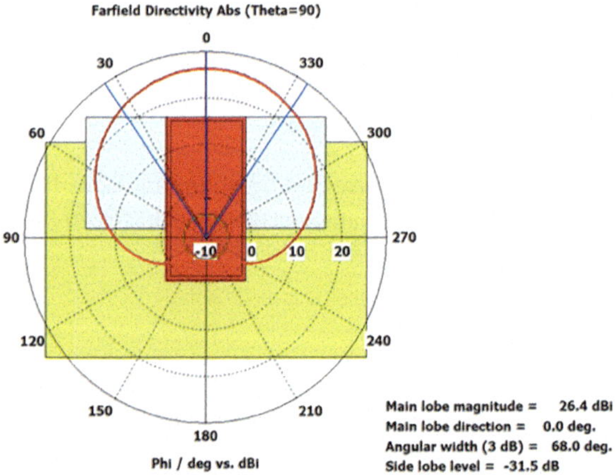

Fig. 4. Farfield results of slotted waveguide antenna element in elevation plane.

elevation angle of 0.4° and 2°, respectively, have been obtained from the simulation with optimized dimensions. The simulation results for farfield radiation pattern in azimuth and elevation plane are shown in Figs.6 and 7 respectively. The air filled WR-284 rectangular waveguide can be used for RF power flow up to 3.5 MW and therefore, the proposed antenna array can be used safely up to 100 MW of RF power flow without electric field breakdown.

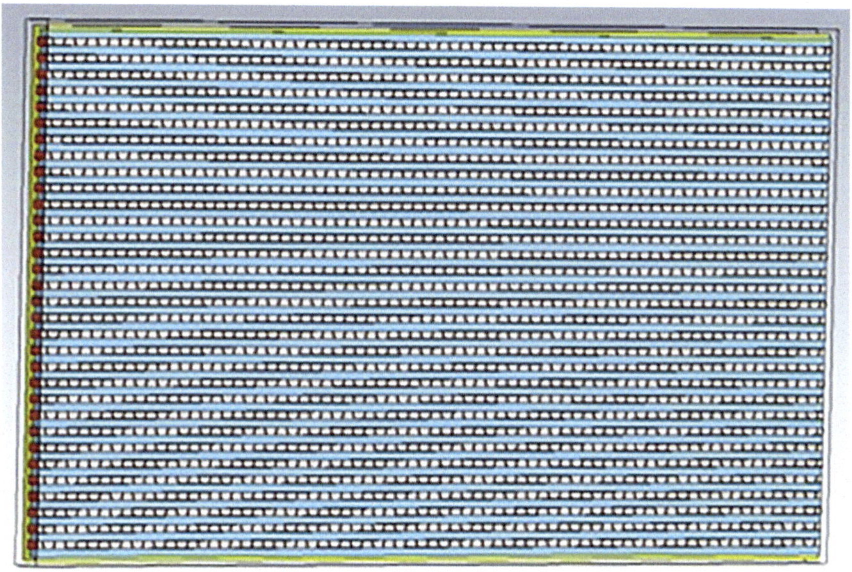

Fig. 5. 3D model of slotted waveguide antenna array with 32 elements.

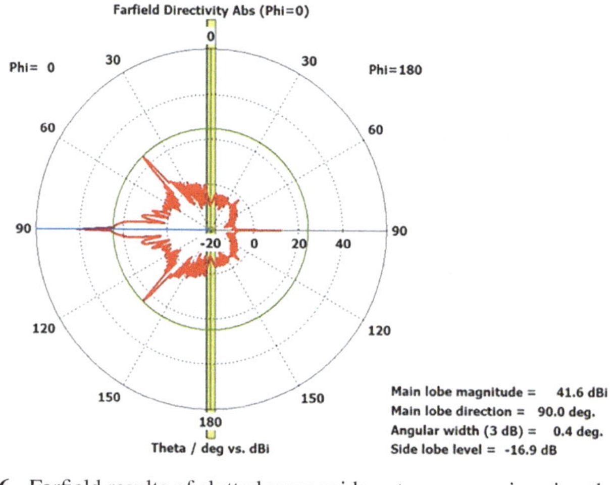

Fig. 6. Farfield results of slotted waveguide antenna array in azimuth plane

4 Conclusions

Design of a high gain and high power slotted waveguide antenna element and slotted waveguide antenna array operating in S-band has been presented. Modeling and simulation have been carried out on CST studio. Slotted waveguide antenna element with 36 pairs of slots could provide 26.4 dBi gain. Further, slotted waveguide antenna array of 32 elements has been optimized and a gain of 41.6 dBi, beamwidth of 0.4° and 2°

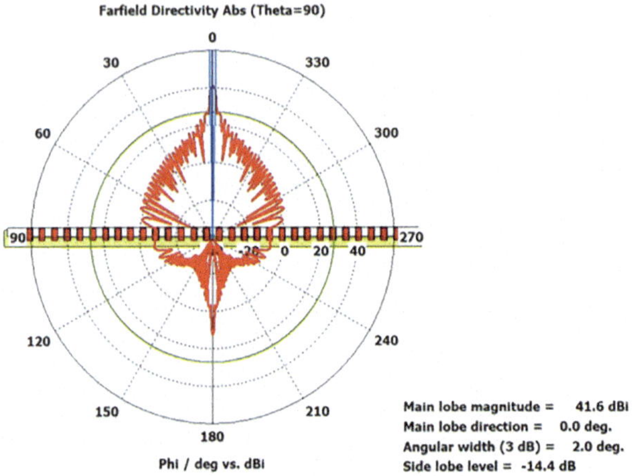

Fig. 7. Farfield results of slotted waveguide antenna array in elevation plane

in azimuth and elevation, respectively and side lobe level of −16.9 dB and −14.4 dB in azimuth and elevation, respectively, could be achieved. The proposed slotted array antenna can be used for RF power of 100 MW and can be used for detection and tracking of airborne targets as it provides narrow beamwidth.

Acknowledgement. The authors would wish to thank the Centre Head of Microwave Tube Research and Development Centre (MTRDC) for allowing us to carry out this research work and publish the same.

References

1. Wahyu, Y., Paramayudha, K.: Design of an edge-slotted waveguide antenna array for ground-controlled interception radar. In: 2021 International Conference on Radar, Antenna, Microwave, Electronics, and Telecommunications (ICRAMET), pp. 17–20
2. Monish Raj, D., Rajesh Kumar V.: Design of slotted waveguide antenna for wireless application. Int. Res. J. Eng. Technol. (IRJET) **9**(5), 1462–1470 (2022)
3. Rahayu, Y., Kurniawan, B., Saputera, Y.P.: Narrow wall Beamwidth slotted waveguide antenna for portable coastal radar communication. J. Eng. Sci. Technol. ICIST2022, **8**(1), 60–70
4. Constantine, A.: Balanis, Antenna Theory Analysis and Design, 2nd edn. Wiley, New York (2003)
5. Liao, S.Y.: Microwave Devices and Circuits. Prentice-Hall, Englewood Cliffs, NJ, USA (1985)
6. Gilbert, R.A.: Antenna Engineering Handbook, Chatpter: Waveguide Slot Antenna Arrays. McGraw-Hill (2007)
7. CST Studio Reference Manual, Germany (2019)

Design of an Interaction Circuit for a High-Power Multiple-Beam Klystron

Ashok Bansiwal(✉), P. Shalini, V. Nallasamy, and S. K. Datta

Microwave Tube Research and Development Centre, Ministry of Defence, DRDO, Bangalore, India
ashokbansiwal@gmail.com

Abstract. Design of an interaction circuit for a high power, S-band Multiple-beam Klystron (MBK) has been presented in this paper. The interaction circuit consists of a six numbers of coaxial double reentrant cavities operating in TM_{020} mode. Cold circuit parameters of coaxial reentrant cavities have been computed using microwave studio and RF performance has been obtained through Particle-In-Cell (PIC) simulation. An output power of 15 MW over 4 MHz bandwidth with a power gain of more than 50dB could be achieved through the simulations.

Keywords: Cavity interaction circuit · Coaxial double reentrant cavity · Higher order mode cavity · High power Klystron · Multiple-beam Klystron · PIC simulation

1 Introduction

High power S-Band Klystrons find their wide application in linear electron accelerators for medical as well as industrial applications [1]. Conventional Single-beam Klystron (SBK) as well as Multiple-beam Klystrons (MBK), both type of klystrons are being used in linear accelerators. MBKs having several advantages over the SBKs like; low operating voltage, permanent magnet focusing system, compact and light weight device, low phase noise, etc. [2, 3].

High power MBKs are being developed mainly in Russia and Sweden. An attempt has been made to design a cavity interaction circuit for a high power S-Band MBK. A coaxial double reentrant cavity operating in TM_{020} mode has been used for the interaction circuit [4, 5]. The analytical design of the cavity has been done using equivalent circuit analysis in [6] and the cavity dimensions optimized for required parameters using Eigen-mode solver of CST studio [7]. Couplers have been designed with standard WR-284 waveguide as input and output ports [8, 9]. A six cavity interaction circuit has been designed for the high power MBK. A 40 number of electron beams carrying a total beam current of 470A and having beam voltage 80kV have been used. Simulation has been carried out using CST particle studio. A 15 MW of output power over 4 MHz bandwidth has been obtained through PIC simulations.

The design and simulation of interaction circuits has been discussed in Sect. 2. PIC simulation and results have been discussed in Sect. 3. The conclusions are drawn in Sect. 4.

S. K. Ghosh et al. (Eds.): VEDA 2024, SPPHY 424, pp. 31–38, 2025.
https://doi.org/10.1007/978-981-96-5174-0_6

2 Design and Simulation of Interaction Circuit

Interaction circuit for high power S-band Multiple-beam Klystron (MBK) has been designed using coaxial double reentrant cavity. Coaxial double reentrant cavity operates in TM_{020} mode and has 40 number of beam holes. The modeling of coaxial double reentrant cavity is done on CST Microwave Studio and cavity parameters have been obtained using Eigen mode solver. Cavity dimensions such as drift gap, outer and inner radii of cavity, height of the cavity and outer and inner radii of the ferrule are optimized for the required resonant frequency (f_0), lower unloaded quality factor (Q_0) and higher characteristic shunt impedance (R_{sh}/Q_0). The 3D model of coaxial double reentrant cavity and its cross sectional view are shown in Fig. 1(a). Simulations has been carried out for finding out the effect of loading the cavity with lossy material on unloaded quality factor of the coaxial double reentrant cavity. The height and radius of the lossy buttons are varied to obtain the desired quality factor which will help in carrying out the PIC simulation and to achieve the controlled growth of output RF power. The 3D model of coaxial double reentrant cavity with four lossy buttons and its cross sectional view is shown in Fig. 1(b). The optimized coaxial double reentrant cavity parameters are given in Table.1.

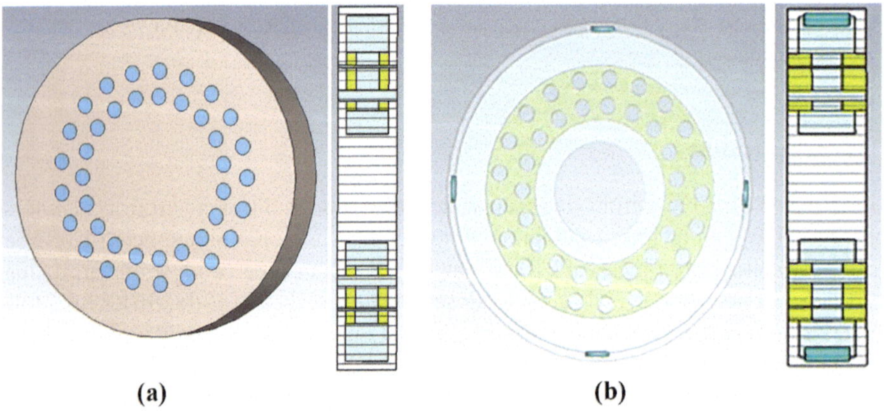

Fig. 1. 3D simulation model of (a) coaxial double reentrant cavity (b) coaxial double reentrant cavity with 4 lossy buttons.

Waveguide coupler with rectangular aperture has been chosen for the input and output coupling circuit. A standard waveguide WR-284 has been chosen as input and output port. A step impedance transformer has been used for impedance matching. The dimensions of the coupling circuit has been optimized to achieve the VSWR of nearly 1:1. The model of the input coupler is shown in Fig. 2 and the optimized VSWR plot is shown in Fig. 3.

Output coupling circuit determines the overall bandwidth of the device and therefore, loaded quality factor should be sufficiently low to support the desired operating band. Output coupling circuit model is shown in Fig. 4 and dimensions are optimized, for compensating the change in resonant frequency due to wider aperture as well as to

Table 1. Optimized main dimensions and parameters of the coaxial double reentrant cavity.

Sl. No	Parameters	Value
1	Operating frequency	S-Band
2	Drift gap	8.2 mm
3	Cavity outer radius	60 mm
4	Beam hole radius	3.25 mm
5	Number of beam holes	40 Nos
6	R_{sh}/Q_0 at one beam hole	18.7 Ω
7	Q_0 of the cavity	5705

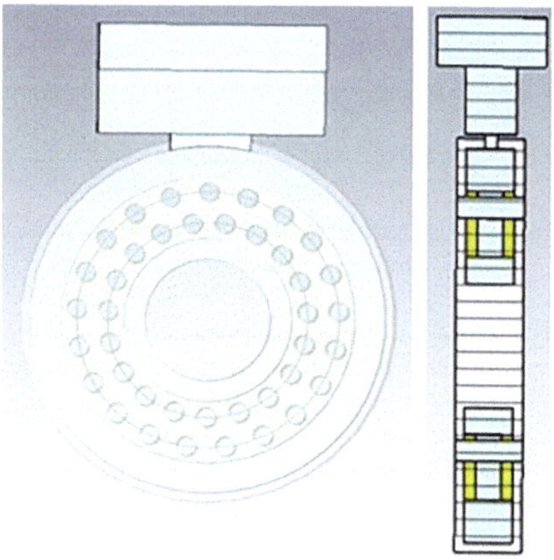

Fig. 2. 3D simulation model of input waveguide coupler with cavity.

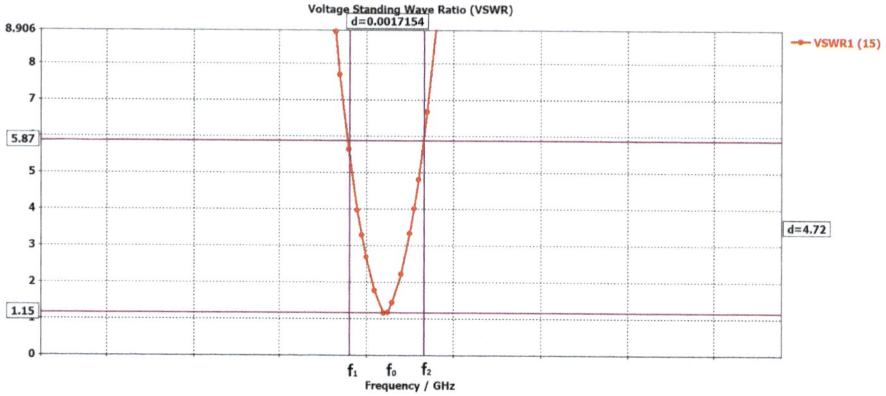

Fig. 3. VSWR plot of input waveguide coupler with cavity.

achieve the desired loaded quality factor required for 4 MHz band of operation. VSWR plot of the optimized coupling circuit is shown in Fig. 5.

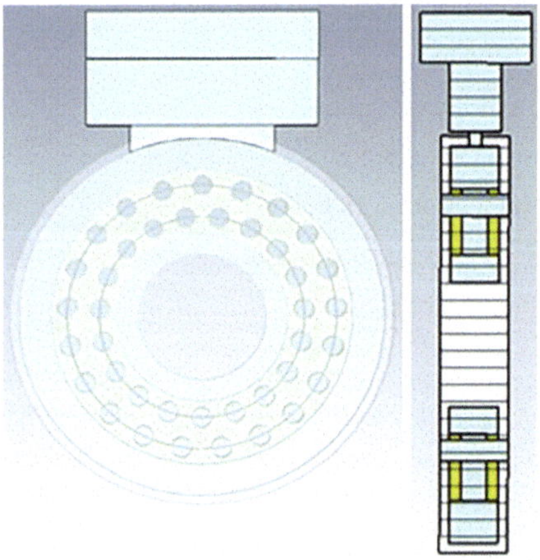

Fig. 4. 3D simulation model of output waveguide coupler with cavity.

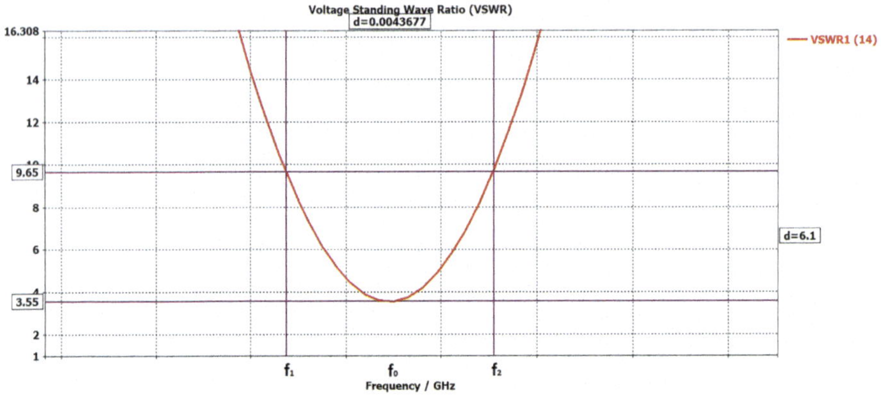

Fig. 5. VSWR plot of output waveguide coupler with cavity.

3 PIC Simulation and Results

A six cavity interaction circuit has been designed. The interaction circuit consists of 6 number of coaxial double reentrant cavities including first and last cavities as input coupling circuit and output coupling circuit, respectively. Simulation for the RF output power from the interaction circuit has been carried out using the PIC simulations. The 3D model of multiple beam klystron with six cavities is as shown in Fig. 6. Initially, all the cavities has been simulated in synchronous condition and the penultimate cavity frequency has been raised to increase the output power. The interaction circuit parameters such as RF drive level, drift gap of each cavity, spacing between the cavities were varied to obtain the desired output power and gain of more than 50 dB. The input and output coupling circuits are placed apart by an angle of 90° as it helps in extracting the RF output power more efficiently without any back streaming from collector end of interaction circuit. The second, third and fourth cavity are loaded with lossy buttons.

PIC simulation of six cavity interaction circuit has been carried out with a beam voltage of 80 kV, beam current of 470A distributed among 40 number of beamlets and magnetic field of 0.15T for an interaction duration of 400 ns. An output power of 15.7 MW is achieved for the input drive level of 27 W which corresponds to the power gain of 58 dB and electronic efficiency of 42%. The output power plot is shown in Fig. 7. The Fourier transform of the output signal is shown in Fig. 8 which indicates that the output signal is having single frequency component and is same as the input frequency. The electron beam trajectory plot is shown in Fig. 9 and it can be noted that no back streaming of electrons observed throughout the simulation duration of 400 ns.

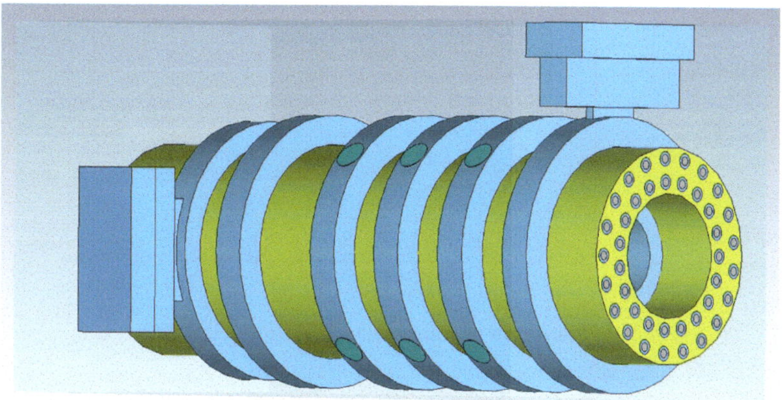

Fig. 6. 3D model of six cavity interaction circuit of MBK.

In order to verify 4 MHz bandwidth of the interaction circuit, simulations have been carried out by varying the frequency of RF input signal. The output power plot for resonant frequency of $f_0 - 2$ MHz and $f_0 + 2$ MHz are as shown in Figs. 10 and 11, respectively. It can be noted that 15 MW (minimum) RF output power over 4 MHz of bandwidth has been obtained from the designed interaction circuit of high power S-band Multiple-beam Klystron.

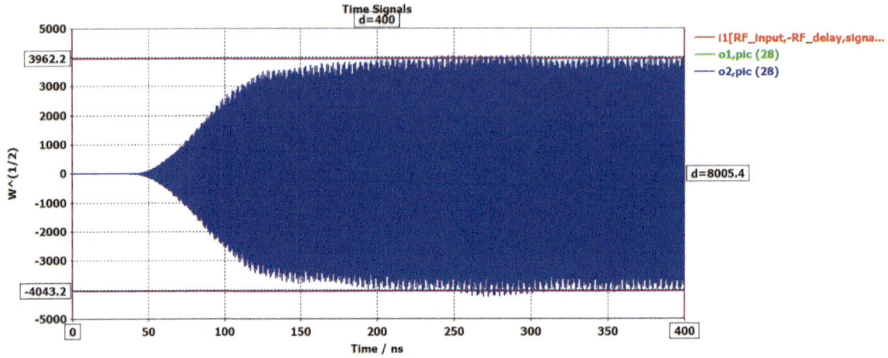

Fig. 7. 15.7 MW of RF output power plot of interaction circuit at f_0.

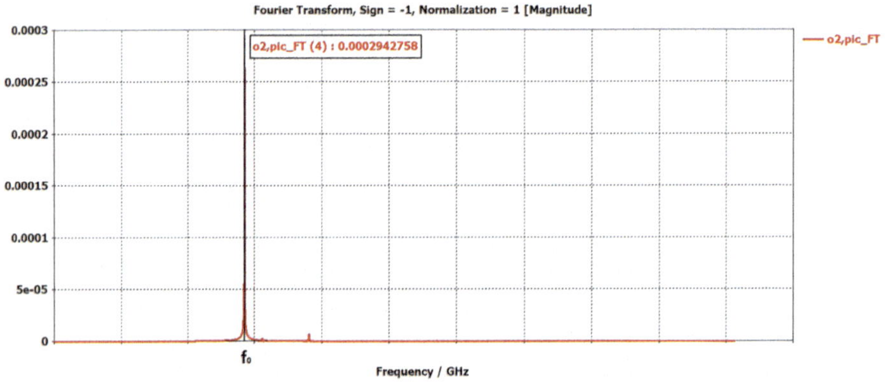

Fig. 8. Fourier transform of output signal from interaction circuit at f_0.

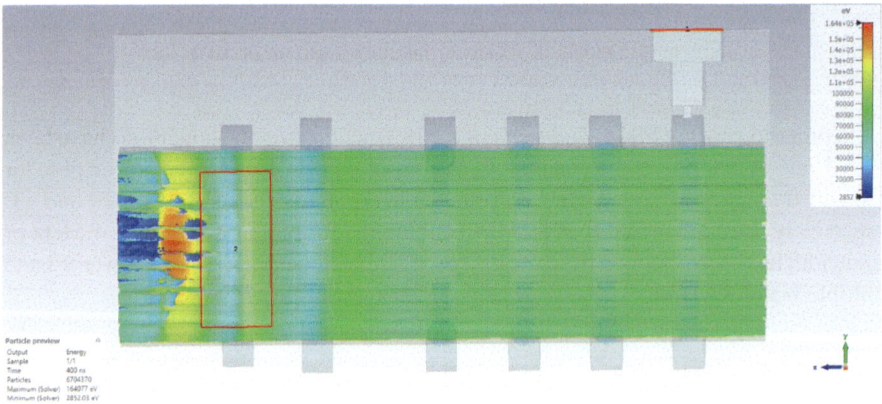

Fig. 9. Particle trajectory plot from PIC simulation at 400 ns.

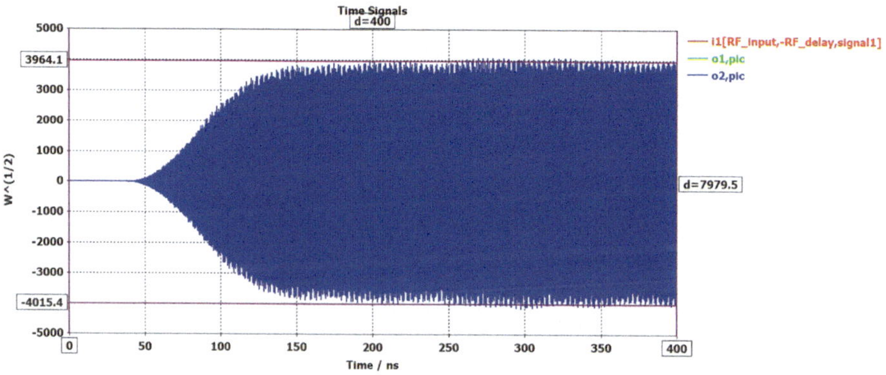

Fig. 10. 15.7MW RF output power plot of interaction circuit at f_0 − 2MHz.

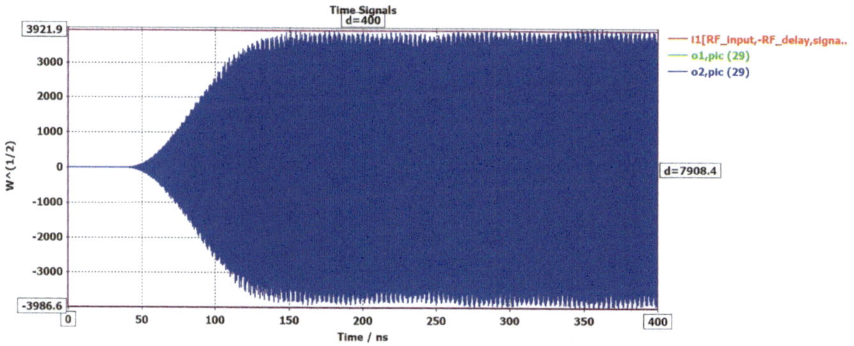

Fig. 11. 15.4MW RF output power plot of interaction circuit at f_0 + 2MHz.

4 Conclusion

Design of a cavity interaction circuit along with coupling circuit for a high power S-band Multiple-beam Klystron has been proposed. A coaxial double reentrant cavity operating in TM_{020} mode has been used for interaction circuit. Coupling circuit with WR-284 waveguide ports and aperture coupling has been used for input and output coupling circuit. Cold circuit parameter of cavity and coupling circuit have been simulated using Eigen mode and frequency domain solver of CST Studio. The RF performance of the interaction circuit has been simulated using Particle-In-Cell module of CST studio. An output power of 15 MW over 4 MHz of bandwidth with a gain and electronic efficiency of 58 dB and 42%, respectively, could be achieved. Efforts are being made towards increasing the electronic efficiency more than 60%.

Acknowledgement. The authors would wish to thank the Centre Head of Microwave Tube Research and Development Centre (MTRDC) for allowing us to carry out this research work and publish the same.

References

1. Komarov, D.A., Yekushkin, E.P., Paramonov, Yu.N., Darmaev, A.N., Melnichuk, G.V.: Development of a C-Band high power amplifier klystron for linear electron accelerators. IEEE (2017)
2. Korolyov, A.N., Gelvich, E.A., Zhary, Y.V., Zakurdayev, A.D., Poognin, V.I.: Multiple-beam klystron amplifiers: performance parameters and development trends. IEEE Trans. Plasma Sci. **32**(3), 1109–1118 (2004)
3. Gilmour, A.S.: Klystrons, traveling wave tubes, magnetrons, crossed-field amplifiers, and gyrotrons. Artech House, Norwood (MA) (2011)
4. Lin, X., Zhang, R., Zuo, L., Chao, Q., Xie, B., Wang, Z., Geng, Z., Liao, Y., Yang, X.: Design and simulation of an S-band compact multi-beam klystron. AIP Adv. **13**, 115023 (2023). https://doi.org/10.1063/5.0157492
5. Lin, X., Zhang, R., Chao, Q., Xie, B., Geng, Z., Liao, Y., Yang, X.: Multi-beam gun design for an S-band klystron. AIP Advances **12**, 085026 (2022). https://doi.org/10.1063/5.0100499
6. Bansiwal, A., Shalini, P.: An equivalent circuit analysis of a coaxial double-reentrant cavity for klystrons. In: National Conference on Emerging Trends in Vacuum Electronic Devices & Applications, Bangalore (2023)
7. CST Studio Reference Manual, Germany (2019)
8. Carter, R.G.: Microwave and RF Vacuum Electronic Power Sources. Cambridge Univ. Press, Cambridge, U.K. (2018)
9. Liao, S.Y.: Microwave Devices and Circuits. Prentice-Hall, Englewood Cliffs, NJ, USA (1985)

Design of a Pill-Box Window for a High-Power Multiple-beam Klystron

Biji Mathew(✉), P. Shalini, Ashok Bansiwal, V. Nallasamy, and S. K. Datta

Microwave Tube Research and Development Centre, Ministry of Defence, Near BEL North Gate, Jalahalli PO, Bangalore, Karnataka 560013, India
bijimathew991@gmail.com

Abstract. This paper describes about the design of a pill-box window for the input and output coupling circuit of a high power Multiple-beam Klystron (MBK). The modeling and simulation of the pill-box window operating in S-band is carried out using CST Microwave Studio. A standard rectangular waveguide WR-284 is used as input and output ports of pill-box window assembly. The window dimensions have been optimized and VSWR less than 1.05, insertion loss less than 0.05dB could be achieved at the desired frequency f_0 through simulation results. The simulation results for VSWR and insertion loss are found to be in good agreement with the measured results. The multipaction analysis has also been performed using Particle-In-Cell solver in CST Microwave Studio, to ensure safe operation of window assembly at rated power.

Keywords: High power window · Multiple-beam Klystron · Multipaction · Pill-box window

1 Introduction

High power Multiple-beam Klystrons (MBKs) are widely being used in electron linear accelerators for industrial application due to lower operating voltage and compact size. The RF windows plays a critical role in isolating the outside atmosphere with vacuum internal to the klystron device as well as allowing RF signal to pass through it with minimal RF loss.

A pill-box type RF window assembly is most widely used in the high power klystrons and design of the same has been proposed here for the use in the input and output coupling circuit of a high power MBK having operating frequency in the S-band [1, 2]. The proposed pill-box window assembly consists of a circular window and a metallic sleeve with both the ends having standard rectangular waveguide WR-284 sections. The circular window ceramic material is chosen as alumina, to withstand both mechanical and electrical stress and has low dielectric loss [3, 4]. Modeling and simulation have been carried out using CST microwave studio [5] and cold circuit parameters like resonant frequency, VSWR and insertion loss have been computed. In order to verify the simulation results, a measurement setup has been made to measure the window assembly parameters and found to be in good agreement.

S. K. Ghosh et al. (Eds.): VEDA 2024, SPPHY 424, pp. 39–45, 2025.
https://doi.org/10.1007/978-981-96-5174-0_7

For high power MBKs, the maximum power handling capability is restricted by RF breakdown caused by multipaction effect. The multipaction is a process of emission of secondary electrons in resonance with an alternating electric field which leads to exponential electron multipaction [6, 7]. This results in excess of heat and may lead to damage of window assembly. Therefore, Multipaction analysis has been carried out using Particle-In-Cell simulation of CST studio. Simulation results confirms the safe operation of the RF window assembly at rated 6 MW of power.

In this paper, design and simulation of a pill-box window assembly have been described in Sect. 2. Multipaction analysis has been described in Sect. 3 and finally, conclusions are drawn in Sect. 4.

2 Design and Simulation of a Pill-Box Window Assembly

The schematic view of a pill-box window is given in the Fig. 1. A circular disc ceramic of Al_2O_3 material has been used for the window assembly. A 96% pure Al_2O_3 ceramic has dielectric constant (ε_r) of 9.4 and loss tangent (tanδ) of 0.0004. The thickness of ceramic is chosen as 3.2 mm based on the details available in the open literature for similar and high power klystron [1, 2].

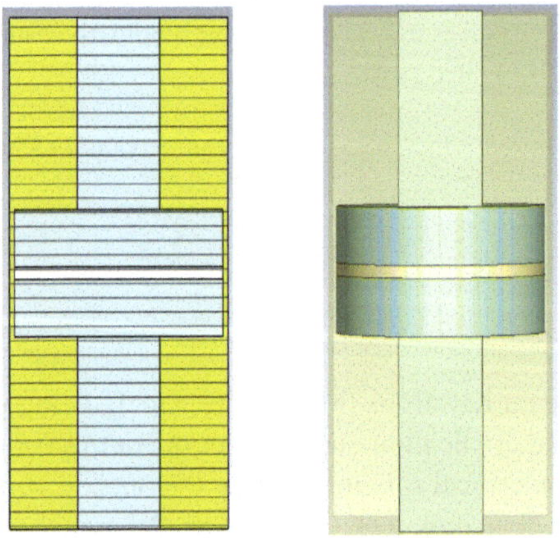

Fig. 1. 3D model of pill-box window assembly

Modeling of the pill-box window assembly has been carried out on CST microwave studio [5]. One end of the rectangular waveguide is considered as the input port where excitation is applied and other end is considered as the output port to extract the power. Tetrahedral mesh has been optimized to obtain the converged results. In the pill-box window, TE_{10} mode of rectangular waveguide converts to TE_{11} mode of cylindrical waveguide and then again converts to TE_{10} mode of rectangular waveguide. The electric field distribution in the pill-box window in transverse and longitudinal plane operating at the desired frequency f_0 are shown in Figs. 2 and 3 respectively.

The dimensions of the pill-box window assembly are optimized for achieving the smooth transition between the modes and achieving VSWR as low as possible. Using

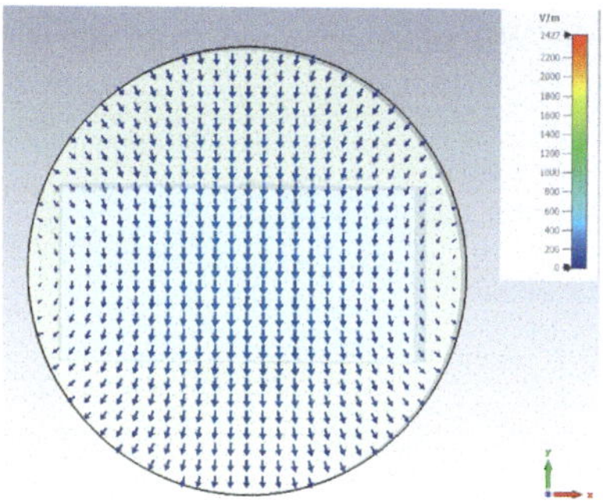

Fig. 2. Electric field in transverse plane of pill-box window.

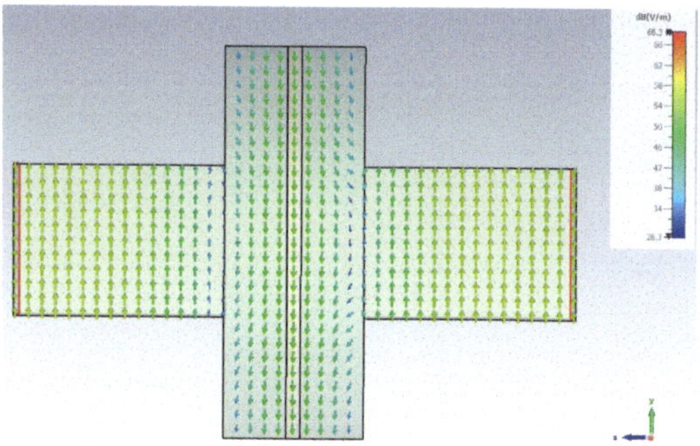

Fig. 3. Electric field in longitudinal plane of pill-box window.

frequency domain solver, the S-parameters and VSWR of the pill-box window assembly have been simulated. The VSWR less than 1.05 and insertion loss of less than 0.05 dB could be achieved through simulations over the desired bandwidth of 100 MHz in S-band after the optimization of window dimensions. The optimized dimensions of the pill-box window is tabulated in Table. 1.

Table 1. Optimized dimensions of pill-box window.

Sl. No.	Parameters	Value
1	Operating frequency	S-Band
2	Diameter of circular window	86 mm
3	Height of circular window	32.05 mm
4	Thickness of ceramic disk	3.2 mm
5	Width of rectangular waveguide	72.14 mm
6	Height of rectangular waveguide	34.04 mm

The S-parameter and VSWR plots obtained from the simulation are as given in Figs. 4 and 5 respectively.

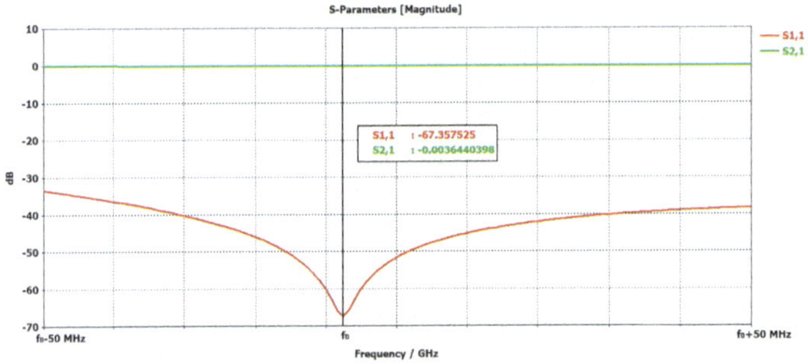

Fig. 4. S-parameter plot of pill-box window assembly obtained from simulation.

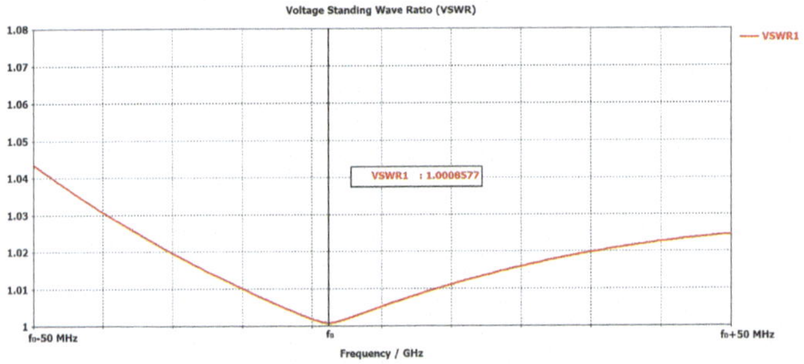

Fig. 5. VSWR plot of pill-box window assembly obtained from simulation.

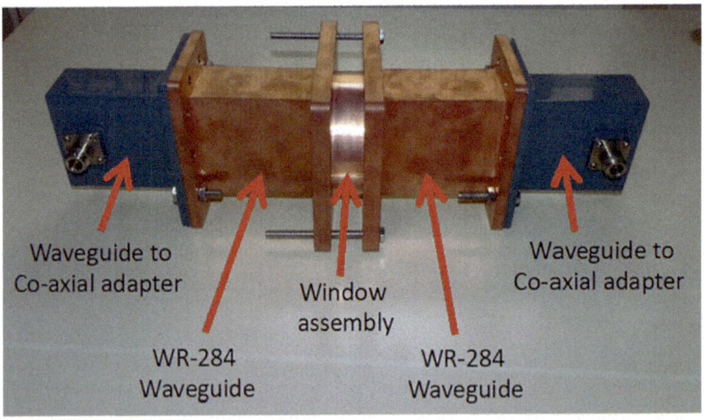

Fig. 6. Measurement setup of pill-box window assembly.

As per the proposed design, a pill-box window assembly has been made where the WR-284 standard waveguide section and sleeve have been made of OFHC copper and an alumina ceramic disc has been used. A back to back measurement setup has been made as shown in Fig. 6. Window assembly parameters have been measured using network analyzer. The measured VSWR is found to be less than 1.05 and insertion loss (S_{21}) less than 0.05 dB at the desired frequency f_0 as shown in Figs. 7 and 8, respectively. The measurement results are in good agreement with design.

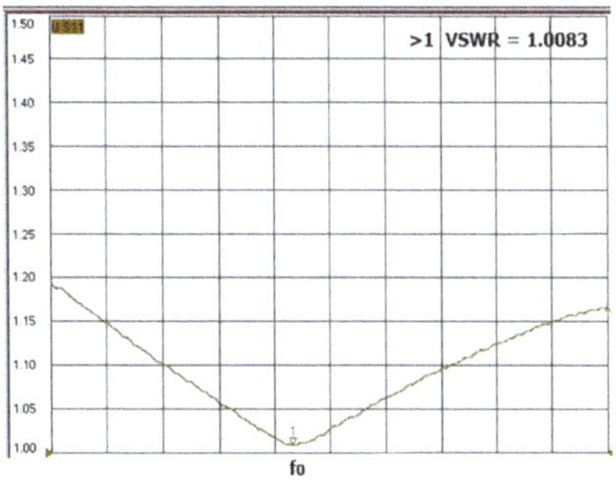

Fig. 7. VSWR plot of pill-box window obtained from RF measurements

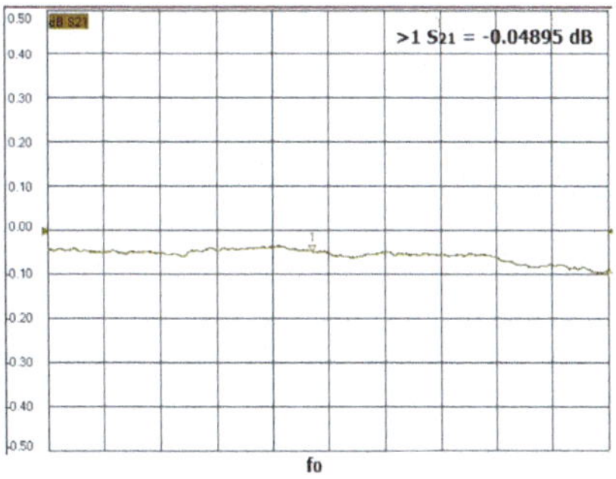

Fig. 8. Insertion loss (S_{21}) plot of pill-box window obtained from RF measurements

3 Multipaction Analysis

The RF performance of pill-box window can be severely affected by multipaction. The multipaction analysis has been performed using Particle-In-Cell solver of CST Microwave Studio. The electric and magnetic field data of the proposed pill-box window has been imported from frequency domain solver of CST Microwave Studio. In the multipaction simulation, 10000 Maxwellian source particles were chosen and the material has been characterized by Secondary Electron Yield (SEY) by Vaughan model. The simulation has been carried out for 10 micro second with 6 MW of power flow. The multipaction could not be observed as the electron gradually decays with time. The particle number versus time obtained through PIC simulation for multipaction analysis is shown in Fig. 9. The multipaction simulation results confirms that the designed pill-box window assembly is safe for 6 MW of rated power flow in high power Multiple-beam Klystron.

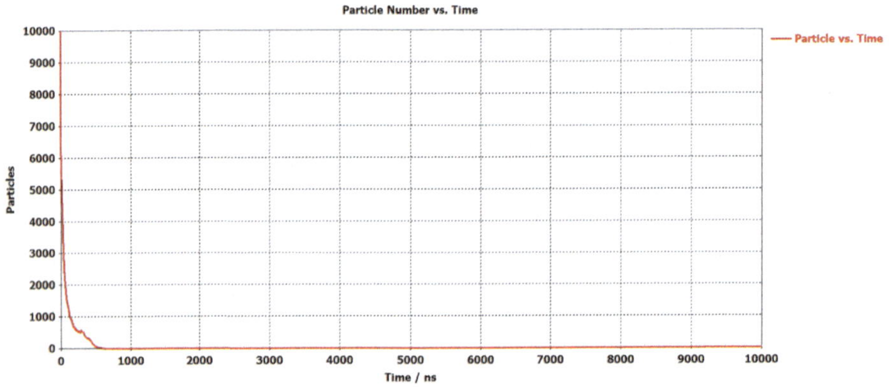

Fig. 9. Particle number vs Time plot for multipaction analysis with 6 MW of power flow.

4 Conclusions

A Pill-box window assembly design has been carried out for the use in couplers of an S-band, high power MBK. Modeling and simulation of pill-box window has been performed using CST Microwave Studio. WR-284 rectangular waveguide operating in TE_{10} mode is used as input and output ports of the pill-box window. The dimensions of the pill-box window has been optimized by varying the thickness of the ceramic disc, height and diameter of window sleeve. The performance of the designed pill-box window is found to be within the desired range of VSWR less than 1.05 and insertion loss less than 0.05 dB. The simulation results are also found to be in good agreement with the measured results. The multipaction analysis of proposed pill-box window has been carried out and the designed pill-box window assembly found safe for 6 MW of power flow at the desired frequency in S-band.

Acknowledgement. The authors would wish to thank the Centre Head of Microwave Tube Research and Development Centre (MTRDC) for allowing us to carry out this research work and publish the same.

References

1. Lamba, O.S., Saxena, V.K., Pal, D., Joshi, L.M., Kant, D., Bandhopadhyay, A.: Electromagnetic analysis of RF window for S-band high power klystron. IJPRET **4**(7), 42–47 (2016)
2. Miura, A., Matsumoto, H.: Development of an S-band high power pillbox-type RF window. In: Proceedings of the 1992 Linear Accelerator Conference, pp. 124–126 (1992)

3. Lamba, O.S., Pal, D., Kant, D., Giri, T., Verma, M.K., Nangru, S.C., Joshi, L.M.: Design and optimization of asymmetric RF window for 6 MW pulse power S-band Klystron. In: Proceedings of International Conference on Microwave, pp. 44–46 (2008)
4. Rizzi, P.A.: Microwave engineering (passive circuits). Prentice Hall International Inc, USA (1988)
5. CST Studio Reference Manual, Germany (2019)
6. Joshi, M.K., Tiwari, T., Bhattacharjee, R.: Design and multipactor analysis of a high power RF window. In: International Vacuum Electronics Conference (IVEC) (2019)
7. Alaria, M.K., Bhomia, V., Parikh, K.S., Sinha, A.K., Srivastava, V.: Multipactor analysis in coaxial RF window for space TWT. Int. J. Microw. Opt. Technol. **6**(5), 255–258 (2011)

Multi-physics Studies and Design Optimisation of Re-entrant RF Cavity for Klystron

Elina Mishra(✉), Tadi Dinesh Kiran, Praveen K. Rai, Janvin Itteera, Kumud Singh, Rajesh Chimurkar, Sampada Sawant, D. H. Darvesh, and Sanjay Malhotra

Electromagnetic Application and Instrumentation Division, Bhabha Atomic Research Centre, Mumbai, Trombay, India
elinam@barc.gov.in

Abstract. As the electron beam travels through the klystron, the RF cavities in klystron bunches and provides effective velocity modulation to the beam. Interaction between the electron beam and the RF fields occurring within the RF section of the klystron is responsible for the overall gain of the amplifier. Design of the RF cavity involves optimization of several figures of merit including resonant frequency, shunt impedance, quality factor, power dissipation, R/Q ratio amongst others. The re-entrant feature in the cavity aids in concentrating electric field in the RF gap region and hence enhancing the velocity modulation of electron beam. Various designs of the RF cavity for an S-band klystron have been modelled and analysed to optimise the figures of merit, while minimising the peak electric field within Kilpatrick limit. Study of tuner efficacy in individual cavities has been carried out for determining the overall tuning range. Error sensitivity analysis has been done to study effect of individual geometrical parameter on the resonant frequency. Temperature rise and heat load on the RF cavities were calculated. This serves as an input for the cooling mechanism design. In this paper, a detailed multi-physics coupled study and design optimisation of the RF section of klystron has been presented.

1 Introduction

Klystron is a vacuum electron device used as RF power amplifier for variety of high frequency applications such as particle accelerators, microwave radars systems, ultra-high frequency transmission amongst others [1]. For indigenous program, Bhabha Atomic Research Centre is developing an S-band, 6 MW (peak), 24 kW (avg.) power klystron for industrial linac development. Operating frequency of the klystron is 2.856 GHz. A summarised technical specification of the klystron under development is given below in Table 1.

The major sub-components of the klystron include an electron gun, RF section, electromagnet and collector. Electron gun generates the electron beam of desired beam emittance and perveance. The electron beam passes through the RF section of the klystron where velocity modulation of the beam bunch takes place. A multi coil electromagnet provides axial magnetic field of a desired profile for confining and focusing the electron beam throughout its trajectory. At the ultimate cavity, the electron beam bunch induces

S. K. Ghosh et al. (Eds.): VEDA 2024, SPPHY 424, pp. 46–58, 2025.
https://doi.org/10.1007/978-981-96-5174-0_8

Table 1. Technical specifications of S-band klystron

S. No.	Parameter	Value	Units
1	Frequency	2856	MHz
2	Peak output	6	MW
3	Average output	24	kW
4	Pulse width	10	μs
5	PRF (max)	400	
6	Gain	45	dB
7	Efficiency	45	%
8	−1 dB bandwidth	± 4	MHz
9	Beam voltage	130	kV
10	Beam current	95	A

electromagnetic fields in the cavity which is then is transmitted through a waveguide. The electron beam bunch is then safely dumped in the collector region of the klystron.

The number of RF cavities in the RF section of klystron is decided by the effective bunching and overall velocity modulation achieved in the electron beam [1, 2]. The cavities are interconnected with each other via drift tubes. The velocity of the beam changes depending on the interaction between RF phase in the cavity gap and the beam. The beam passing the gap through positive half of RF cycle will get accelerated and those passing in the negative RF cycle will get decelerated. Gradually, depending on the drift spaces between the RF gaps the electron beam will get bunched due to velocity modulation over the RF gaps in the cavities. Each RF cavity helps in improving this bunching phenomena and thus increasing the beam current density. At the output cavity, the highly bunched electron beam then induces strong RF fields whose energy is then transmitted to the load through suitable waveguide system.

2 Theory

The efficiency in which the RF cavity modulates the velocity of the electron beam depends on its electromagnetic design. The mode of interest for klystron RF cavity is TM_{010} where the electric field is in axial direction and magnetic field in the azimuthal direction. The axial electric field interacts with the incoming electron beam to modulate its velocity and eventually bunch it. Optimisation of the following key design parameters need to be carried out during electromagnetic design of RF cavity:

1. Resonant frequency
2. Shunt impedance
3. R/Q
4. Quality factor
5. Transit time factor

All these above parameters are a function of cavity geometry mainly:

1. Diameter of the cavity
2. Beam aperture
3. Gap distance between nose cone
4. Nose cone dimensions
5. Tuner position and dimension

Resonant Frequency of an RF cavity is purely a function of cavity dimensions. An RF cavity can have 'n' number of resonant modes corresponding to 'n' discrete frequencies. The desired resonant mode for our application is TM_{010} mode corresponding to 2.856 GHz. At this frequency, maximum power transfer will occur between the RF cavity and electron beam. For a typical pillbox cavity,

$$\omega_0 = 2.405\frac{c}{r} \tag{1}$$

where ω_0: resonant frequency

r: radius of RF cavity

For re-entrant cavity, this simple analytical equation needs to be modified to achieve accurate resonant frequency. Resonant frequency is a function of cavity dimensions. In klystron, the different cavities are staggered tuned to maximize the gain-bandwidth product over the entire section.

Quality Factor (Q) of an RF cavity describes how long (in terms of RF cycles) it takes the cavity to completely lose the stored energy in form of power dissipation. Q depends largely on the surface material properties. It can be improved by accurate machining, surface preparation and cleaning mechanisms and seamless joining and sealing of RF sub-components.

$$Q_0 = \frac{\omega_0 U}{P_c} \tag{2}$$

Surface resistivity Rs describes the RF resistance of the RF cavity and depends on the frequency and material properties.

$$R_s = \frac{1}{\delta\sigma} \tag{3}$$

where δ: skin depth

σ: conductivity of the cavity material

Shunt impedance quantifies the efficacy of a cavity to generate accelerating voltage (V) for a given power (Pc) dissipation per cycle.

$$R = \frac{V^2}{Pc} \tag{4}$$

R/Q quantifies the efficiency of an RF cavity in terms of geometrical property and is material independent. This is a very crucial figure of merit and signifies the gain-bandwidth product of the RF cavity.

$$\frac{R}{Q} = \frac{V^2}{\omega_0 U} \tag{5}$$

Transit Time Factor (TTF) quantifies the measure of the spatial and temporal effect of acceleration due to RF fields when compared to DC field.

$$\Delta W = qV_0 Tcos\Phi \tag{6}$$

$$T = \frac{sin\left(\frac{\pi L}{\beta\lambda}\right)}{\frac{\pi L}{\beta\lambda}}$$

3 Design and Analysis of RF Cavity

A. Electromagnetic Design

Dedicated 3D multi-physics codes from CST microwave studio [3] were used for the electromagnetic design, analysis and optimization of the klystron cavities. Figure 1 shows the electromagnetic model of the cavity. The internal details of the cavity consisting of re-entrant nose cones and tuner system is shown in Fig. 2.

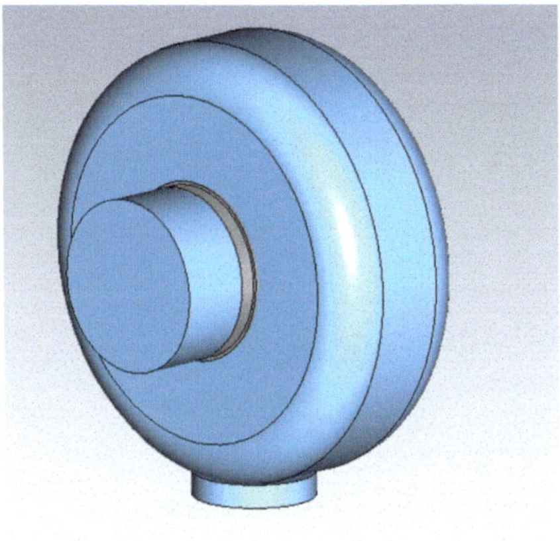

Fig. 1. Electromagnetic model of klystron RF cavity

Eigen mode solver was used to determine the eigen modes of cavity. The mode that is our mode of interest is TM_{010} where the E-field is in the axial direction and B-field is in the azimuthal direction. Figures 3 and 4 show the E-field and B-field patterns in the cavity.

The resonant frequency of TM_{010} mode is 2.886 GHz as per the simulation. Figures 5 and 6 gives the electric and magnetic field energy density in the klystron cavity. The electric energy density is maximum near the tips of the nose cone where the electric field density is highest and the magnetic energy density is highest near the cavity equator region as seen in the tuner rod portion.

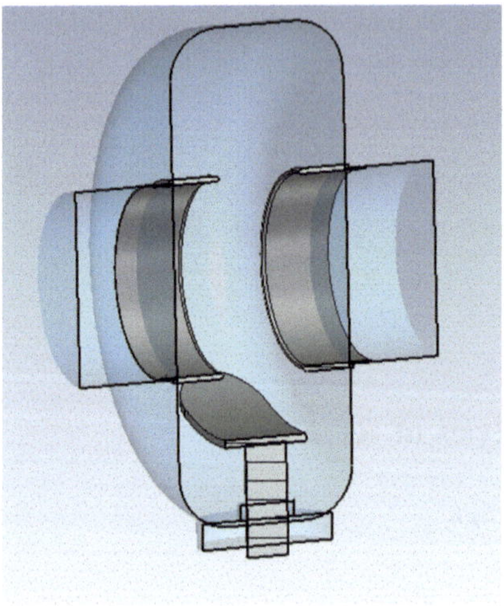

Fig. 2. Internal details of the RF cavity

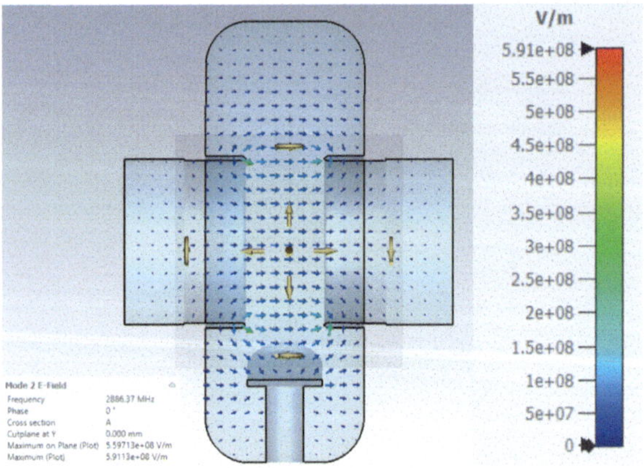

Fig. 3. E-field patterns in the cavity

Plot 7 shows the axial electric field (R = 0 mm) and plot 8 shows the axial electric field at R = 12 mm. From the plots, it can be inferred that the electric fields peak in the re-entrant nose cone area hence enhancing the voltage gain across the RF gap (Fig. 7 and 8).

As detailed in the section II, the figures of merit of the RF cavity were calculated. In CST microwave studio, the eigen mode analysis is done considering total energy stored in the cavity as 1 J. Accordingly, voltage gain across the RF gap is calculated. From 1-D AJDISK simulation code, electron beam and RF field interaction along the complete length of the klystron has been studied. For obtaining maximum output power from the last cavity, optimisation studies have been carried out. The figures of merit in CST studio

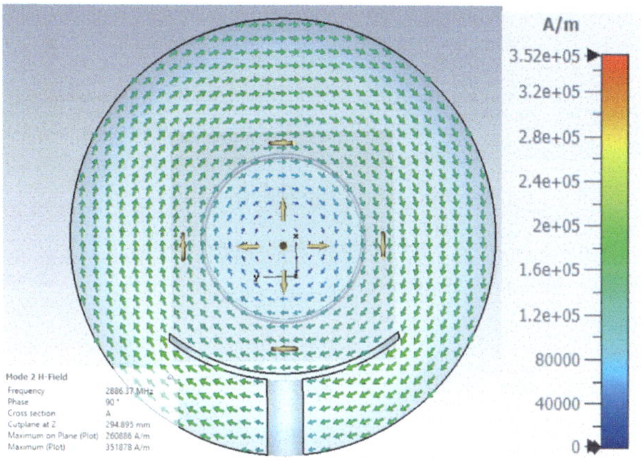

Fig. 4. B-field patterns in the cavity

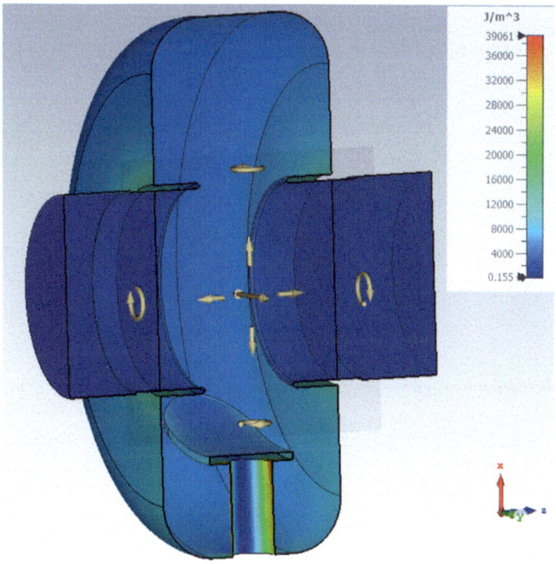

Fig. 5. E-field energy density in klystron cavity

are determined by scaling the voltage across the RF gap. Table 2 summarised the figures of merit of the designed cavity.

Design of RF cavity is an iterative process, which takes inputs from AJDISK to determine various output parameters, which in turn is re-simulated in AJDISK to determine the output power. According to this, five RF cavities of the klystron tube were designed.

B. **Tuner Efficacy**

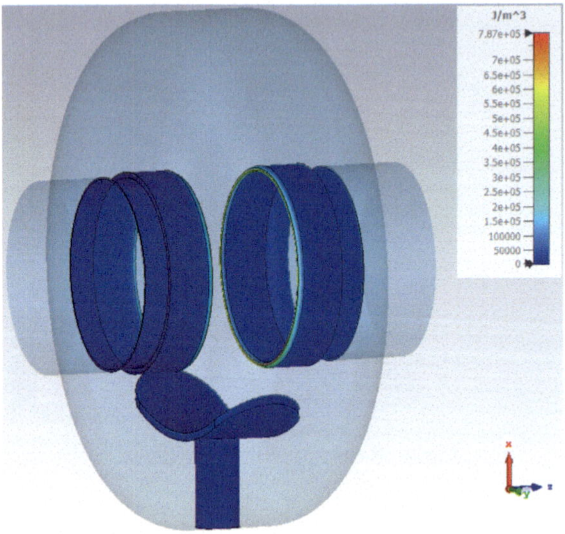

Fig. 6. B-field energy density in klystron cavity

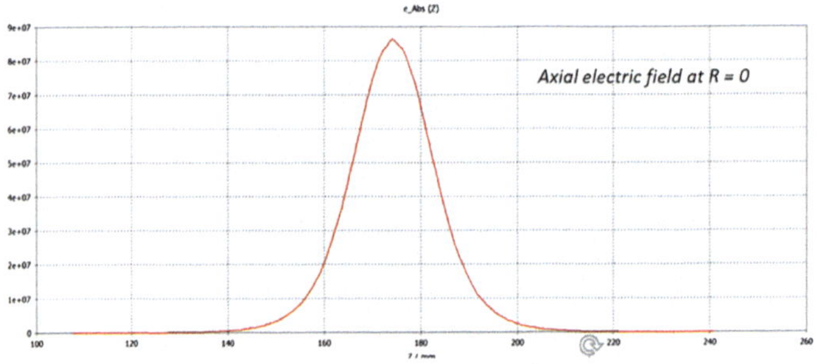

Fig. 7. Electric field along axis of cavity (R = 0 mm)

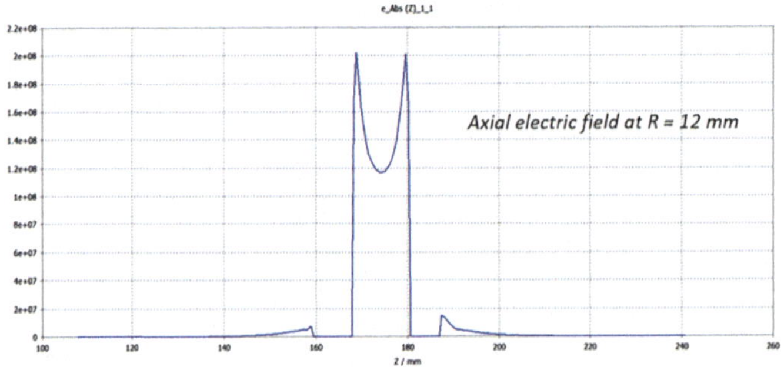

Fig. 8. Electric field along axis of cavity (R = 12 mm)

Table 2. Figures of merit of designed cavity

S. No.	Parameter	Value	Units
1	Resonant frequency	2886	MHz
2	Quality factor	9233	
3	Shunt impedance	1.968	MΩ
4	R/Q (Geometrical)	213	MΩ
5	Voltage across RF gap (data from AJDISK)	~ 46	kΩ

While development and fabrication, due to several engineering challenges, the resonant frequency of the developed cavity may not exactly match with that of the designed cavity (in simulation). This may be due to manufacturing errors, assembly errors and temperature variation amongst others. There needs to be a provision for fine-tuning the resonant frequency of the cavity [4]. A tuner arrangement consisting of tuner rod and a pad is used to tune the cavity during commissioning as well as operation stages. Plot 9 shows the variation in resonant frequency of the cavity with tuner position inside the cavity (Fig. 9).

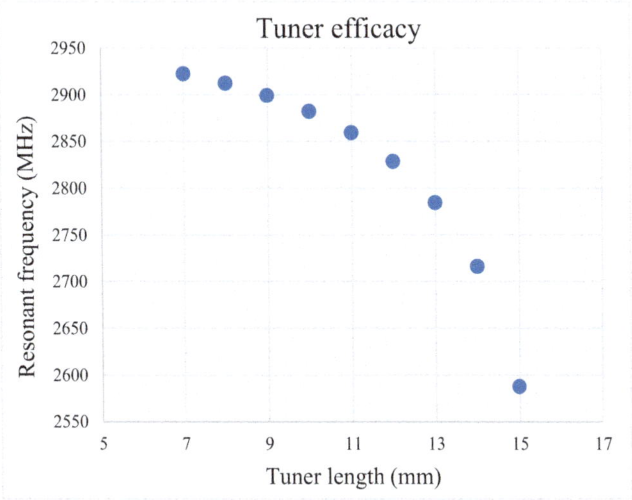

Fig. 9. Change in resonant frequency with tuner position

From the figure, it can be concluded that effective tuning range of the tuner is ~ 90 MHz. This is where the change in resonant frequency with the tuner is almost linear. Beyond this range if the tuner is inserted, in enters the dominant electric field zone and the resonant frequency falls sharply.

C. **Error Sensitivity Analysis**

While drafting engineering and manufacturing drawings for the development of the cavity, it is crucial to specify the fabrication tolerances in the drawings. The tolerances which needs to be specified in individual sub-parts of the cavity needs to be established in connection to the change in resonant frequency pertaining to change in dimension of the sub-parts. Hence, error sensitivity analysis was carried out to ascertain the effect of dimensional variation of individual sub-part on resonant frequency. Table 3 summarises the effect of change in cavity dimensions with resonant frequency.

Table 3. Effect of change in cavity dimensions with resonant frequency

S. No.	Parameter	Sensitivity (MHz/mm)
1	Cavity diameter	91
2	Cavity length	82
3	Nose cone outer diameter	61
4	Nose cone inner diameter	52.8
5	Nose cone blend	58
6	RF gap length	91.7
7	Tuner rod diameter	10.1
8	Tuner pad thickness	7.9

From the above analysis, suitable fabrication tolerances have to be specified keeping in consideration that the net change in resonant frequency should remain within the tuning range of the tuner system (including manufacturing and assembly errors).

D. **Thermal Analysis**

Coupled multi-physics analysis was carried out to determine the rise in temperature and heat load on individual RF cavity. Input to the thermal solver was taken from the Eigen mode solver. Figure 10 shows the rise in temperature distribution of the 2nd cavity and Fig. 11 shows the heat load on the cavity surface.

4 Development and RF Characterisation

Two number of RF cavities were developed and fabricated for validation of the RF design and low power RF characterisation was done. Figure 12 shows the photograph of the developed RF cavity. Low power RF characterisation was carried out using VNA to meausre the resonant frequency, quality factor and tuner efficacy. The different parts of the cavity were not brazed but push-fit to close the cavity. Figure 13 shows the photograph of the measurement setup.

Figure 14 shows the S21 parameter of the developed cavity and the unloaded quality factor. It can be seen that the resonant frequency of the developed cavity is 2.849 GHz and the quality factor is ~ 2000. The difference in simulated and measured resonant

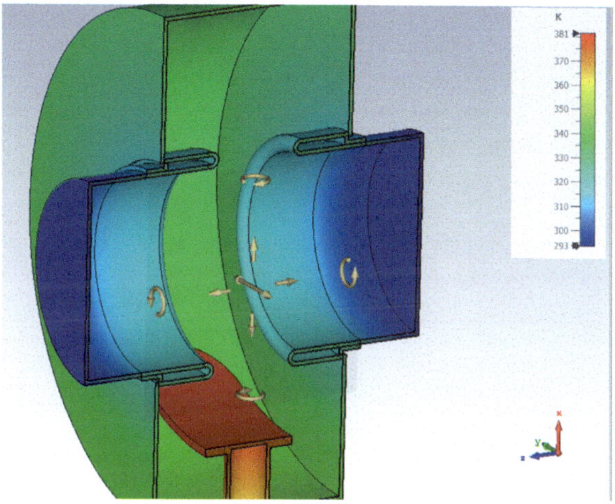

Fig. 10. Temperature distribution on RF cavity surface

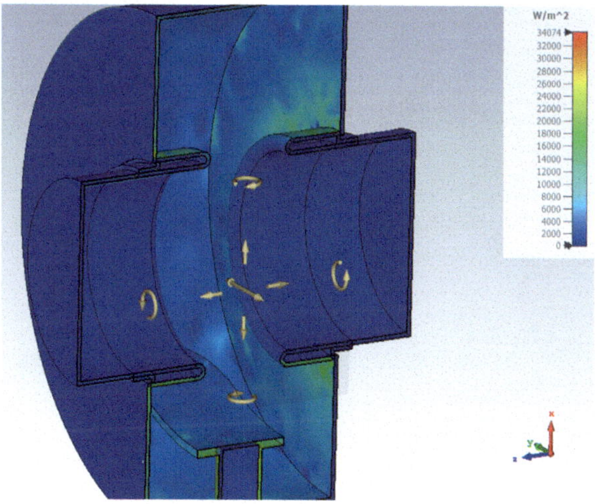

Fig. 11. Heat load on the cavity surface

frequency can be attributed to manufacturing and assembly errors and that the cavity is closed by push-fitting. Brazing the cavity components will also improve the quality factor.

Tuner was moved over the entire possible range of operation and to determine the tuning efficacy of the system. Plot 15 shows the change in resonant frequency with change in tuner position. From the above plot, the effective tuning range of the tuner over nearly linear range of operation is ~ 72 MHz.

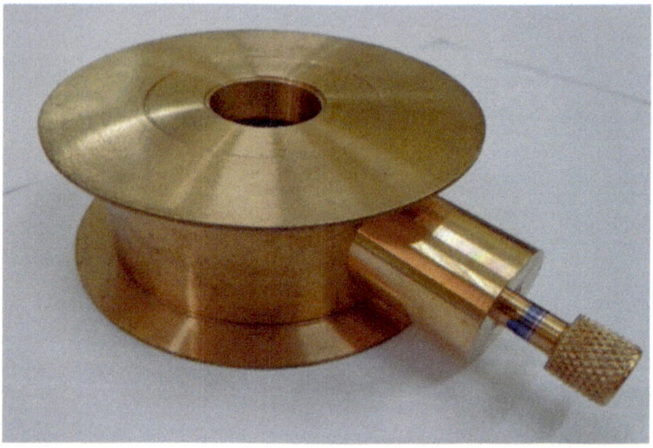

Fig. 12. Developed klystron cavity

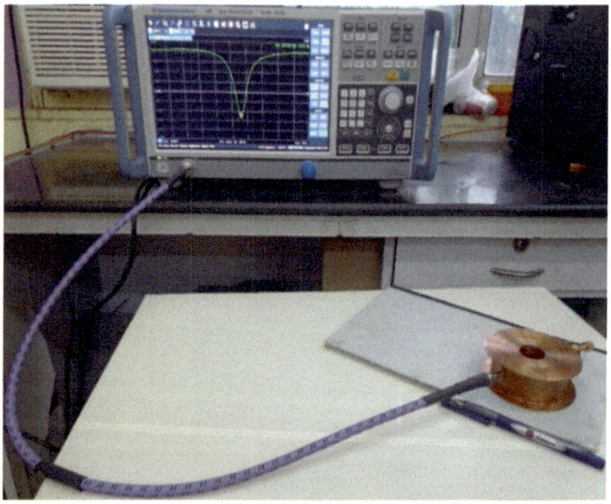

Fig. 13. Photograph of the cavity under test using VNA

5 Summary and Conclusion

Electromagnetic analysis, thermal and error sensitivity studies for klystron RF cavity were carried out. Two number of the cavities were developed and low power RF measurements were carried out. Tuner efficacy of the cavity were also determined. The measurement results closely match with those of the simulated values (Fig. 15).

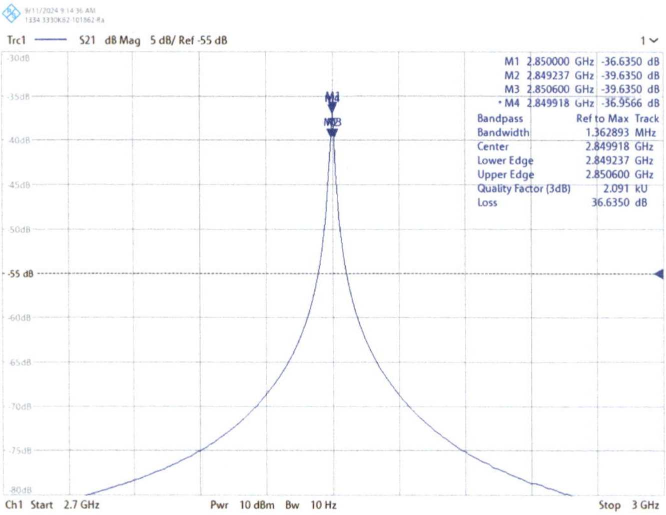

Fig. 14. Transmission parameter (S21) of the develope cavity

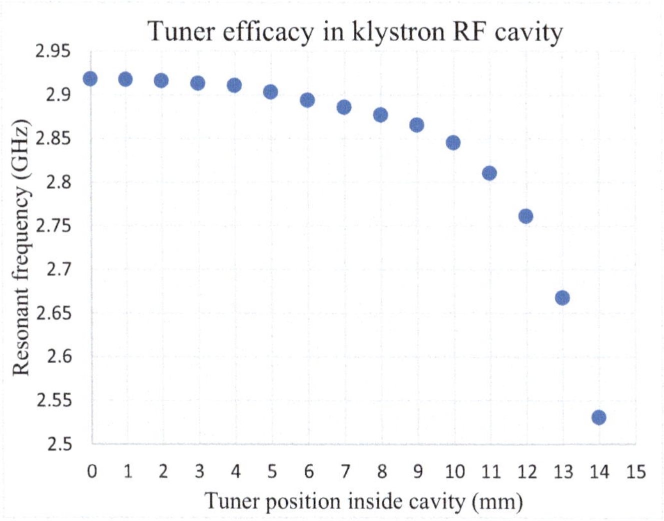

Fig. 15. Measured change in resonant frequency with tuner position

Acknowledgement. Authors would like to thank Dr. S.K. Ghosh, Chief Scientist & Group Head, Vacuum Electron Device Group, Dr. Deependra Kant, Senior Principal Scientist, Vacuum Electron Device Group & Dr. Ayan Kumar Bandyopadhyay, Senior Principal Scientist & Group Head, High-Frequency Devices and System Group (HFDSG), CSIR-CEERI for their reference inputs and technical interaction on experiment results on the developed S-Band Klystron tube at CSIR-CEERI.

References

1. Joshi, L.M., et al.: Simulation of RF section and beam-wave interaction in a 6 MW peak power S-band klystron. In: 2008 IEEE International Vacuum Electronics Conference, Monterey, CA, USA, pp. 276–277 (2008). https://doi.org/10.1109/IVELEC.2008.4556510
2. Sprehn, D., et al.: 150-MW S-Band Klystron Program at the Stanford Linear Accelerator Center. Pulsed RF Sources for Linear Colliders, Hayama, Japan, April 8–12 (1996)
3. Le Floch, J.M., et al.: Rigorous analysis of highly tunable cylindrical Transverse Magnetic mode re-entrant cavities. arxiv, 1308.2755
4. Holzbauer, J.: "RF Theory and Design" USPAS - Grand Rapids June 2012
5. CST studio suite 3D user guide

Electromagnetic Design and Optimization Studies for Multi-coil Focusing Lens for Indigenous 6MW S-Band Klystron Tube

Kumud Singh(✉), Himanshu Bisht, Mahima, Janvin Itteera, Vikas Tiwari, Elina Mishra, Prashant Kumar, Sanjay Malhotra, Rajesh Chimurkar, R. R. Singh, and R. K. Jalan

Electromagnetic Applications and Instrumentation Division, Bhabha Atomic Research Centre, Mumbai, India

Kumuds@barc.gov.in

Abstract. Electron beam focussing and transmission in the klystron tube is guided by the magnetic field and it has implications on the output power, bandwidth, gain, volume, and mass of the vacuum tube. High frequency Klystrons with high pulse repetition factor and high efficiency are operated in immersive flow beam confinement mode. In this mode electron beam has strong beam stiffness in the beam–wave interaction region but it necessitates at least 2–4 times the Brillouin magnetic field strength thus necessitating a stronger magnetic field for beam focusing. Waveguide and tuner aperture ports introduce asymmetries in the magnet design which can interfere with the uniform solenoidal coil configuration. Spatial constraints introducing asymmetric field configuration is more severe in the output cavity region where the magnetic field reduces due to presence of output waveguide port. On the contrary, magnetic field requirement in the output cavity beam-wave interaction region is higher than the first feeding cavity, buncher cavities and penultimate cavity to reduce beam perturbations of current modulated bunched beam. A water cooled tunable multi-coil electromagnet assembly has been designed for 6-MW peak power S-Band Klystron tube in order to meet 0.11T field requirement in the output cavity region. Electromagnetic analysis and simulation studies were done to optimize the axial magnetic flux density in the good field region to meet beam optics requirements with given spatial constraints using three dimensional Finite element method (FEM). Due to the multi-coil design of the magnetic system, easy tuning is possible to tune the beam filling ratio within a reasonable range by adjusting the coil currents to evaluate tube performance during RF conditioning and trials. This paper discusses the electromagnetic design and optimization studies for tuning the magnetic field profile [1] and the peak field location in the vicinity of output cavity region of the S-Band Klystron tube.

1 Introduction

The fields are assumed to be axially symmetric for analysis of electron beam [2]. Axial magnetic field B is used to balance the space charge forces in the beam. The quantities that are assumed to be constant in the radial direction is Charge density, ρ, and the axial velocity, u_0, while it varies in axial direction,

S. K. Ghosh et al. (Eds.): VEDA 2024, SPPHY 424, pp. 59–69, 2025.
https://doi.org/10.1007/978-981-96-5174-0_9

Motion of electrons can be depicted from the force equation,

$$\frac{d^2r}{dt^2} - r\left(\frac{d\theta}{dt}\right)^2 = -\eta(E_r + Br\frac{d\theta}{dt}) \tag{1}$$

where, $\frac{d\theta}{dt}$ is the term estimated by Busch's theorem, which states that charged particle beams attains a kinetic angular momentum as it exits the axial field solenoid field to conserve the canonical angular momentum, established when the beam is inside the solenoid field [2].

For the beam spread analysis the electric field is calculated by using Gauss's law. At the radial boundaries of beam where radius = b, the electric field is can be written as $E_r(b)$,

$$E_r(b) = -\frac{I}{2\pi\ b\ u_0\varepsilon_0} \tag{2}$$

For calculating the motion of electrons on the outer boundary of the beam is obtained by combining Eq. (1) and (2):

$$\frac{d^2b}{dt^2} + b\omega_l^2\left[1 - \left(\frac{B_c}{B}\frac{b_c^2}{b^2}\right)\right] - \frac{\eta I}{2\pi\ bu_0\varepsilon_0} = 0 \tag{3}$$

where, $\omega_L = \eta B/2$ term represents the Larmor frequency, being equal to 0.5 times of the cyclotron frequency. Hence, we obtain the beam equation. For simplification we define,

$$a = \frac{1}{B}\left(\frac{2I}{2\pi\ bu_0\varepsilon_0}\right)^{1/2} \tag{4}$$

Now, the above constant "a" is the equilibrium radius for a beam subjected to Brillouin field level. The magnetic force increases as magnetic field level B rises, and the equilibrium radius being inversely proportional to the field reduces. Equilibrium radii and the space charge force is large for higher beam current and higher charge density in the the beam. As beam expands due to increase in velocity, the space charge forces decreases, in turn decreasing the value of "a". Using the above relation, beam equation modifies as,

$$\frac{1}{a}\frac{d^2b}{dt^2} + \omega_l^2\left[\frac{b}{a}\left[1 - \left(\frac{B_c}{B}\frac{b_c^2}{b^2}\right)\right] - \frac{a}{b}\right] = 0 \tag{5}$$

The radial force on the electrons is related to the radial acceleration on the electrons which is given by the first term. The overall radial force becomes zero, when the coefficient value in square brackets reduces to zero. (This occurs when the space charge and the centrifugal forces which cause electron beam to grow or diverge balances the focusing forces generated by the magnetic field to constrict or focus the beam) When the force F_r in radial direction becomes zero, electron beam attains equilibrium and does not further grows or compresses. If this situation can be preserved, then beam diameter will remain uniform and a cylindrical beam will be established. For estimating this value of radius at which the radial force is not present (b_e), the coefficient of the second term in Eq. (5) would need to be equated to zero,

$$\frac{b_e}{a}\left[1-\left(\frac{B_c}{B}\frac{b_c^2}{b_e^2}\right)^2\right]-\frac{a}{b_e}=0 \tag{6}$$

This can further be made further simple by evaluating the interdependency between Magnetic field B and Magnetic field at Cathode B_c. This value of B can be written in terms of magnetic flux density at cathode and the Brillouin flux density B_B. The radial acceleration will reduce to zero at equilibrium radius and thus, we can write

$$\eta E_r = -\eta B_B a\frac{d\theta_B}{dt} + a\frac{d^2\theta_B}{dt^2} \tag{7}$$

where, $d\theta_B/dt$ is the value of $d\theta/dt$ for B_B and for immersed flow it can be written down as,

$$\eta E_r = -\eta B\, b_e\frac{d\theta}{dt} + b_e\frac{d^2\theta}{dt^2} \tag{8}$$

Beam radii is same for Brillouin or immersed flow and Electric field E_r is also same, so the Eqs. (7) and (8) can be combined to find

$$B^2 = B_B^2 + B_c^2\frac{A_c^2}{A_e^2} \tag{9}$$

where, Ac $\propto b_c^2$ is cross-sectional area of the cathode surface and $A_e \propto b_e^2$ is the cross-sectional area of the beam at equilibrium radius. The fraction A_c/A_e is called area compression ratio of the electron gun. Since A_c/A_e is of the order of 25 or more. It is often easy to reference the actual flux density, B, to B_B, by the confinement factor m, which is defined by

$$B = mB_B \tag{10}$$

Now, the equilibrium beam radii can be written as:

$$\frac{b_e}{a} = m \tag{11}$$

Hence, for immersed flow, the ratio of the equilibrium beam radii to the Brillouin beam radii is the confinement factor. We can obtain the solution for beam equation by writing beam radius in incremental form,

$$b = b_e(1+\delta) \tag{12}$$

where $\delta \ll 1$ and neglecting the higher-order term, and equation for finding beam radius can be written as

$$\frac{d^2\delta}{dt^2} = \omega_s^2\delta = 0 \tag{13}$$

$$\omega_s = \sqrt{2}\left(2 - \frac{1}{m^2}\right)^{1/2} \tag{14}$$

Solution to above equation is of form

$$\delta = A \sin \omega_s t + B \cos\omega_s t \tag{15}$$

Since, $b = b_e(1 + \delta)$ ant $t = z/u_0$,

$$b = b_e\left(1 + A \sin \frac{\omega_s}{u_0} z + B \cos \frac{\omega_s}{u_0} z\right) \tag{16}$$

Or

$$b = b_e\left[\left(1 + C \sin(\frac{\omega_s}{u_0} z + \Phi\right)\right] \tag{17}$$

It indicates that the beam radius oscillates (scallops) about the equilibrium radii b_e with a frequency ω_s. And period, of the scallop is obtained by the wavelength of scallop that is given by,

$$\lambda_s = \frac{u_0}{f_s} = \frac{0.030 \times 10^{-3}}{B}\left(\frac{V}{\left(2 - \frac{1}{m^2}\right)}\right)^{1/2} \text{m} \tag{18}$$

Values of constant C and phase φ can be determined by exploring the conditions at the entrance of the beam at $z = 0$ into the main magnetic focusing field. Brillouin field for the present S-Band configuration comes out to be ~ 513 Gauss, and for confinement factor $m = 1.559$, B field is kept at ~ 800 G in the input, buncher cavities and Penultimate cavities [3]. Field in the output cavity is slightly higher to keep the current modulated beam focused. Figure 1, shows the desired magnetic field profile derived from the beam dynamics simulations.

2 Electromagnetic Design

Analytical as well as Parametric studies using FEM simulations were carried out for the geometry of the main and the auxillary coils for field enhancement in output cavity region. Electromagnetic design for these magnets was done using low frequency Magnetostatics solver of Opera TOSCA and CST Microwave studio [4]. Table 1 lists down the main functional requirement of magnet derived from beam optics simulations. Figure 1 shows the cut view section of 3D Electromagnetic model of magnet assembly integrated with Klystron tube.

Table 1 lists down the major specifications of the magnet assembly including good field region and minimum clear aperture for facilitating klystron tube assembly with the magnet.

Main coil is designed to provide the solenoidal field for focusing, while auxillary & additional coils are placed in output cavity region to enhance the field in output cavity

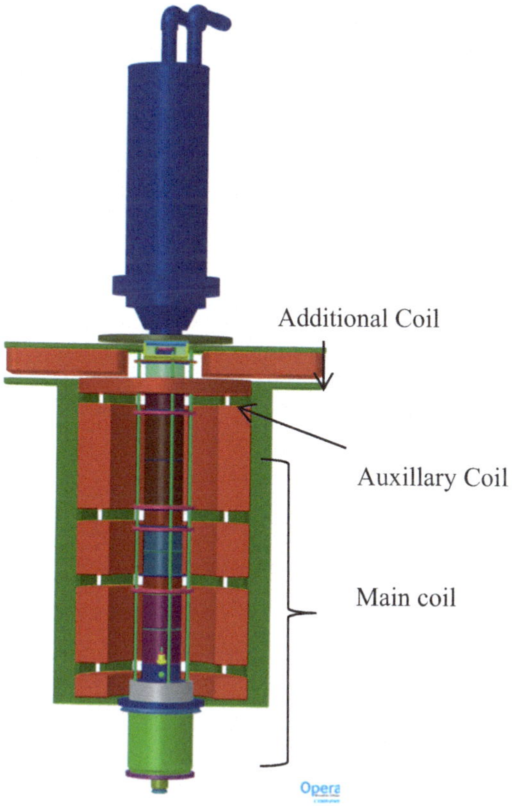

Fig. 1. Magnet 3D electromagnetic model integrated with 3D S-Band klystron tube model

Table 1. Magnet specifications

Sr. No.	Parameter	Specified
1	Magnetic field in output cavity Magnetic field in cavity 1,2,3,4	0.11 T 0.08 T
2	Field homogeneity in GFR	< 2.5%
3	Good field region	Ø 20 × 700 mm
4	Minimum clear aperture	170 mm
5	Mode of heat dissipation	Water-cooling

region. Multiple coil design has been selected to provide tunability and the space between the coils is used for accessing the cavity tuners.

Figure 2 shows the desired axial magnetic field profile given as functional input for 2D analysis of klystron tube which is required for focusing of the current modulated electron beam in beam wave interaction region. Figure 3 shows the axial magnetic field profile along the tube length obtained in the electromagnetic design. Inherent gun pole piece and collector pole piece are marked to show the field levels at the entrance and departure out of the beam-wave interaction region (RF section). Designed magnetic field profile matches the beam optics requirements. Figure 4 shows the cut plane view of 3D electromagnetic model of the magnet assembly. The peak field levels in the output cavity region can be adjusted the input current fed to auxillary and additional coil.

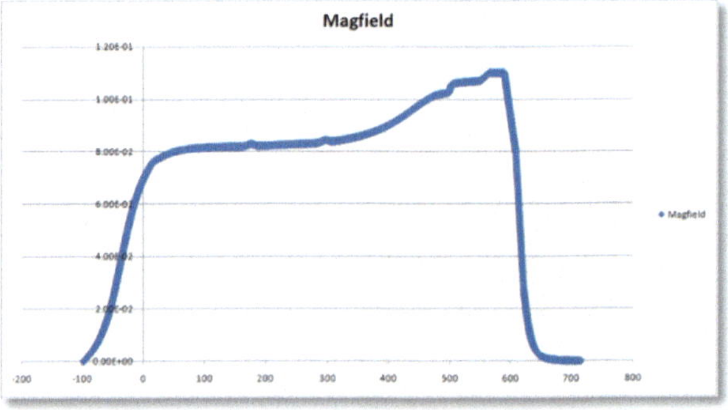

Fig. 2. Desired axial magnetic field profile from beam dynamics simulations where firs cavity location has been taken as "0" location

Radial field distribution

Radial field uniformity in the output cavity region is observed to be 1.84% in the design simulations owing to non axi-symmetric coil arrangement, whereas in input, bunching and penultimate cavity it is 0.04%. The radial field non-uniformity will introduce asymmetry in beam profile in the output cavity. Multiphysics coupled design studies shall be performed to evaluate its effect on tube's RF performance and collector heat flux distribution. Figure 5 shows the zonal field plot of magnetic flux density at axial locations corresponding to center of the cavity gap of cavity 1, 2, 3 and 4. Figure 6 shows the zonal field plot of magnetic flux density in the output cavity.

Fringe fields

Fringe magnetic field levels in collector region and in the peripheral region outside the electromagnetic yoke is essential to evaluate the requirement of shielding structures in the collector region and shielding of stepper motors required for tuning of the RF cavities [5]. Figure 7 the fringe field levels outside the electromagnet yoke.

3 Particle Trajectory Studies

Particle trajectory studies were performed to evaluate beam size using a parallel beam at the entrance of magnet assembly in the post processor after field evaluation for an accelerating voltage of 145 KV, and current of 90 A. For the purpose of particle trajectory calculations in simplified manner it is longitudinal velocity, u_0 and the charge distribution (density), ρ, are assumed to be uniform in the radial direction.

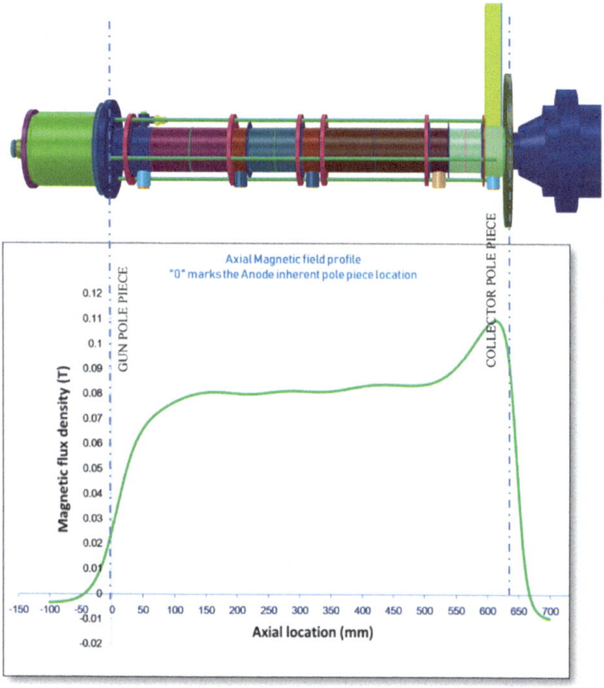

Fig. 3. Axial Magnetic field profile obtained with electromagnetic simulation studies. Gun pole piece is taken as the "0" axial location for EM simulations.

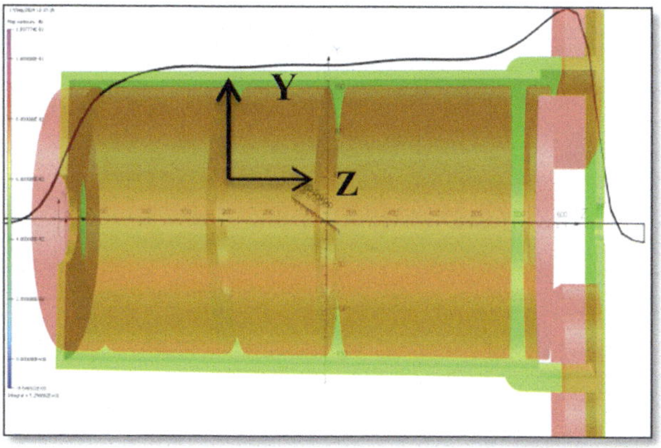

Fig. 4. Minimum field location optimization for different bucking coil configuration

Figure 8 shows the particle trajectory in X and Y transverse dimensions for axial magnetic field profile without additional coils and Fig. 9 shows the particle trajectory in X and Y transverse dimensions for the magnet assembly with additional coils powered on. The non-axisymmetric coil arrangement for the latter case in the output cavity region

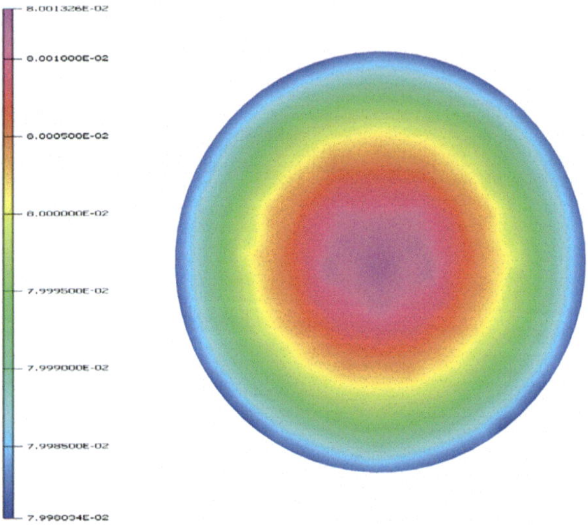

Fig. 5. Radial field distribution in Cavity gap 1, 2, 3 & 4

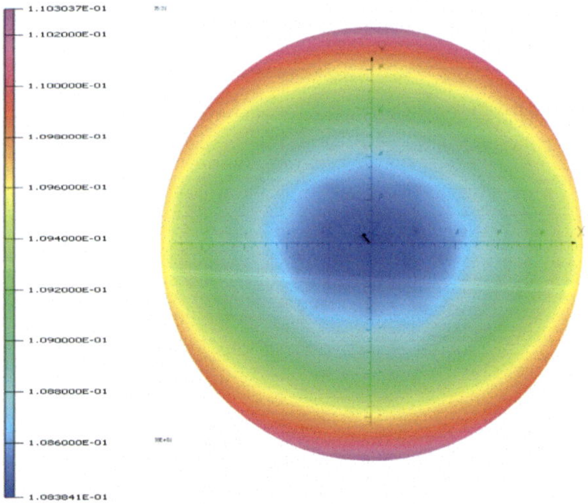

Fig. 6: Radial magnetic field distribution in output cavity gap

will introduce asymmetry in the beam, which is clearly shown in the particle trajectory beam profile mismatch in two transverse dimensions.

4 Engineering Design

Engineering design of the coils ascertain that dipole components are minimized due to in and out current conductor length so as not to introduce any steering field in the solenoidal field. Magnetic steel used as electromagnet yoke is designed to be in discreet bar forms

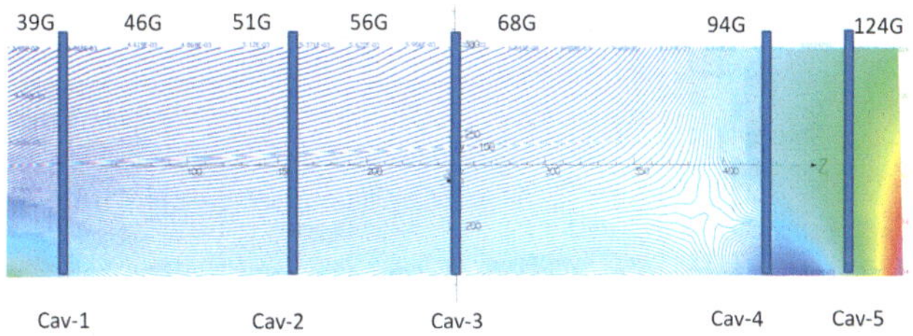

Fig. 7. Fringe field levels at a radial distance of 10 mm from the outside diameter of the electromagnet yoke.

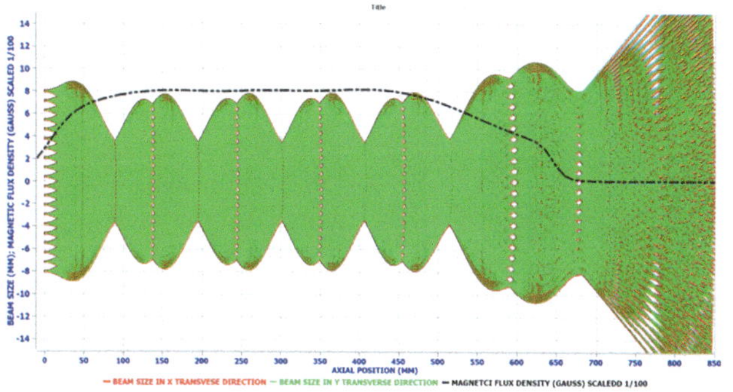

Fig. 8. Particle trajectory in the beam-wave interaction region with only main and auxiliary coils powered on

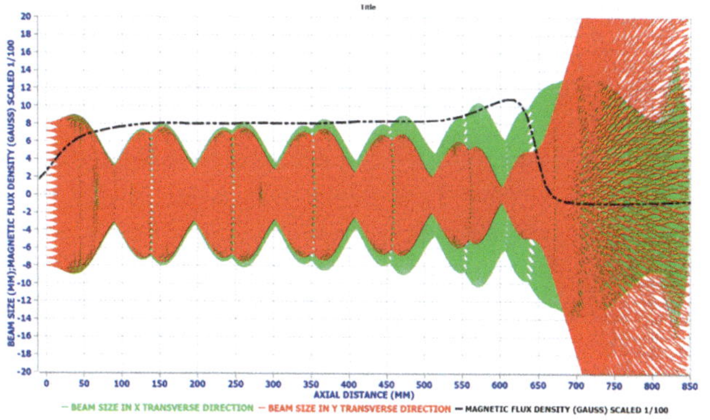

Fig. 9. Particle trajectory in the beam-wave interaction region with main, auxiliary coils and additional coils powered on

so as to facilitate easy assembly and disassembly of the magnet and additionally provide space to insert mechanical assembly for tuning of the RF cavity during conditioning of the klystron tube. Thermal and hydraulic calculations of the coil have been calculated. Table 2 summarizes the design factors estimated form electromagnetic design of the multi-coil arrangement.

Table 2. Summary of electromagnetic coil design

Sr. No.	Parameter	Designed
1	Topology	Double Pan-cake
2	Design-current	Main coil: 130 A Auxillary coil: 300 A Additional coil: 500 A
3	Power dissipation / coil	1.25 KW in main coil 0.66 KW in Auxillary 2.8 KW in additional coils
4	Conductor cross-section	8 x 8(mm) with Ø5mm coolant channel
5	Coolant material	DM-Water
6	Target coolant flow velocity	1.5 m/s (approximately 4.5 Lpm)
7	Pressure drop	1.5 bar
9	Total no. of turns	276 in main coil 24 in Auxillary coil 80 in Additional coil
10	Max. temperature rise	6^0 C
11	Power supply regulation topology	Constant current source
12	Maximum current	500 A (Additional coil)
13	Power supply stability	Long term (8 h) < 100 ppm

5 Conclusion

The FEM analysis of the magnet is completed. Thermal and hydraulic design of the coil is completed and prototype coil is manufactured and tested for thermal and hydraulic validation. Final engineering details of the magnets have been worked out which includes water-cooled double pancake coils, Low conductivity water system, high permeability soft iron core & a precision power supply. Development of the above magnet is under progress at Electromagnetic Applications & Instrumentation Division BARC. Multi-physics simulations to study the coupled effect of Electrostatic field in gun region, Magnetostatic fields in gun and beam wave interaction region and RF perturbations in the cavity region is under progress to evaluate the optimized field profile for high efficiency of the Klystron tube performance.

Acknowledgement. The authors would like to thank Dr. S.K. Ghosh Chief Scientist & Group Head Vacuum Electron Device Group, Dr. Deependra Kant Senior Principal Scientist Vacuum Electron Device Group & Dr. Ayan Kumar Bandyopadhyay Senior Principal Scientist & Group Head, High-Frequency Devices and System Group (HFDSG), CSIR-Central Electronics Engineering Research for their reference inputs and technical interaction on experiment results on the developed S-Band Klystron tube at CSIR-CEERI.

References

1. Joshi, L.M., et al.: Design and development of a 6 MW peak, 24 kW average power S-band Klystron. In: IEEE Internal Conference on Vacuum Electronics (2011)
2. Gilmour, A.S.: Microwave and MM wave vacuum electron devices: inductive output tubes. Klystrons, traveling wave tubes, magnetrons, crossed-field amplifiers, and gyrotrons. Chapter 7
3. Bandyopadhyay, A.K., et al.: Design and development aspects of Klystrons. In: IEEE Internal Conference on Vacuum Electronics (2021)
4. OPERA 3D, reference guide and user Manual
5. Singh, K., Itteera, J., et al.: Design and development of permanent magnet based focusing lens for J-band klystron. In: 2014 International Symposium on Discharges and Electrical Insulation in Vacuum (ISDEIV)

Synthesis and Characterization of Scandia Doped Nano Tungsten Powders

U. Devarajan, V. Deepa, K. Santhosh Kumar(✉), M. Vijay Kumar, P. Sidharthan, and S. K. Datta

Microwave Tube Research and Development Centre, DRDO, Ministry of Defence, Bengaluru 560013, India
ksantosh.mtrdc@gov.in

Abstract. Scandate Cathodes have the potential to replace conventional M-Type dispenser cathodes due to their improved emission characteristics such as high emission current density (Top Layer-Laser Ablation Deposition: 1–400 A/cm^2) [1], lower operational temperatures (750–950 °C) [2], and longer lifetime (10,000 h). In the present work, Scandate doped Tungsten nano powders are synthesized by Liquid-Liquid method using Sol-Gel chemical route process [3]. Synthesized powders are characterized for particle size, shape, crystal structure and elemental composition. Powder X-ray diffraction (XRD) is employed to analyze the crystal structure, phase formation and average particle size. Transmission Electron Microscope (TEM) is used to analyse nano particle structure and crystalline size. The powders are pressed into 7 mm diameter pellets using hydraulic press followed by sintering. Scanning Electron Microscope (SEM) is used to capture surface morphological images like pore size, pore distribution and distribution of Scandium oxide on Tungsten particles. Uniform distribution of the ingredients on the pellet surface is confirmed by Energy Dispersive X-ray analysis (EDAX). These analysis helped to understand the properties of synthesized powders and to optimize the synthesis process.

Keywords: Dispenser cathodes · Scandate · Sol-Gel · Nano particles

1 Introduction

Thermionic Dispenser Cathodes are source of electron beam for vacuum electron devices (VED's) such as Travelling Wave Tubes (TWT's) , Linear accelerators, Magnetrons and Backward Wave Oscillators, etc. M-type cathodes with maximum zero field current density of 15 A/cm^2 at 1050 °C are commonly used in these VED's. However, for the devices being developed at higher frequency range (THz) which needs compact size cathodes with higher zero field current densities (>100 A/cm^2). Scandate cathodes are attracted significantly since more than five decades due to their higher electron emission at low temperature. Scandate cathodes are developed by adding the scandium oxide in the impregnation mixture [4, 5] or by top layer method [1, 6], or by doping with tungsten powder [7]. Scandium oxide doped Tungsten powders are prepared by two

S. K. Ghosh et al. (Eds.): VEDA 2024, SPPHY 424, pp. 70–76, 2025.
https://doi.org/10.1007/978-981-96-5174-0_10

methods. One by is by liquid-solid (L-S) [7–9] and other by liquid-liquid (L-L) method [10, 11]. From L-S method, Wang et al. [2] repeatedly achieved space charge-limited (SCL) current densities of over 30 A/cm^2 at 850°C$_b$. Wang et al. [12] also reported that the Sc-in-W cathode developed by L-L method delivered SCL current density of 52 A/cm^2 at 850 °C$_b$. The resulting uniform active layer above the underlying pellet microstructure from L-L method resulted in its excellent observed emitting properties and that small, near spherical W grains in the pellet contributed to the formation of this layer. The uniform, equiaxed sub-micrometer W grains coated with nano meter size particles of Scandium/Barium/Oxygen (Sc/Ba/O) represent an optimal microstructure for high current density thermionic emission [12, 13]. Based on all above mentioned reports and the extensive literature survey report [14], we have chosen the L-L method to produces Scandium doped W powders along with Barium Calcium aluminates to get above mentioned microstructure.

2 Experimental

The experimental part covers the synthesis of Nanoparticles using Sol-Gel method (L-L) and the techniques used for powder characterization. It also covers the pellet fabrication and its characterization.

2.1 Synthesis of Nanoparticles

The Scandia doped Tungsten nano powders are prepared by Liquid-Liquid method using Sol-gel technique [3]. The aqueous solutions of all the ingredients (Sc, Ba, Calcium and Aluminum) are taken in their nitrate forms and Tungsten is in ammonium meta tungstate form. The work flow of sol-gel process is as follows (Fig. 1).

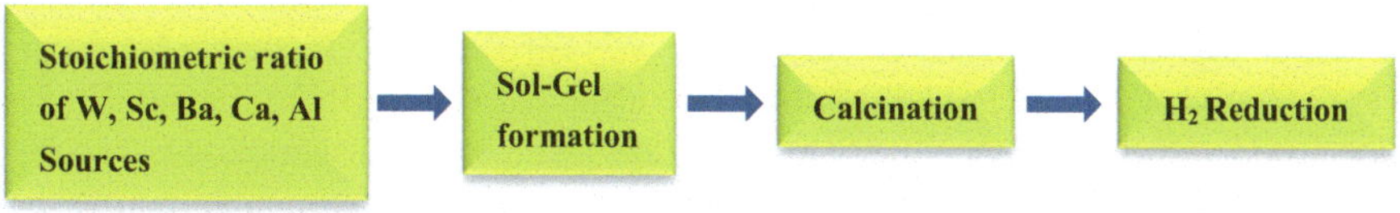

Fig. 1. Schematic flow chart of So-Gel synthesis process for making Scandate powders.

Reduction of oxide powders resulted in composite oxide doped tungsten powders which are subjected to low energy ball milling. Small portion of the powders are used for characterization purpose and remaining are used for pellet fabrication.

2.2 Characterizations Techniques

2.2.1 Powder X-Ray Diffraction (XRD)

The XRD pattern (Brucker's) is recorded from 10° to 90° range (2θ). The phase formation is identified by comparing experimental data with theoretical Crystallographic Data Base

(ICDB) values. The lattice parameter is calculated and the average crystalline size is calculated using Debye-sherrer formula's

$$L = \frac{(0 \cdot 9\lambda)}{\beta \cos \theta} \tag{1}$$

where, L = Average crystalline size (nm); λ = wavelength of X-ray (1.547 A°).
β = Full width half maximum of high intensity peaks; θ = Angle of Diffraction.

2.2.2 Transmission Electron Microscopy (HR-TEM)

JEOL 2010F TEM is used for measuring particle size as well as its morphology. Dark-field imaging is performed using EM 24140 detector.

2.2.3 Scanning Electron Microscope (SEM)

SEM (Carl Zeiss, EVO-18) coupled with Oxford Inca Energy Dispersive Analysis of X-rays (EDAX) is used to study the microstructure and elemental composition of both synthesized powder and pressed pellets.

2.2.4 Fabrication of Pellets

Die and Punch is fabricated using SS 304 for pressing the pellets (pellet diameter = 7.7 mm) using hydraulic pressing machine (1–10 Ton). The pellets were sintered at 1200 °C in vacuum atmosphere using molybdenum heater baskets.

3 Results and Discussion

3.1 Structural Studies

Figure 2 (a) shows X-Ray diffraction pattern of the oxide form of WBaScAl compound. The major high intense peaks correspond to W (1 1 0), W (2 0 0) W (2 1 1) and W (2 2 0) planes which crystallises as Body Centered Cubic (BCC) having Fm3-m space group with lattice parameters a = 8.375A°. These similar structure is reported previously [15, 16]. The other less intese peaks represents mixed double perovskites ($Ba_3Sc_4O_9$, $Ba_2Sc_2O_5$, Ba_2ScAlO_5, $Ba_2Sc_2O_5$) crystalize with tetragonal (space group: I4/m) structure [17]. The calculated average crystalline size changes drastically in W planes than other complex structures. The average crystalline size of Ba- Ca-Aluminates found to be 8–22 nm as per calculation and that of W particles were in sub-micron level. Figure 2(b).

3.2 Transmission Internal Structures (HR-TEM Images)

Figure 3 (a-e) depicts TEM images of synthesized powders at various resolutions (20, 50, 100 nm) and Fig. 3f shows the electron diffraction pattern. It is found that W powders are homogeneously coated with Sc_2O_3. The narrow spherical type particles size varies from 20 to 100 nm. The bright-field (BF), dark-field (DF) and phase-contrast imaging are observed from Fig. 3 (f).

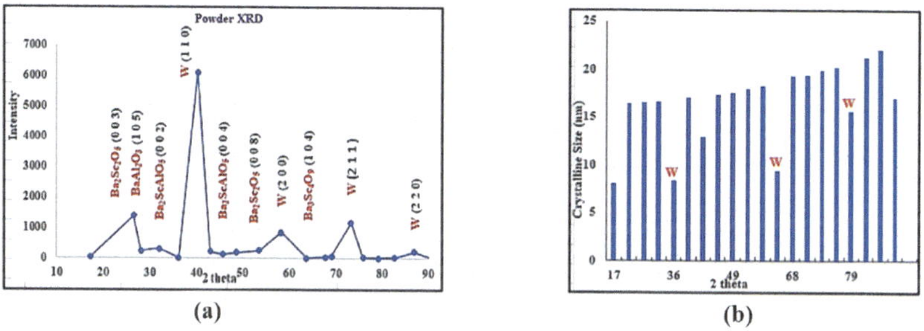

Fig. 2. (**a**) Powder XRD Pattern of sample. (**b**) Crystalline or particle size is in nm range in the bar chart.

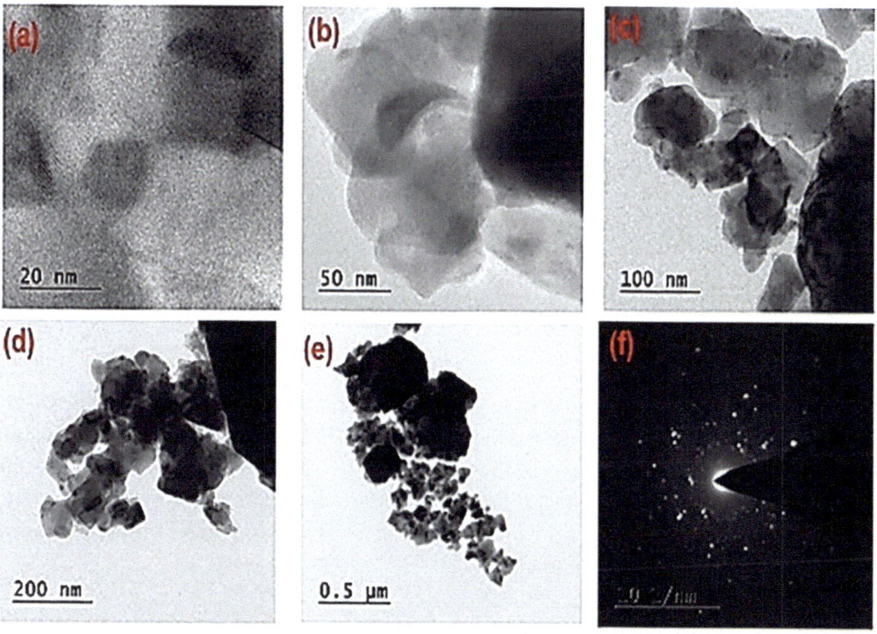

Fig. 3. (a-e) HR-TEM images of Scandia doped W particles (f) Electron diffraction of sample.

3.3 Surface Topography (HR-SEM Images)

Pellets are fabricated from synthesized powders by die-punch pressing process and the green pellet sintering at 1200 ^{O}C for 30 min in vacuum atmosphere. The sintered pellet's surface micrographs are taken using SEM which are as shown in Fig. 4 (a-d).

It is observed that there is a homogeneous coating of nano sized Scandium oxide and Barium Calcium Aluminate particles (Fig. 4 a, b) over the near spherical, sub-micron sized W particles. Some large W particles (~ 1 micro meter dia.) are seen without any attached scandium particles. But on further magnification of these W particles, a rectangular shaped grains with 200 to 400 nm in size, comprising of $Sc_2W_3O_{12}$ as reported by Wang et al. [9] are seen. As the sintering of the pellet was carried out at

Fig. 4. SEM images of the pellet surface (a-d)

lesser temperature, larger inter particle gaps are seen between the W particles (Fig. 4 b). This gap will be reduced by sintering the pellet at 1450 ^{O}C. The distribution of elements (in weight %) on the four regions of the pellet (Fig. 5 a, f) are studied using EDAX and the results are given in Table 1. All the elements are distributed uniformly in the four regions of pellet except scandium which is varying from 1.33 to 2.91%. Figure 5 b shows the EDAX spectra of the pellet matrix.

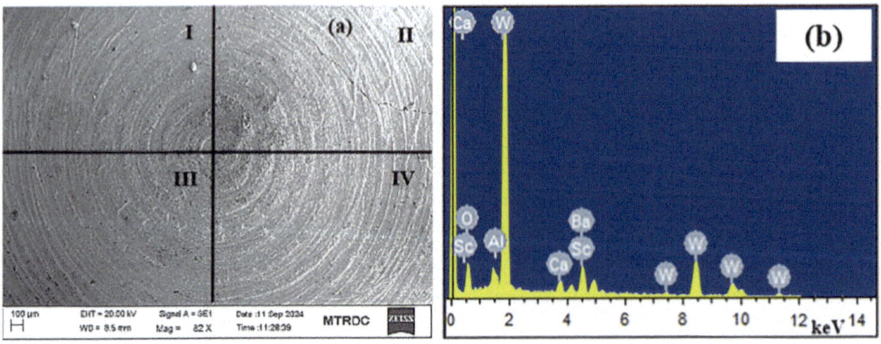

Fig. 5. (a) SEM image of the pellet surface showing four regions of elemental analysis, (b) Elemental EDAX spectra of the pellet matrix.

Table 1. Elemental weight % on four regions of the pellet

Element	Weight (%) in Four regions on pellet surface				Average Weight (%)
	I	II	III	IV	
Oxygen	10.43	10.50	9.28	9.45	9.91
Aluminum	1.59	1.56	1.65	1.34	1.54
Calcium	1.45	1.46	1.73	1.50	1.54
Scandium	1.33	2.14	1.67	2.91	2.01
Barium	13.36	12.61	13.14	13.06	13.04
Tungsten	71.85	71.73	72.52	71.74	71.96
Total	100.00	100.00	100.00	100.00	100.00

Summary: Nano sized Scandate doped W powders along with Ba-Ca-Aluminates are synthesized by sol-gel method. Scandium oxide particles are found adhering to the W particles more uniformly along with other ingredients when viewed in SEM. Cubic phase of W and others tetragonal structures are confirmed from powder XRD. The average particle size of Ba- Ca-Aluminates found to be 8–22 nm and that of W particles were in sub-micron level. The pellets has to be sintered at higher temperature to get further compaction. Future work is to develop cathodes form this powder and to check its emission performance.

Acknowledgement. Authors are thankful to Centre Head, MTRDC and DRDO for encouraging to carry to this research work.

References

1. Gartner, G., Geittner, P., Lydtin, H., Ritz, A.: Emission properties of top-layer Scandate cathodes prepared by LAD. Appl. Surf. Sci. **111**, 11–17 (1997)
2. Wang, Y., Wang, J., Liu, W., Zhang, K., Li, J.: Development of high current-density cathodes with Scandia-doped tungsten powders. IEEE Trans. Electron Devices **54**(5), 1061–1070, May (2007)
3. Wang, J., Liu, W., Zhou, M., Wang, Y., Li, H., Zuo, T.: Method of Manufacturing a pressed Scandate Dispenser Cathode. US 2008/0025864 A1 United States Patent Application, Irvine CA 92618 (US), pp. 1–5, January (2008)
4. Koppius, O.G.: Impregnants for dispenser cathodes. U.S. Patent 3719856, Mar 6, (1973)
5. van Stratum, A.J.A., van Os, J.G., Blatter, J.R., Zalm, P.: Barium-Aluminum-Scandate dispenser cathodes. U.S. Patent 4007393, Feb, 8, (1977)
6. Hasker, J., Van Esdonk, J., Crombeen, J.E.: Properties and manufacture of top-layer scandate cathodes. Appl. Surf. **26**(2), 173–195 (1986)
7. Wang, J., Wang, Y., Tao, S., Li, H., Yang, J., Zhou, M.: Scandia doped tungsten bodies for Sc- type cathodes. Appl. Surf. Sci. **215**(1–4), 38–48, Jun. (2003)
8. Tao, S., et al.: Microstructure and emission properties of Sc- Type dispenser cathodes with scandia doped Tungeston matrix. Vac. Electron **247**(6), 15–20 (2003)

9. Barik, R.K., Singh, A.K., Shukla, S.K., Singh, T.P., Raju, R.S., Gun-Sik Park.: Development of nanoparticle-based high current density cathode for THz devices application. IEEE Trans. on Electron Devices **63**(4), 1715–1721, Apr. (2016)
10. Wang, J., Liu, W., Li, L., Wang, Y., Wang, Y., Zhou, M.: A study of Scandia-doped pressed cathodes. IEEE Trans. Electron Devices **56**(5), 799–804, May. (2009)
11. Zao, J., et al.: High current density and long life nano composite scandate dispenser cathode fabrication. IEEE Trans. Electron Devices **58**(4), 1221–1228 (2011)
12. Wang, J., Li, L., Liu, W., Wang, Y., Wang, Y., Zhou, M.: Sc_2O_3 –W matrix impregnated cathode with spherical grains. J. Phy. Chem. Solids **69**(8), 2103–2108, Aug. (2008)
13. Cui, Y., Wang, J., Liu, W., Wang, Y., Zhou, M.: Characterization of Scandia doped pressed cathode fabricated by spray drying method. Appl. Surf. Sci. **258**(1), 327–332, Oct. (2011)
14. Seif, M.N., Zhou, Q., Liu, X., Balk, T.J., Beck, M.J.: A review of Sc-containing scandate thermionic cathodes. *Cond-mat.mtrl-SCI* (2022)
15. Liu, W., Wang, J., Wang, Y., Cui, Y., Zhang, X., Zhou, M.: XRD Analysis of Scandate Cathodes. IEEE 978-1-4244-8660-1/11 (2011)
16. Zhang, Z., Zhou, Y., Gong, T., Xu, L., Zhou, X., Fu, H.: Formation of Ba3Sc2WO9 on the Ba2ScAlO5 Therminoic Cathode Surface. In: 24th International Vacuum Electronics Conference (IVEC) | 978-1-6654-7321-7/23/$31.00 ©2023 IEEE (2023)
17. Day, B.E., et al.: Structures of ordered tungsten- or molybdenum-containing quaternary perovskite oxides. J. Solid State Chem. **185**, 107–116 (2012)

Simulation Studies of Magnetically Insulated Line Oscillator

Shivabhagya M. S.(✉), D. Senthil Kumar, and S. K. Datta

Microwave Tube Research and Development Centre, Bengaluru, India
Shivabhagya24@gmail.com, Senthil.mtrdc@gov.in

Abstract. MILO development has garnered increased attention due to its ability to generate high power microwaves without relying on an external magnetic field. Instead, it utilizes self-generated magnetic fields. The oscillator's anode which is part of slow wave structures, composed of thin conductive vanes creating cavities between them. The anode and cathode are separated by a gap, where cross field magnetic fields serve to insulate and confine the electron flow. This paper presents modelling and simulations carried out using CST software, exploring the effects of voltage and electrode gap in S-band MILO structure. The proposed design is simulated for maximum output power of 2.3 GW with beam voltage of −600 kV, and beam current of 45 kA, with specified gap of 14 mm in S Band frequency. And also, the simulation was carried out for the variation in voltage from −200 kV to −600 kV and electrode switch gap parametric from 5 mm to 15 mm.

Keywords: High power microwaves · Electrode switch gap · Magnetically insulated line oscillator

1 Introduction

In recent years, there has been a notable resurgence in focused research and development (R&D) in the domain of high power microwaves (HPM) for both military and civilian uses. Primary goal of these efforts is to disrupt or disable electronic equipment intentionally, without causing harm to critical infrastructure or personnel [1]. The first experimental MILO was reported using 195 kV applied voltage and beam current of 47 kA. This setup achieved an efficiency of 10.8%, delivering 2.1 GW of power at 50ns duty cycle. Numerous studies by various authors have since documented various types of MILOs, employing particle in cell (PIC) simulations with diverse simulation software [2, 3]. Another reported instance is an L band MILO, which demonstrated efficiency of 12.3% and generated RF output power of 2.1 GW at a 130 ns pulse width. This MILO operates at a 500 kV applied voltage and uses a 45 kA beam current. Notably, it features a reduced quantity of cavities, which effectively removes azimuthally symmetric

S. K. Ghosh et al. (Eds.): VEDA 2024, SPPHY 424, pp. 77–83, 2025.
https://doi.org/10.1007/978-981-96-5174-0_11

mode competition. Electrons are released from both sections, that is cathode surface and the end electrode surfaces, allowing the device to oscillate over a broad voltage range near the desired frequency [4, 5]. Many applications demand High Power Microwaves (HPM) sources, in consideration with compactness, total weight and form factor. The MILO holds an advantage over other devices because it operates without any externally magnetic systems. Another critical aspect in the development of High Power Microwaves (HPM) devices is the selection of pulsed power used to drive the source. Traditionally, Marx generators are preferred as better choice for MILO in laboratory applications [6, 7].

Publication of the peer research group has achieved the maximum power of 2.2 GW, with current of 42 kA, with applied voltage of 405 kV. Considering ideal material properties and parametric studies of MILO components, the present work is carried out for optimization. In MILOs, prime input power distributes mainly for two purposes, one for energizing generating the RF power and another for establishing the insulating magnetic field. As, voltage increases, there is a risk of electrical breakdown prior to the operation of the device. Present study aims to optimize the required voltage for the generation of maximum RF power from the designed structure [8]. In MILO, instead of external coils, it itself generates the magnetic field by the flow of electrons. This removes the need to synchronize the voltage applied with the magnetic field. Since the accelerating voltage and insulating magnetic field are produced by the same source, electrical breakdown cannot be experienced for the device to be operated at high voltages. [9]. The emission occurs with length of the cathode, when high voltage pulse is given. This implies that prior to microwave generation, electrons travel aligning to the cathode axis beneath the slow wave structure. As a result, the electron flow interacts with an electromagnetic wave, defined by the slow wave structure, and transfers energy from the electron beam to the wave [10].

2 Design and Analysis of MILO

Numerical simulations have been employed to explore the nonlinear behavior in a load constrained system focusing particularly on the power conversion efficiency. The MILO consists of six annular disks, spaced 2.15 cm apart and each 0.5 cm in thickness, which is referred as vanes. 4.5 cm and 4.9 cm are the inner and outer radius of the vanes. Final vane in the slow wave structure (SWS) also serves as the extractor, with an inner radius of 5.2 cm. This slightly larger radius ensures a proper match between the gap (electric field). The load region includes the cathode, which has a radius of 5.48 cm. In simulation, the cathode surface is defined as particle emission but in experiments, the cathodes surface within the emission region is defined with a material such as velvet, which has a much lower emission threshold. Another crucial design feature was the focusing electrode, which directed the electron beam, thereby improving the efficiency of the beam-wave interaction. Additionally, the focusing electrode helps in controlling the beam current density, which in turn minimizes the formation of anode plasma.

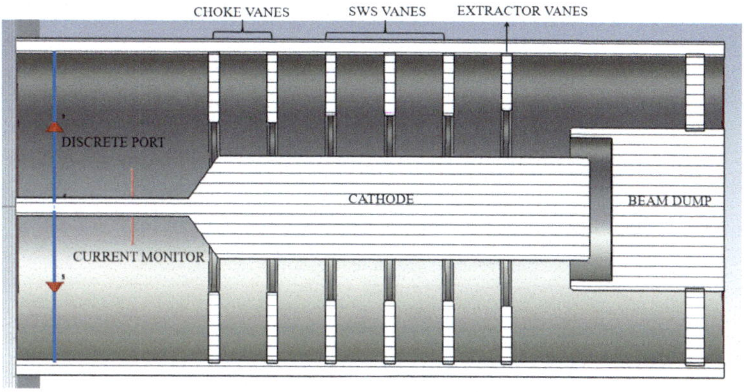

Fig. 1. Model of the MILO

Figure 1 depicts the complete model of the MILO. The discrete port is introduced and it is varied to optimize the results. Monitors positioned at input port measure the total injected current. Explosive electron emission occurs from the cylindrical section of the cathode. Most of the current flowing in the device is captured by beam dump, which produces an axial current that generates a radial magnetic field. The beam collector or the dump is grounded via four conducting rods placed at specific azimuthal locations, forming a one fourth wave structure that remains transparent to microwaves. Current and voltage measurements are used to determine the efficiency.

3 Simulation Results

Based on the injected voltage (ranging from −200 kV to −600 kV), the results were well aligned for voltages below −600 kV. However, in simulation it tended to overestimate the output power at higher applied voltages. At low voltages, the CST parametric sweep did not show any operational mode in the slow wave structure (SWS). This is assumed to be because, at lower electric fields, the axial beam velocity predicted by the simulation is too low to synchronize with an SWS mode. Voltage was applied at the input, ranging from −200 kV to −600 kV with a total time of 100 ns, with 5 ns rise and fall time.

Figure 2(a) and Figure 2(b) explains the output results for varied voltage and the current. The length of the electrode switch gap (IR) was treated as an independent variable. Adjusting IR influences the downstream diode current, as a result, the magnetic field in the cavity region is modified. The radius of cathode is varied such that the axial length (IR) should be lesser than the radial gap.

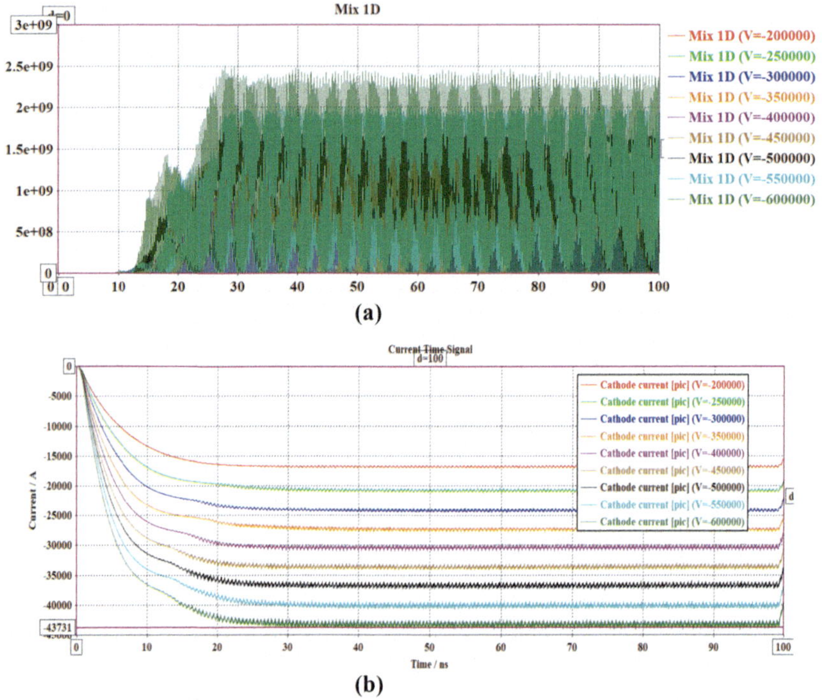

Fig. 2. Plot of output power and current for various input voltage

Figure 3a and b illustrates the output power and current for various IR values varying from 5 mm to 15 mm. In the simulations, IR = 14 mm performed significantly better than IR = 10 mm. However, future designs will take additional parameters into account to mitigate field enhancement in this region. A comparison of Tables 1 and 2 presents the simulated values for output power and current.

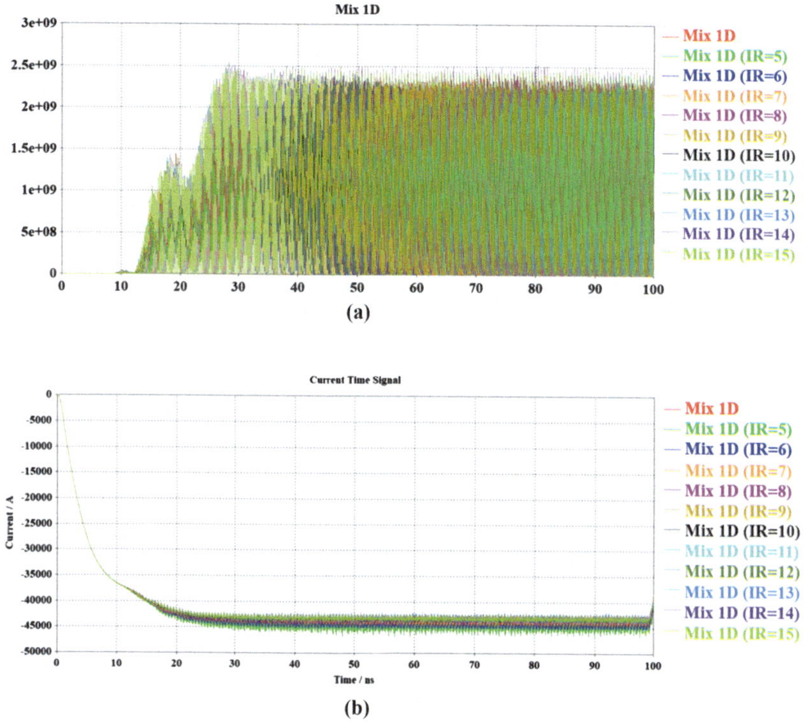

Fig. 3. Output power and current for various IR values ranging from 5 mm to 15 mm.

Table 1. Output results for various voltages.

Voltage (kV)	Power (MW)	Current (kA)	Efficiency (%)
200	43	16.9	1.27
250	231	20.9	4.42
300	500	24.3	6.85
350	828	27.7	8.54
400	1130	30.8	9.17
450	1141	34.1	7.43
500	1630	37.3	8.73
550	1970	40.7	8.80
600	2300	43.7	8.77

Table 2. Output results for various electrode switch gap

Electrode switch gap (mm)	Power (MW)	Current (kA)	Efficiency (%)
5	2000	46.1	7.23
6	2230	44.9	8.27
7	2260	44.6	8.44
8	2250	44.1	8.50
9	2250	44.1	8.50
10	2160	43.9	8.20
11	2290	43.9	8.69
12	2290	43.9	8.69
13	2240	43.8	8.52
14	2370	43.7	9.03
15	2320	43.7	8.84

4 Conclusion

In this study, we discuss the results of modifying the input voltage and electrode switch gap. The simulation data clearly indicates that the maximum output is achieved at a specific voltage, which enhances overall efficiency. Rather than modifying the downstream radius, varying the downstream cathodes axial length can control the current supplied to the downstream diode. Additionally, to enhance electron flow within the cavity region, changes can be applied at the launch point and upstream of the resonant cavities. Thus, it is observed that the output microwave frequency predominantly falls within the S band, with the output power reaching 2.3 GW with beam current of 45 kA at −600 kV.

References

1. Benford, J., Swegle, J.A., Schamiloglu, E.: High power microwaves, 3rd edn. CRC Press, Taylor & Francis Group, New York (NY) (2016)
2. Kim, D.H., Jung, H.C., Min, S.H., Shin, S.H., Park, G.S., Yim, D.W., Kim, C.H.: Experimental investigation of giga-watts magnetically insulated transmission line oscillator (MILO). In: Proc. IEEE Int. Vacuum Electron. Conf., pp. 561–562, Sep. (2006)
3. Kim, D.H., Jung. H.C., Min, S.H., Shin, S.H., Park, G.S.: Dynamics of mode competition in a gigawatt-class magnetically insulated line oscillator. Appl. Phys. Lett. **90**(12), 124103–1–124103–3, Mar. (2007)
4. Yanhua, G., Zhikai, F., Hu, H.: Simulation investigation of C-band magnetically insulated line oscillator. In: Proc. IEEE Interact. Symp. Microw., pp. 1339–1344, Aug. (2007)
5. Cousin, R., Larour, J., Gouard, P., Raymond, P.: Evidence of the 3pi/4 interaction mode in a compact magnetically insulated line oscillator process. J. Appl. Phys. **100**(8), 084512–1–084512–5, Oct. (2006)
6. McBride, R.D., et al.: A primer on pulsed power and linear transformer drivers for high energy density physics applications. IEEE Trans. Plasma Sci. **46**(11), 3928–3967, Nov. (2018)

7. Koval'chuk, B.M., et al.: Fast primary storage device utilizing a linear pulse transformer. Russian Phys. J. **40**(12), 1142–1153, Dec. (1997)
8. Wang, X., et al.: Experimental study on the frequency tunability of a magnetically insulated transmission line oscillator. 0018–9383. IEEE Trans. Electron Devices, © (2023)
9. Clark, M.C., Marder, B.M., Bacon, L.D.: Magnetically insulated transmission line oscillator. Appl. Phys. Lett. **52**(1), 78–80, Jan. (1988)
10. Calico, S.E., Clark, M.C., Lemke, R.W., Scott, M.C.: Experimental and theoretical investigations of a magnetically insulated line oscillator (MILO). Proc. SPIE **2557**, 50–59 (1995)

Design Study of Helical Slow Wave Structure for a High Frequency (K-band) Space TWT

Amitavo Roy Choudhury(✉), Pawan Pareek, Subhradeep Chakraborty, and S. K. Ghosh

VED Group, CSIR-Central Electronics Engineering Research Institute, Pilani 333031, India
arc@ceeri.res.in

Abstract. A helix TWT for K-band 40W has been designed using the conventional design approach for the slow wave structure (SWS). For meeting the required communication parameters such as Output power (> 40W), gain (> 45dB) and electronic efficiency (>12%), A helical slow wave structure (SWS) has been finalized using CST-MWS software and in house developed SUNRAY-1D software.

Keywords: Interaction structure · Gyro-TWT · Azimuthal and axial periodic structure · Dispersion characteristics · Simulation

1 Introduction

Helix-TWT (Fig. 1) is a widely used broad band power amplifier for ground, air borne and space application. TWTs used for space communications must have a long life, high efficiency, high reliability along with small size and weight [1, 2]. The Slow Wave Structure (SWS) is crucial for determining the strict requirements of space Traveling Wave Tubes (TWTs). A helical structure is commonly employed as the Slow Wave Structure in TWTs for satellite communication because of its simplicity, compact size, light weight, and low dispersion characteristic.

The design and development activities for helix TWTs have been conducted at CSIR-CEERI for over a decade. This paper presents the design philosophy [3] of helix SWS for K-band 40 W helix SWS for space TWTs.

2 Design Methodology

In this manuscript, the authors has goal to obtain a helix TWT operating at K-band having high value of linearity as well as efficiency. These design objective are the largest part of the crucial goals which are required to obtain using simulation and modeling procedure. These are essential properties of any vacuum amplifiers especially for our modern satellite communication systems. Thus in order to obtain these properties we require to have a rigorous helix design having different helix profiles.

S. K. Ghosh et al. (Eds.): VEDA 2024, SPPHY 424, pp. 84–89, 2025.
https://doi.org/10.1007/978-981-96-5174-0_12

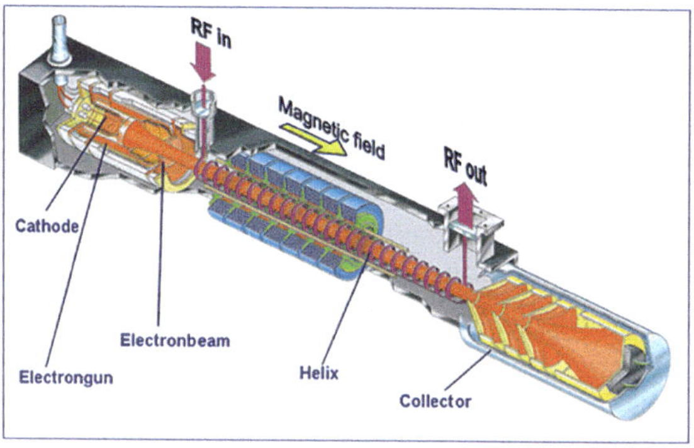

Fig. 1. Schematic of high efficiency Helix TWT.

To achieve constant output power across the required bandwidth, one should maintain perfect beam-wave synchronism in a linear beam tube. However, for satellite communication, a more linear condition is preferred. To meet this requirement, the design of the helix needed to be altered, which would reduce the bandwidth and efficiency of the traveling wave tube (TWT). Thus a careful trade-off between these conditions is therefore required to be maintained [4].

For the desired performance of the space TWT, we optimize the pitch profile of the helical structure methodically along the length of the slow wave structure. The final helix design was simulated and analyzed across the required frequency band.

In the helix TWT design process, optimizing the interaction process throughout the interaction region (SWS) involves establishing a growing wave, followed by electron bunching, and finally extracting power from the bunched electron beam. The tapered output end of the helix pitch profile enables the bunched electron beam to stay in the decelerating field for a longer axial distance. This is due to the reduced phase velocity, which in turn allows for more power extraction from the helix. As a result, this enhances the conversion efficiency of the helix Traveling Wave Tube (TWT).

In this design process, the design methodology was primarily based on the fundamental process happening within the interaction structure, these are: RF wave growth on the helix, electron bunching along the length of the tube and in return the power was extracted from the electron beam. The pith profile must be incorporate the above fundamental process as they happen along the interaction structure.

The helix pitch profile is designed using a single uniform section, with the following optimizations:

1. Pitch p0 is optimized for maximum small signal gain.
2. Pitch P1 is optimized for maximum bunching (maximum RF current).
3. Pitch p2 is optimized for maximum output power or efficiency.

Additionally, the input section length z0 along with pitch p0 is optimized for maximum RF current or electron bunched intensity. With a positive step taper, z1 along with pitch p1 is then adjusted to achieve maximum forward RF power at the end of the tube. Subsequently, considering a negative step in the helix profile (with pitch p2), the taper length z2 is then in tune for getting utmost to the fore power at the output coupler end.

Thus by adjusting the helix pitch values (p_0, p_1 & p_2), we can finally optimize the RF axial phase velocity along the length of the interaction structure to maintain the beam-wave synchronization. To optimize the RF axial phase velocity along the interaction structure and maintain beam-wave synchronization, we adjust the helix pitch values (p_0, p_1, and p_2). Using CST software, we calculate the axial propagation constant (β) and interaction impedance (K) for each helix pitch value. These values are then utilized in the LSM program to optimize the entire helical structure. Figure 3 shows that the CST simulated interaction impedance (K) and axial propagation constant (β) with the different helix pitch values (Fig. 2).

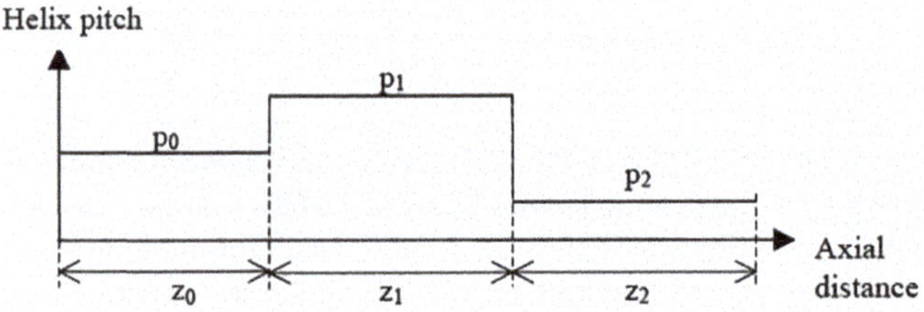

Fig. 2. Stepped helix pitch profile for optimized performance.

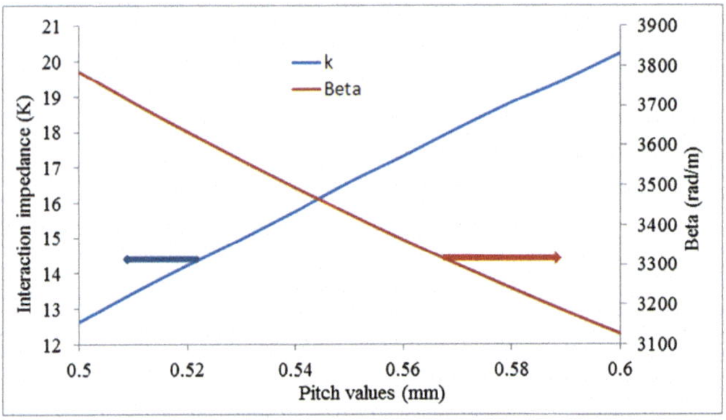

Fig. 3. Dependence of helix pitches on K and β.

The loss profile over dielectric support structure considered in the helix design was truly based on the previous design experience. In the helix design, loss profile varies along dielectric rod length and is maximum at sever end to reduce the signal reflections at the RF coupler port. In the design process, some parameters will remain constant, such as the DC beam power, beam perveance, and helix radius. As we already know the required output power, assuming an electronic efficiency of around 12% will allow us to determine the DC beam power. The beam voltage of 6.7 kV and beam current of 74 mA are determined based on the desired beam perveance. The helix radius (a) is determined by considering the approximation $(\beta_e)a \approx 1.25$ (where $\beta_e = \omega/u_0$ and $u_0 =$

electron beam velocity). Here 50% beam filling factor was considered and thus beam radius will be ~ 0.22 mm. Analytically initial helix pitch value was calculated by using $p = 2\pi a * u_0/C$.

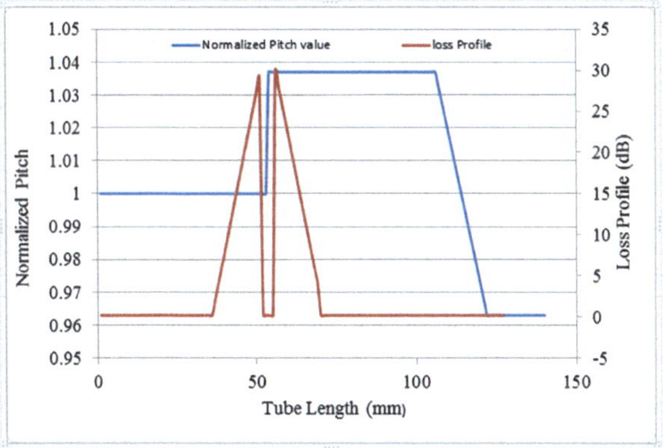

Fig. 4. Normalized pitch and loss profile of 40 W K-band space TWT

3 Slow Wave Structure (SWS) Design

An in-house developed code SUNRAY-1D and commercially available CST-MWS have been used for design of SWS for K-band 40W space TWT. SUNRAY-1D [4–6] The TWT's various output parameters, such as output power, gain, efficiency, phase shift, etc., were calculated using a large signal multi-signal code. The code utilizes a Lagrangian formulation for beam wave interaction. The propagation constant (b) and impedance (K) on the helix axis were determined using CST-MWS with a single-turn approach. Figure 4 displays the TWT's helix assembly with a single-turn helix modeled in CST-MWS [7–9], complete with appropriate boundary conditions (periodic boundary with phase shift). Table 1 outlines the optimized dimensions of the helix SWS assembly.

After calculating all electrical parameters (cold) from CST-MWS, SUNRAY-1D code is used to design SWS for required power, gain and efficiency. In this case helix SWS was in two sections: first section consist uniform pitch and pitch of this section is optimized for maximum small signal gain, second section consist a pitch correspond to maximum bunching followed by a uniform negative tapering for maximum efficiency. Each section consist a total tip loss of 30 db on length of 15 mm. Thus finally all the three pitch values were optimized using the above mentioned design methodology and was plotted in Fig. 5 (normalized w.r.t. p_0). Proper helix pitch profile design was used to determine the proper RF performance of the helix TWTs. 2-secting helix pitch profile used in this case study for the space communication TWT is shown in Fig. 5. Using this helix pitch profile, the RF performance of the TWT has been simulated.

Figure 6 show the saturated output power and gain respectively simulated using in-house code SUNRAY-ID (Both Saturated and small signal gain) against normalized

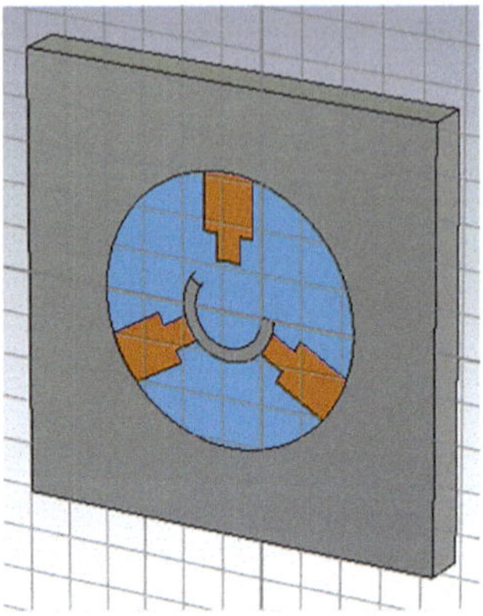

Fig. 5. CST-MWS model of helix TWT.

Table 1. Dimension of helix SWS assembly.

Name	Value (mm)	Description
Barrel id	1.57	Barrel radius
Helix id	0.42	Helix radius
Rod 1 height	0.70	Rod to barrel
Rod 1thickness	0.60	Rod touching to barrel
Rod 2 height	0.30	Rod touching to helix
Rod 2 thickness	0.30	Rod touching to helix
Tape width	0.30	Helix tape width
Tape thickness	0.15	Helix tape thickness

frequency at operating voltage of 6.7 kV and beam current of 74 mA. From Fig. 6 it is clear that simulated saturated output power is more i.e. around 57–60 W against the required output power of 40 W. Also Fig. 6 shows that both gain at saturation and small signal gain are almost flat within the require frequency band.

4 Conclusion

By using this simplified design approach, An accurate design of helix SWS has been developed for space communication within K-band. This delivers over 40W of power, along with high gain and electronic efficiency. The design aims to achieve the desired saturated power and gain at saturation. Further optimization is required to fine-tune the helix design, and we are currently working on PIC simulation, which will be presented in the near future.

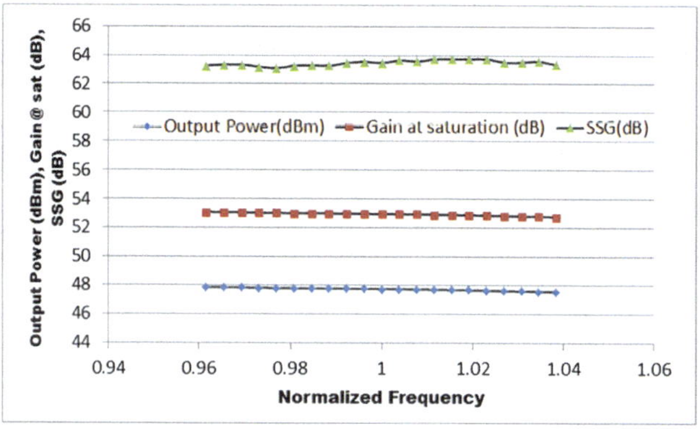

Fig. 6. Simulated saturated output power, saturated gain & small signal gain vs. normalized frequency (normalizing w.r.t center frequency).

Acknowledgment. The authors are grateful to Director CEERI, Pilani, for permission to publish this paper. Thanks to team member of Space TWT project at CSIR-CEERI, CSIR for their co-operation.

References

1. Gilmour, A.S. Jr.: Principles of Traveling Wave Tube. Artech House, Boston, London (1994)
2. Choudhury, A.R., Sharma, R.K., Bera, A., Sharma, S.M., Srivastava, V.: Design and Development of Ku-Band 140W Space TWT. Proc. IVEC 2008, p. 171
3. Srivastava, V., Carter, R.G., Ravinder, B., Sinha, A.K., Joshi, S.N.: Design of helix slow-wave structures for high efficiency TWTs. IEEE Trans. Electron Devices **47**(12), December (2000)
4. Srivastava, V., Joshi, S.N.: One-dimensional nonlinear model for helix TWTs. IETE Tech. Rev. **6**(6), 500–507 (1989)
5. Srivastava, V., Joshi, S.N.: Improved nonlinear model for multilingual analysis of helix TWTs. IEE Proc.-H, **7**(12), 129–134, Apr. (1992)
6. Bera, A., Choudhury, A.R., Srivastava, V.: SUNRAY-1D code for accurate and fast large signal analysis of a helix TWT. Proc. IVEC 2008, p. 115
7. Kory, C.L.: Validation of an accurate three dimensional helical slow wave circuit model. NASA Tech report March 1997, Report no. NASA contractor report 4766
8. Kory, C.L., Dayton, J.A.: Accurate cold-test model of helical TWT slow-wave circuits. IEEE Trans. Electron Devices **45**(4), 966–971, April (1998)
9. Aloisio, M., Waller, P.: Analysis of helical Slow-wave structures for Space TWTs using 3-D Electromagnetic simulators. IEEE Trans. Electron Devices **52**(5), 749–754 (2005)

Effect of Cathode Tilting on the Electron Beam Dynamics

Suneeta Arya(✉), Narendra Singh, A. R. Choudhury, Pawan Pareek, Rajesh Meena, Bijendra Kumar, Chirag Mistry, and S. K. Ghosh

Central Electronics Engineering Research Institute, Pilani, India
suneeta@ceeri.res.in

Abstract. The design and simulation of electron gun with required beam dynamics is done by using CST-PS software package. Geometry and position, namely axial, radial, concentricity of cathode and electrodes are the key parameters for required beam optics. The concentricity of the beam focusing electrode (BFE) is optimized with respect to the position of the cathode. Both radial and axial distances between BFE and cathode is fixed, typically at 0.2 mm for 130 mA beam current at 7 kV cathode voltage. However, it has been observed for some cases that during development of electron gun, the cathode gets tilted at its axis subsequently followed by offset of radial and axial distances of BFE. This paper presents the effect of cathode tilting on beam dynamics of the electron gun with different cathode position with respect to the BFE position and also determines its effect on helix interception current during testing of TWTs.

Keywords: Dual-anode electron gun · Helix space TWT

1 Introduction

Helix travelling wave tubes (TWTs) are the key component in various applications such as, satellite communications, electronic countermeasures systems, broadcasting and high-power radar systems etc. A TWT consists of three main parts, namely, an electron gun, slow-wave structure, and collector. The electron gun plays a significant role in space TWTs. It is used for its prime characteristics such as long life (typically > 15 years), high laminarity, reliability, stability and electronic efficiency at the desired output power and gain [1].

The Dual-anode electron gun designed and developed for high-reliable and long-life applications (Fig. 1). In this electron gun, there are electrically isolated beam focusing electrode (BFE), a beam controlling anode, an positive ion reflector anode and a zero potential anode. An M-type dispenser cathode is invariably used as the same is unique choice for space applications TWTs. To meet these characteristics a Pierce type convergent electron gun is designed using CST-PS software [2].

S. K. Ghosh et al. (Eds.): VEDA 2024, SPPHY 424, pp. 90–96, 2025.
https://doi.org/10.1007/978-981-96-5174-0_13

Fig.1. Developed dual anode electron gun

2 Electron Gun Design

A Pierce type dual anode electron bean source (gun)modelled for ~130 mA current (I_0) at 7 kV beam voltage (V_0) with 0.4 mm beam waist radius (r_w). The line diagram of dual anode electron gun is shown in Fig. 2. The electron beam sourcecontains dispenser cathode with 3.1 mm diameter with low emission loading (~1.0 Amp/cm^2), low cathode operating temperature ~950 °C at heater power around 5.0 W, an electrically isolated beam focusing electrode (BFE), a control anode (A_1) and one additional anode i.e., ion control anode (A_2) and accelerating anode (A_3). The control anode is employed in order to enhance the beam current with time without changing the accelerating ground potential.

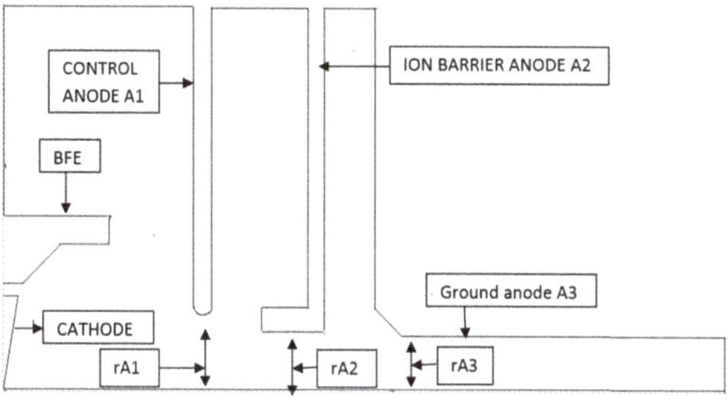

Fig. 2. Line diagram of a dual-anode electron gun

The control anode is used when emission current decreases with time—with additional nominal voltage emission current increases and hence, life. In this design of electron gun, the control anode potential is kept at 2.8 kV with respect to the cathode. Hence, we found that the control anode potential plays an important role in controlling the desired beam perveance and enhancing the life of a space TWT. Similarly, the ion barrier anode (A2) is kept at +100 to 300 V with respect to ground anode (A3) to avoid cathode poisoning from hitting of positive ions generated in the interaction region, enhancing the life and reliability of the cathode. In addition, BFE is negatively biased with respect to cathode of the order of −7 to −10 V which helps to control the beam optics and beam transmission in the tube >99% with the help of magnet focusing scheme.

Initial estimation of gun geometry is obtained using Vaughan's synthesis [3] approach for the calculation of desired beam parameter in terms of potential, current, beam minimum radius and cathode loading. Finally, design of double anode beam source with respect to geometry and potential is obtained through iteration. Through the iteration process, geometry of electrodes and inter electrode distances (both axial and radial) and voltages are optimized to achieve the required beam optics and beam current [4]. The parameters such as radial and axial distances and concentricity of the cathode w.r.t electrodes are important for the required beam dynamics. In this design both the radial and axial distances between BFE and cathode is fixed (keeping the margin of fabrication tolerances—obtained through rigorous sensitivity analysis) and these are typically kept ~0.2 mm for the required beam current. Development of the electron gun using suitable jigging is a completely mechanical process and it has been observed that in some cases the cathode gets tilted/offset about its axis—may be due to cumulative effect of fabrication tolerances of all parts and jigs. The offset of the cathode deforms beam optics causing higher BFE current, spot heating, high helix interception current, etc. and finally lead to damage the device (Fig. 3). Thus, effect of beam current on various parameters, namely, geometry of the electrodes and their distances and effective potentials have been studied, simulated and analyzed in CST-PS followed by sensitivity study to take care the fabrication tolerances for practical relevance [5, 6].

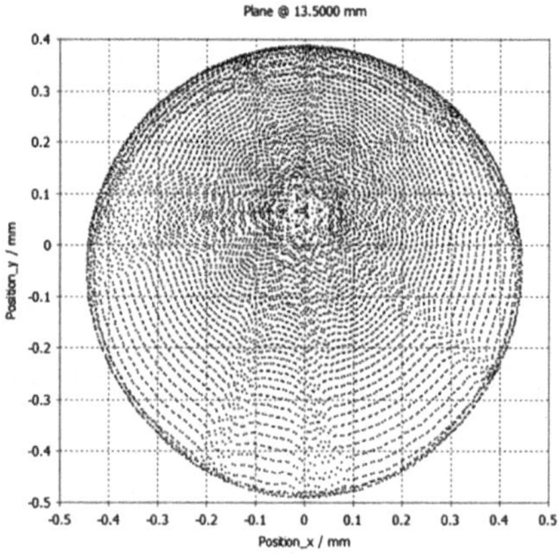

Fig. 3. Effect of cathode offset on beam optics

3 Development Procedure of Electron Gun

The developed dual electron gun (Fig. 1) consists of an M-type dispenser cathode, BFE, control anode(A_1), ion barrier anode (A_2), ground accelerating anode (A_3). The dual anode electron gun with designed values has been developed using suitable jigs and fixtures to achieve beam current ~130 mA at 7.0 kV cathode potential. The development of electron gun is completely manual process. In this process, after development of electron gun assembly—brazing of metal and non-metal components, the first step is, the post machining of electron gun with closer tolerances according to BFE and cathode outer dimensions. After laser welding of BFE, cathode laser welding is done using jigs/fixture. In order to achieve the radial and axial concentricity of cathode w.r.t BFE a special jigs/fixture is designed as shown in Figs. 4 and 5. In this fixture arrangement a plunger is used to force the cathode base along with alignment of the cathode axis. The cathode holding fixture is tightening with the screw, at the top of the electron gun. Now, cathode laser welding is done from the top. Hence, with the suitable cathode holding fixture arrangement the tilting of cathode is reduced and improves the yield.

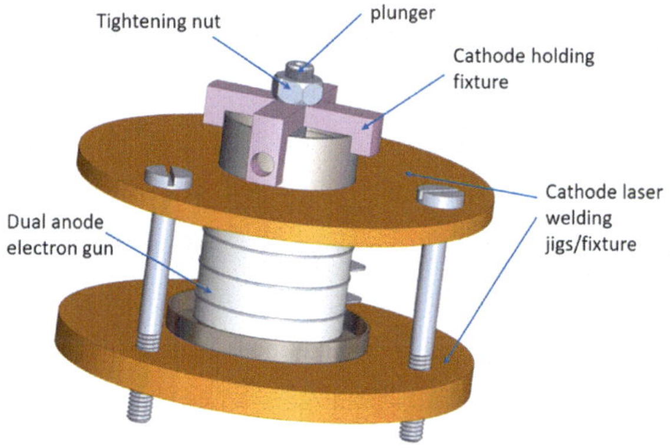

Fig. 4. Dual anode electron gun with cathode holding fixture

4 Results and Discussion

This paper presents the development procedure of a double-anode electron beam source for satellite applications a developed electron gun without using cathode holding jigs/fixture is given in Fig. 6 and measured data is given in Table 1. In order to reduce this measurement variation, a suitable jigs/fixtures are used to develop a perfect dual anode electron gun, with the proper tolerances/dimensions.

Using these proper jigs/fixtures the tilting of cathode is reduced, which is given in Fig. 7. Table 2 shows the measurement result, as the concentricity of the cathode and BFE is within the acceptable tolerances. It is also found that the electron gun dimensional measurement results and the simulation results are almost in similar agreement.

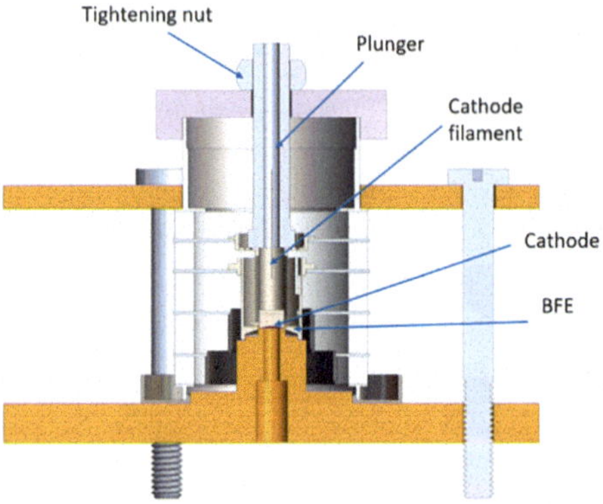

Fig. 5. Cross-section view of dual anode electron gun with cathode and jigs/fixtures

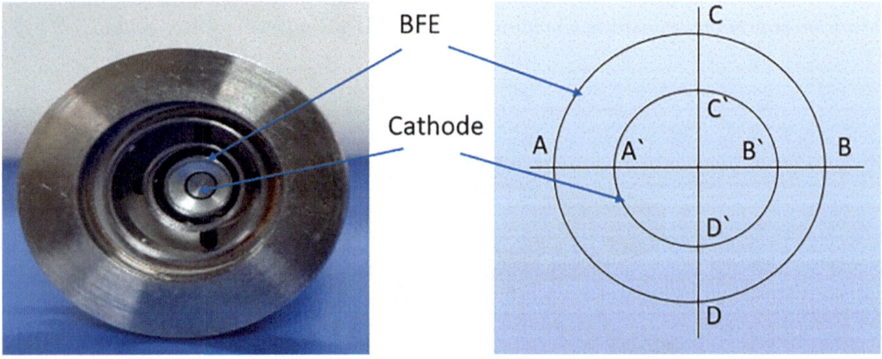

Fig. 6. Cathode welding after without using cathode holding jigs/fixture

Table 1: Concentricity measurement of cathode with respect to BFE

S.No	Distance	Actual value	Measured value
1	AA′	0.200	0.222
2	BB′	0.200	0.212
3	CC′	0.200	0.217
4	DD′	0.200	0.224

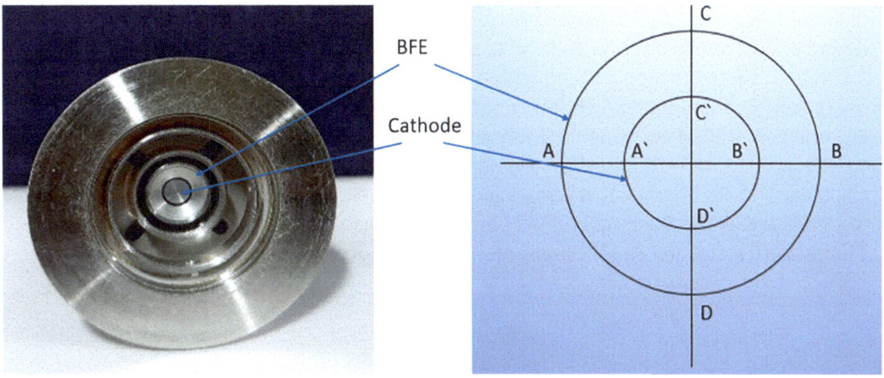

Fig. 7. Cathode welding with precised cathode holding jigs/fixture

Table 2: Concentricity measurement of cathode with respect to BFE using cathode holding jigs/fixture

S.No	Distance	Actual value	Measured value
1	AA′	0.200	0.201
2	BB′	0.200	0.202
3	CC′	0.200	0.204
4	DD′	0.200	0.206

5 Conclusion

Development of a dual-anode electron gun with precision jigs & fixtures for potential application in a space TWT has been presented in this paper. It is shown that to improve the yield, the gun needs to be developed with precision jigging. The precision jigging improves concentricity of both BFE and cathode about their axis, which finally improves beam optics and its propagation through the interaction structure. In this process, one gets better agreement between simulation and measured beam parameters.

Acknowledgement. The authors are thankful to the Director, CSIR–CEERI for giving permission to publish thispaper and also thankfulto othercolleagues for their support.

References

1. Gilmour, A.S., Jr.: Klystrons, traveling wave tubes, magnetrons, crossed-field amplifiers, and gyrotrons. Artech House, Norwood (2011)
2. CST Particle Studio. Computer Simulation Technologies. [Online]. https://www.cst.com/Products/CSTPS
3. Vaughan, M.: Synthesis of pierce gun. IEEE Trans. Electron Devices **28**, 37–41 (1981)

4. Sharma, R.K., Sharma, P., Choudhury, A.R., Sharma, S.M., Arya, S., Srivastava, V.: P2–19: design of dual anode electron gun and beam focusing for Ku-band 140 W short length space TWT. In: IEEE International Vacuum Electronics Conference, pp. 259–260, May 18–20 (2010)
5. Sharma, R.K., Choudhury, A.R., Sharma, S.M., Arya, S., Bera, A., Srivastava, V.: Dual anode electron gun & PPM focusing for Ka-band TWT. In: IEEE International Vacuum Electronics Conference, pp. 185–186, April 28–30 (2009)
6. Choudhury, A.R., Sharma, R.K., Sharma, S.M., Arya, S., Bera, A., Srivastava, V.: Dual anode electron gun for Ka-band helix TWT. In: International Conference on Recent Advances in Microwave Theory and Applications, pp. 192–195, November 21–24 (2008)

Investigating the Impact of External Port Coupling on Microwave CPW Resonators for Circuit QED Experiments

Deepak Paste, Sumita Sura(✉), Praveen Kumar Suggisetti, M. Y. Dixit, and Gopal Joshi

Accelerator Control Division, Bhabha Atomic Research Centre (BARC), Trombay, India
sumitasura@barc.gov.in

Abstract. Microwave Coplanar Waveguide (CPW) resonators have emerged as key components in the field of Circuit Quantum Electrodynamics (CQED) based quantum information processing. Present study explores the effect of external couplings on resonant frequency and quality factor of half-wavelength (λ/2) CPW resonator. Electromagnetic simulations have been performed to design several λ/2 CPW resonators with fundamental frequencies ranging from 2 to 8 GHz, and quality factors due to coupling ranging from a few hundred to several thousand. Simulations have been carried out for various geometry of input and output coupling capacitors. The results presented in the paper will be helpful in designing CPW resonators with specific operating parameters for CQED experiments.

1 Introduction

Microwave Coplanar Waveguide (CPW) resonators have garnered significant interest over past decade. These resonators are subject of intense study for their use in various applications [1–5]. CPW resonators, known for their high-quality factor, ease of integration, and scalability have emerged as important components for rapidly evolving field of quantum information processing (QIP), particularly within the framework of Circuit Quantum Electrodynamics (CQED) [6–8]. In CQED, the interaction between quantum systems and electromagnetic fields confined within a cavity forms the basis for quantum state control and readout. A critical aspect of designing CPW resonators for QIP applications is the precise control of their resonant frequency and quality factor. These parameters dictate the ability of the resonator to store energy and interaction with quantum systems, directly influencing the performance and fidelity of quantum operations. The resonant frequency sets the operational condition, while the quality factor impacts the coherence time of the quantum states, making these parameters crucial for optimizing CPW resonators for specific quantum experiments [6]. Coupling capacitances are crucial as these govern the strength of interaction between the resonator and qubit, and external circuitry. These capacitances have a significant effect on both the resonant frequency and quality factor.

The present study reports the impact of external couplings on the resonant frequency and quality factor of half-wavelength (λ/2) CPW resonators, a common configuration

S. K. Ghosh et al. (Eds.): VEDA 2024, SPPHY 424, pp. 97–102, 2025.
https://doi.org/10.1007/978-981-96-5174-0_14

in CQED experiments. By conducting detailed electromagnetic simulations, we explore how variations in the input and output coupling capacitors affect the performance over a wide range of fundamental frequencies (2 to 8 GHz). The results presented in this paper are aimed to serve as a valuable resource for researchers and engineers working in the field of QIP, providing guidelines for designing CPW resonators.

2 Resonator Design

Sketch of CPW resonators capacitively coupled to external port via a finger capacitor is shown in Fig. 1. CPW has been sketched on a 500 um thick Silicon substrate. It consists of a central conducting strip, of width Tw, in the middle of two ground planes, on top of substrate with spacing Ts with the ground planes on either side. The values of Tw and Ts are chosen to achieve a 50 Ω characteristic impedance, which is a convenient value. Resonators with various centre-conductor lengths Lf_0 between 8 and 29 mm aiming at fundamental frequencies $f0$ between 2 and 8 GHz were designed. These resonators are coupled via gap or finger capacitors to the input and output transmission lines. CPW resonators has been designed with meander geometry. Simulations have been carried out to analyse the effect of coupling capacitors on fundamental frequency and quality factor of these resonators. The chosen resonator parameters were Tw = 32 um, Ts = 20.2 um to achieve characteristic impedance ~50 Ω.

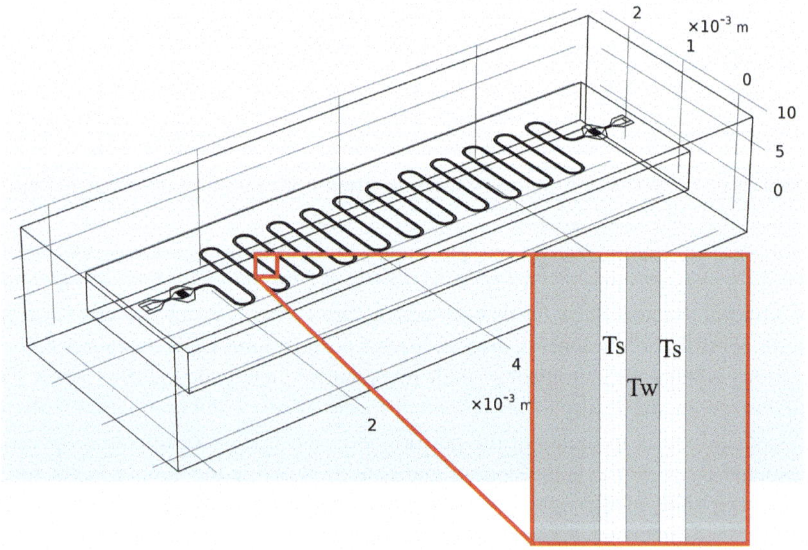

Fig. 1. Representation of the coplanar waveguide resonator. The red square shows a top view of a section of CPW resonator.

3 Simulation Results

To investigate the effect of input and output coupling capacitances, various finger capacitors with different numbers of fingers (A, B, C, D, E, F, G, H) and a gap capacitor (I) were designed. The finger capacitors were designed from one up to eight pairs of fingers with length $lf = 300$ um, width $wf = 10$ um, and separation $sf = 10$ um. The gap capacitor was designed with separation $sf = 3.3$um. The coupling capacitance decreases progressively from A to I. All designs and simulations were carried out using COMSOL software [7], which provided detailed electromagnetic analysis of different geometries. Figure 2 shows few geometrical representations of coupling capacitors.

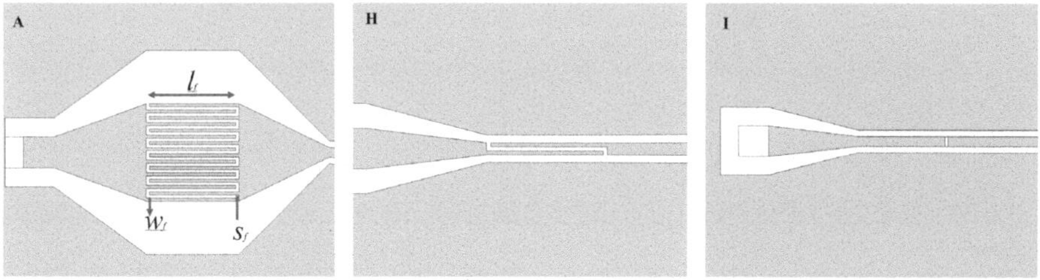

Fig. 2. Representation of 8 + 8 finger capacitor (A), 1 + 1 Finger Capacitor (H) and Gap capacitor (I)

Eigen frequency analysis was performed and Fig. 3 shows electric field distribution for a resonator with $f0$ ~2.1 GHz and 8+8 finger coupling capacitor. The resonant mode was shifted from 2.1 GHz to 1.84 GHz due to coupling effect. The scattering parameter response, particularly the transmission parameter S_{21}, was analyzed for each coupling capacitance. S_{21} is the transmission parameter that gives information about the power delivered from port 1 to port 2. The resonance frequency $f0$ and the quality factor Q_L were extracted by fitting a Lorentzian line shape function to the scattering parameter S_{21}. Table 1. Lists the simulated coupling capacitance for different geometry, fundamental resonant mode and quality factor for 2.1 GHz resonator. It was observed that as the coupling capacitance decreases, the total quality factor increases. Fig. 4(a) illustrates the S_{21} response for various coupling geometries of the 2.1 GHz resonator, showing a clear shift in resonant frequency due to changes in coupling capacitance. This behavior aligns with the theoretical equation provided in Reference [6]. The response of resonance frequency and quality factors for various coupling geometries for 2.1 GHz (Lf_0 =28.453) resonator are shown in Fig. 4(b). The resonant frequency decreases monotonically with coupling capacitances.

Additionally, the same coupling capacitances were analyzed for resonators with different length Lf0 and thus different fundamental resonant frequencies, and the results are plotted in Fig. 5. Consistent with the earlier observations, larger coupling capacitances result in greater shifts in resonant frequency. However, it was also found that the frequency shift is also dependent on the fundamental mode of the resonator. For resonators with shorter lengths and higher fundamental mode, the frequency shift due to the same coupling capacitance is significantly larger compared to longer resonators with lower

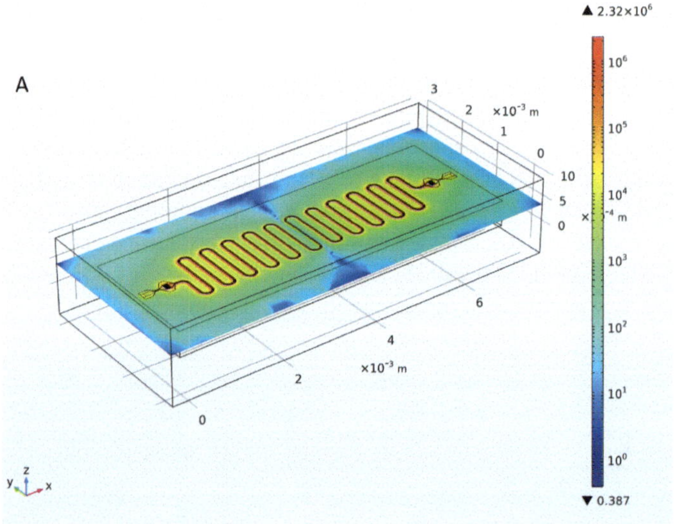

Fig. 3. Planar electric field distribution of 2.1 GHz resonator ($Lf_0 = 28.453$) with 8 + 8 finger capacitor.

Table 1. Parameters for the different simulated coupling capacitances for CPW resonator with fundamental frequency 2.1 GHz ($Lf_0 = 28.453$).

ID	Coupling	Coupling capacitance Ck (fF)	f (GHz)	Quality factor Q
A	8 + 8 finger	396.8	1.8362	23.396
B	7 + 7 finger	348.5	1.8683	30.44
C	6 + 6 finger	303.1	1.8931	40.372
D	5 + 5 finger	250.7	1.9262	55.452
E	4 + 4 finger	208.7	1.9515	88.602
F	3 + 3 finger	157.1	1.9799	158.62
G	2 + 2 finger	108.7	2.0174	428.4
H	1 + 1 finger	60.7	2.047	3670.9
I	10um gap	37.46	2.0813	121120

fundamental mode. For example, using an 8+8 finger capacitor, the frequency shift for a 7.3 GHz resonator was approximately 30%, whereas for a 2.2 GHz resonator, the shift was around 12%.

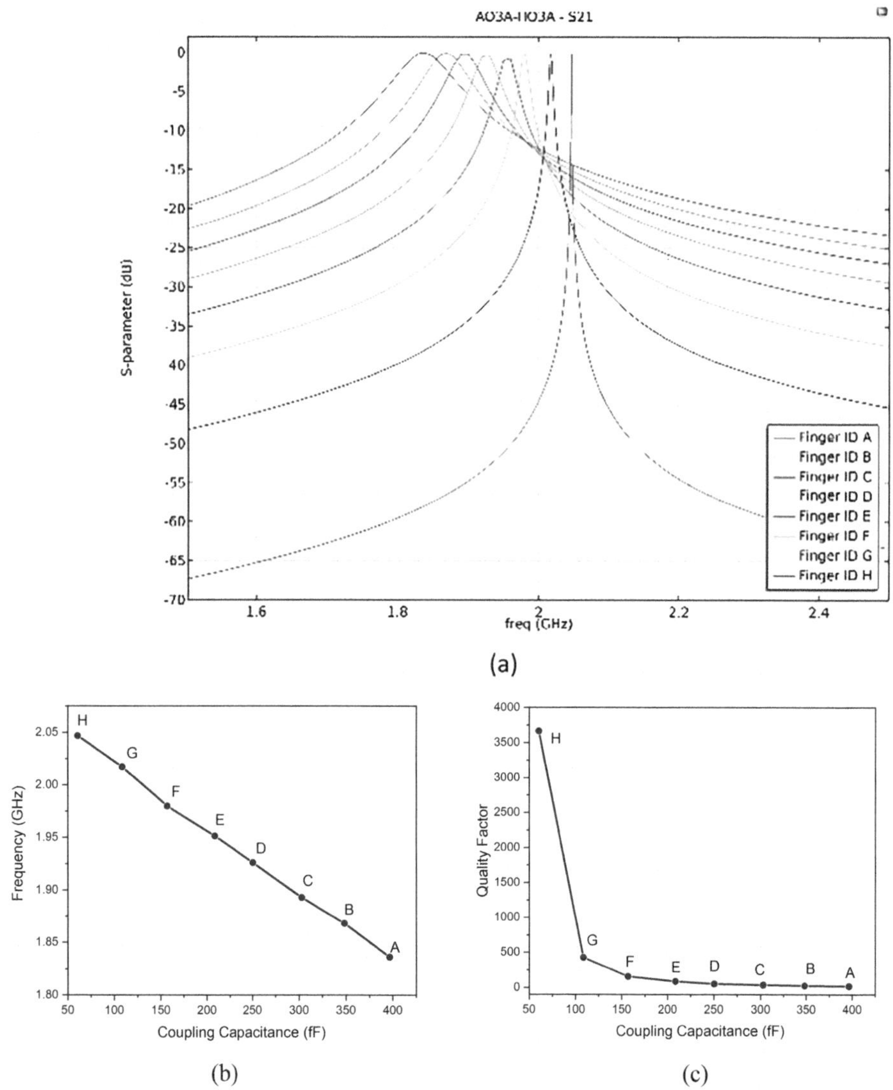

Fig. 4. **(a)** S_{21} vs Frequency **(b)** Frequency vs Coupling Capacitance **(c)** Quality Factor vs Coupling Capacitance.

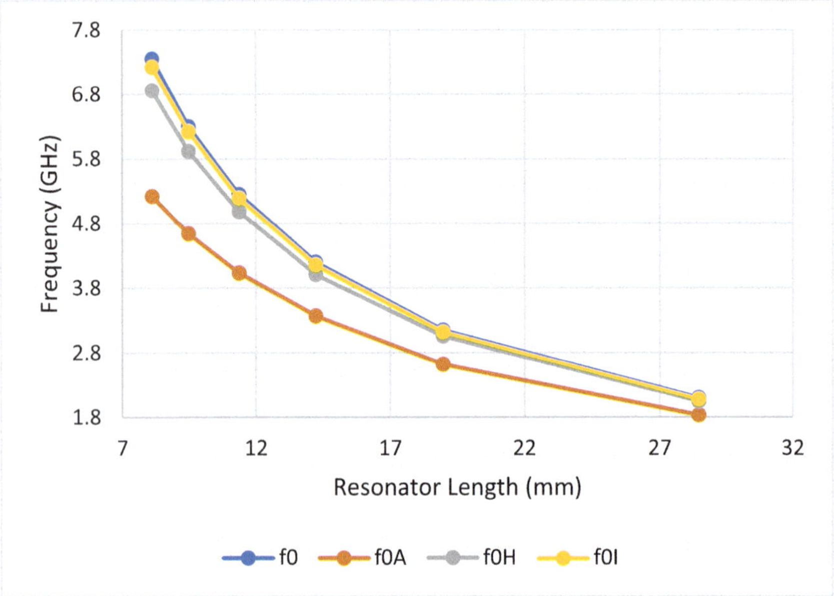

Fig. 5. Variation of resonant frequency with length of resonator for different coupling capacitance geometry.

4 Conclusion

In conclusion, this study investigated the influence of input and output coupling capacitances on the resonant frequency and quality factor of half-wavelength ($\lambda/2$) CPW resonators. Through simulations of various capacitor geometries, it was observed that increasing coupling capacitance reduces the quality factor and causes a significant shift in resonant frequency, with higher shifts occurring in resonators with shorter lengths and higher fundamental mode.

References

1. Mazin, B.A., Day, P.K., LeDuc, H.G., Vayonakis, A., Zmuidzinas, J.: Proc. SPIE **4849**, 283 (2002)
2. Day, P.K., LeDuc, H.G., Mazin, B.A., Vayonakis, A., Zmuidzinas, J.: Nature London **425**, 817 (2003)
3. Zmuidzinas, J., Richards, P.L.: Proc. IEEE **92**, 1597 (2004)
4. Mazin, A., et al.: J. Low Temp. Phys. **151**, 537 (2008)
5. Vardulakis, G., Withington, S., Goldie, D.J., Glowacka, D.M.: Meas. Sci. Technol. **19**, 015509 (2008)
6. Göppl, M., Fragner, A., et al.: J. Appl. Phys. **104**, 113904 (2008)
7. Clauss, C., et al.: J. Phys. Conf. Ser. **592**, 012146 (2015)
8. Salunkhe, K.V., Kundu, S., et al.: Proceedings of the COMSOL Conference, Bangalore (2019)
9. http://www.comsol.com

Topology Optimization of the Radiator to Improve Heat Dissipation by Radiation of TWT

Chirag Mistry[1(✉)], Amitavo Roy Choudhury[1], Subhradeep Chakraborty[2], Suneeta Arya[1], Pawan Pareek[1], and Sanjay Kumar Ghosh[1]

[1] Vacuum Electron Devices Group, CSIR-CEERI, Pilani, India
cpmistry@ceeri.res.in
[2] High-Frequency Devices and System Group, CSIR-CEERI, Pilani, India

Abstract. Radiation cooled helix TWT is most commonly used as a power amplifier for satellite communication for being broadband in nature. In general, the radiator is mounted on the Multistage Depressed Collector (MDC) and the objective of the radiator is to dissipate heat to the space environment. In this paper, the author has presented the design of the radiator to dissipate heat up to 140 W. Further, the topology optimization of the radiator for room temperature and space temperature was carried out to keep the maximum radiator temperature below the allowable temperature limits. In addition, an effort has been made to keep the radiator as lightweight as possible. The radiator is simulated in the ANSYS tool and the simulation results are presented in the paper.

Keywords: Radiator · Space TWT · Thermal simulation

1 Introduction

Radiation cooled Helix TWT is commonly used for satellite communication due to its high efficiency. The function of the radiator of the space TWT is to dissipate waste heat into the environment. The radiator has to mount on the Multistage Depressed Collector (MDC) of the space TWT as shown in the Fig. 1. High energy beams incident on the inner surface of the electrodes of MDC. Some energy has recovered from the electrodes and the remaining produced heat at MDC. In general, TWT has to operate under RF and DC operating conditions. In addition, space TWT has to operate in an environment temperature ranging from −40 °C to 85 °C in a thermo-vacuum chamber for space qualification [1, 2]. Therefore, in this paper thermal simulation of the radiator assembly has been carried out for the RF, DC and at different environment temperatures (−40 °C & 85 °C) under RF case. The temperature profile of the radiator assembly has been estimated using ANSYS tool under different operating cases at different environment temperatures. The simulation results i.e. maximum temperature were validated with the allowable temperature limit criteria of the material. Further, the geometry of the radiator has been optimized to keep the maximum temperature below the allowable temperature

S. K. Ghosh et al. (Eds.): VEDA 2024, SPPHY 424, pp. 103–107, 2025.
https://doi.org/10.1007/978-981-96-5174-0_15

limit of the material. A cross section view of the initial design of the radiator is as shown in the Fig. 2. The initial design of the radiator (D1) has four fins of fin thickness (t) and fin length (L). The heat dissipation capacity of the radiator can be increased by introducing fin structure to increase surface area for heat dissipation. Therefore, to keep the maximum temperature below the allowable limit, the geometry of the radiator is optimized as listed in Table 1. The design iterations from D1 to D4 each has been simulated in the ANYSY for the temperature profile. The thermal simulation presented here is for the final optimized geometry of the radiator (D4). The non-functional space between the fins no 2 to 4 of initial radiator design is utilized to add more fins for the final design of the radiator.

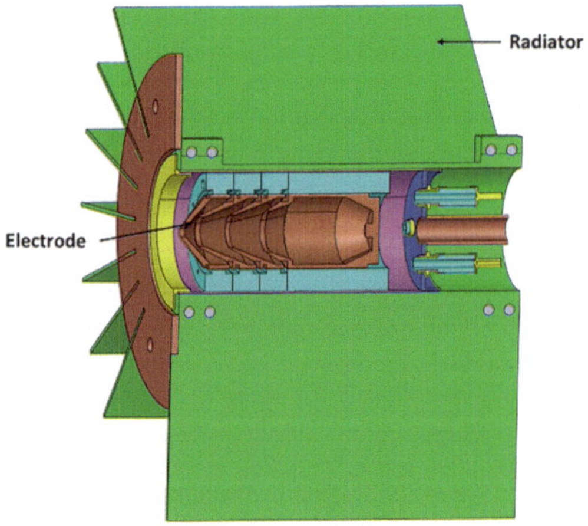

Fig. 1. The radiator mounted on Multistage Depressed Collector (MDC)

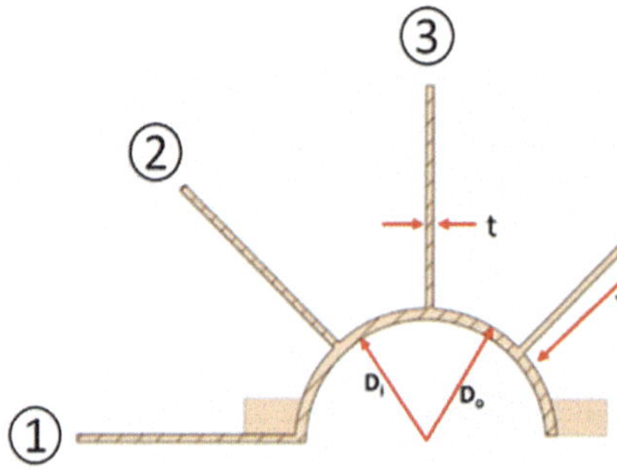

Fig. 2. A cross section view of initial design of the radiator

Table 1. Thermal design optimization iterations

Sr. No	Parameters	D1	D2	D3	D4
1	Inner diameter (Di)	Di	Di	Di	Di
2	Outer diameter (Do)	Do	Do	Do	Do
3	Fins length (L)	**L**	**2L**	L	**2L**
4	Thickness (t)	t	t	t	t
5	No of fins	4	4	**8**	**8**
6	Weight (g)	58	91	91	157

2 Thermal Simulation and Results

The geometry of the radiator assembly has been model in the Solid Works [3] and thermal simulation have been carried out in the ANSYS tool [4]. The heat load on the different electrodes of the MDC are as listed in the Table 2. The heat dissipate through the environment is by means of the natural convection and radiation. Accordingly, thermal boundary condition are applied for the thermal simulation in ANSYS. The final optimized geometry of the radiator (D4) has eight no of fins of double length (2L) compared to initial design as shown in Fig. 3. The weight of the final optimized geometry of the radiator is 157g. There is always tread off between the weight and thermal performance of the radiator when, it was used for the space application. Therefore, based on the thermal performance D4 design of the radiator is best choice. The radiator is made of Al-6061.

Table 2. The heat load on the different electrodes of the MDC

Sr. No	Electrode	RF	DC
1	Electrode 1	19	5
2	Electrode 2	27	5
3	Electrode 3	20	5
4	Electrode 4	74	125

The temperature profile of the radiator for RF case at 85 °C environment temperature is as shown in Fig. 4. The estimated maximum temperature through the simulation is ~141 °C, which is well below the allowable temperature limit of the material of 200 °C. Therefore, the radiator assembly is safe to operate under extreme environmental temperature of 85 °C. The maximum radiator temperature plots for different operating conditions for each design iterations are as shown in Fig. 5. It can be observe from the Fig. 5 that, D4 version of the radiator design has least temperature compared to all other cases. While radiator design, D2 and D3 are also most same for both thermal and weight aspects.

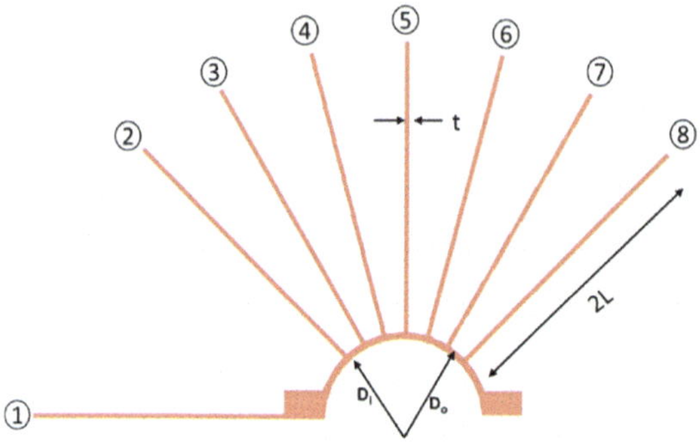

Fig. 3. A cross section view of the final optimized design (D4) of the radiator

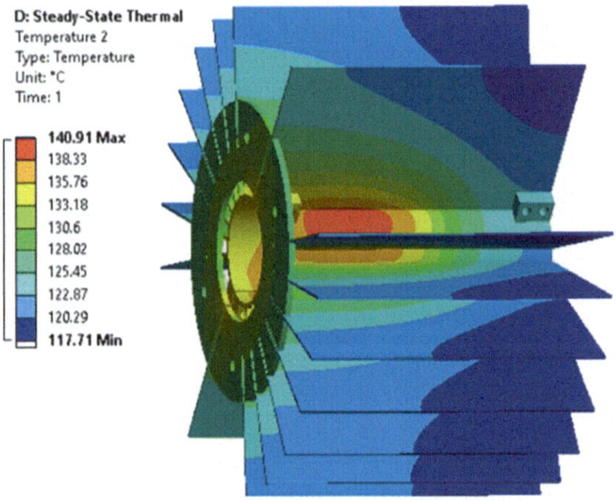

Fig. 4. Temperature profile of the radiator for RF case under 85 °C

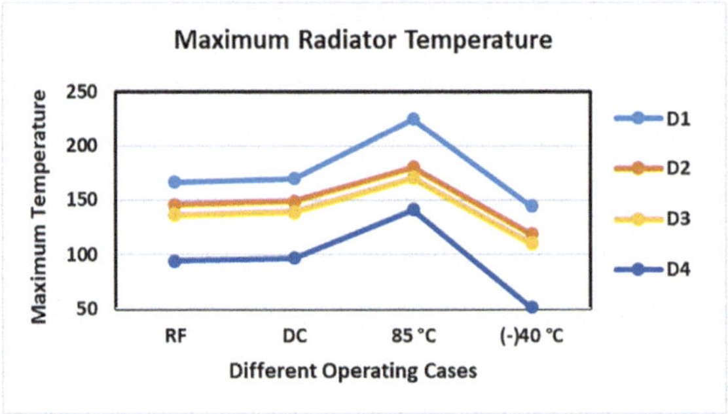

Fig. 5. The maximum radiator temperature plots for different operating cases

3 Conclusion

The thermal design and simulation of the radiator have been carried out to dissipate heat up to 140 W to the environment. The design of the radiator have been optimized to keep the maximum temperature below the allowable limit of the material under extreme environment temperature. Therefore, the proposed design of the radiator is safe to operate under different operating conditions at extreme environmental conditions.

Acknowledgment. The authors would like to thank the Director of the Council of Scientific and Industrial Research–Central Electronics Engineering Research Institute (CSIR–CEERI), for his kind permission to publish this paper.

References

1. Mistry, C., Ghosh, S.K.: Thermal design and analysis of radiation shield cooling assembly for efficient heat dissipation of space traveling-wave tube. IEEE Trans. Electron Devices **71**(1), 374–380 (2024). https://doi.org/10.1109/TED.2023.3338591
2. Chen, P., et al.: Development of Ka-band 100 W Radiation-cooled Space TWT. In: 2023 24th International Vacuum Electronics Conference (IVEC), Chengdu, China, 2023, pp. 1–2 (2023). https://doi.org/10.1109/IVEC56627.2023.10157339
3. SolidWorks Manual, Canonsburg, PA, USA (2011)
4. ANSYS® Academic Research Mechanical, release 19.0, Help System, Static Analysis Guide, ANSYS Inc. (2019)

Metasurface-Based Low-Profile Beam-Steering Antenna for Satellite Applications

Barsha Kumari, Rajan Agrahari, and Manpuran Mahto(✉)

Electronics and Communication Engineering Department, National Institute of Technology Patna, Patna, India
mmahto@nitp.ac.in

Abstract. Beam-steering antennas are becoming increasingly important in wireless communication systems due to their ability to electronically control radiation patterns and beam widths. This article introduces a low-profile, metasurface-based K-band beam-steering antenna specifically designed for satellite communication. The proposed antenna integrates both active and passive meta-cells, offering a compact form factor, enhanced performance, and compatibility with modern devices. The metasurface consists of nine passive meta-cells and four active subarrays, with each active subaaray of 2×2 grid of active meta-cells. These components are fabricated on a low-loss Rogers RT5880 substrate. The active meta-atom within each array features a square patch enclosed within a square ring, interconnected by two PIN diodes. This metasurface is excited by a patch antenna operating at 24 GHz, positioned at a specific height below the metasurface. Real-time beam steering is accomplished by switching the PIN diodes' ON and OFF states. When all diodes are in the ON state, the suggested design attains a gain of 16.3 dBi and a radiation efficiency of 80%. Additionally, it offers a wide beam-steering range of 28°, making it a promising solution for advanced satellite communication systems.

Keywords: Metamaterial · Millimeter-Wave technology · Satellite communication

1 Introduction

Beam steering has seen a surge in research activity recently because of its many applications, which include next-generation and satellite communication. Beam steering is a method used to alter the direction in which the main part of a signal or radiation pattern is focused [1]. It is achieved by adjusting or switching the antenna elements. It reduces interference, saves power, and increases the antenna's gain, directivity, and efficiency [2]. We need a narrow beam to send signals over long distances in some wireless communication setups. This can be done in two ways: making the antenna larger or using an array of multiple antennas. An antenna array is made up of several individual antennas combined [3]. Many radiating elements, each with a phase shifter, make up a phased array antenna. Each element's signal can be phase-shifted to produce either constructive or destructive interference, which enables the beams to be guided in predetermined

S. K. Ghosh et al. (Eds.): VEDA 2024, SPPHY 424, pp. 108–115, 2025.
https://doi.org/10.1007/978-981-96-5174-0_16

directions. In beam steering, issues with phase gradients can be addressed by using metasurfaces.

A metasurface is a two-dimensional periodic structure with periodicity extending in the x and y directions but with an extremely thin thickness in the z direction. This material can effectively control the amplitude, phase, and polarization of electromagnetic waves because of its periodic arrangement of subwavelength unit cells [4, 5]. Applications of metasurface are perfect absorbers, scattering of light, polarization, lens, cloaking devices etc. [6–9].

In the past, a beam steering system with two layers has been proposed in many literatures. In general, the top layer features a partially reflecting surface, both active and passive, which creates a phase gradient from high to low. In [10], the bottom layer contains a microstrip patch antenna operating at 28 GHz. This setup allows for beam steering with a range of $\pm 10^{\circ}$, directing the radiation pattern in both the $\varphi = 0^{\circ}$ and $\varphi = 90^{\circ}$ planes. Similarly [11], a six-layered beam scanning phase shifting surface designed for dual-band operation has been proposed by S. A. Matos et al. that offers a scanning range of 0° to 50°. It utilizes a mechanical feed as the input for one-dimensional beam steering, which is more cost-effective than phased array antennas. However, these designs are bulkier, require more space, and operate at slower speeds [11]. A microfluidic system has been proposed for a one-dimensional focal plane array, which is positioned on the back surface of a hemispherical Rexolite material [12]. This setup introduces beam-steering capabilities with a range of approximately $\pm 30^{\circ}$. A three-layer electronically controlled phase shifting surface, featuring a 20×20 grid of unit cells and a horn antenna positioned at 5.6 λ_0, has been proposed for the Ka-band. By adjusting the PIN diodes, a wide steering range of about $\pm 60^{\circ}$ was achieved [13]. Most of these reported work is limited by their large size, low gain, low beam steering range, and complex fabrication.

This paper proposes a metasurface designed on top of a single microstrip antenna for beam-steering applications. The programmable metasurface for beam steering capability with high gain beam scanning at a fixed frequency has been designed in this paper. A microstrip antenna is used as feeding port and has been designed on a ground plane, and it exhibits resonance at 24 GHz. A metasurface layer consisting of the proposed two PIN diodes (D_1 and D_2) loaded metallic patches deposited on a thin dielectric substrate. Features of metasurface and PIN diode is used for controlling the impedance of the metasurface at a fixed frequency. The PIN diodes when programmed correctly steer the radiation pattern in $\varphi = 0^{\circ}$ planes in the directions $\theta = -12^{\circ}$ and $\theta = 16^{\circ}$.

2 Meta-Cell Design

Figure 1 displays the meta-cells design schematic. The active meta-cells feature an inner square ring with dimensions m = 4.5 mm. The design utilizes a Rogers RT5880 with a thickness of 0.37 mm, characterized by a relative permittivity of 2.2 and a dissipation factor of 0.0009 at 24 GHz. The inner square ring is placed at a specified gap from the edge of the meta-cell (see Fig. 1(a)). Additionally, the periodicity of the meta-cell is defined as 6 mm.

The design of the passive meta-cell, shown in Fig. 1(b), differs from the active meta-cell. In passive meta-cell, the diodes and square ring are removed from the top layer.

The passive meta-cell is designed so that its phase at the operating frequency (24 GHz) is lies approximately at the mean point of the ON phase and OFF phase. In our design, we employed the MA4GP907 PIN diode [10]. Figure 1(c) illustrates the cross-sectional view of the meta-cell design to illustrate the 3-D view of the geometry. Figure 1(d) depicts the PIN diode analogous circuit in both states (OFF and ON). When the diodes are in the ON state, they function as an R-L series circuit ($R_{ON} = 5.2\ \Omega$ and $L_{ON} = 0.05$ nH), whereas in the OFF state, they act as an R-C shunt circuit ($R_{OFF} = 40\ k\Omega$ and $C_{OFF} = 0.025$ pf).

To obtain the frequency response, a plane wave is directed normally onto the designed structure. The reflectance from the design structure is calculated from the CST Microwave Studio 2021. Because of their periodicity, unit-cell boundary conditions are used in two directions. In contrast, a perfectly matching condition is utilized in the z-direction to reduce reflection from that direction.

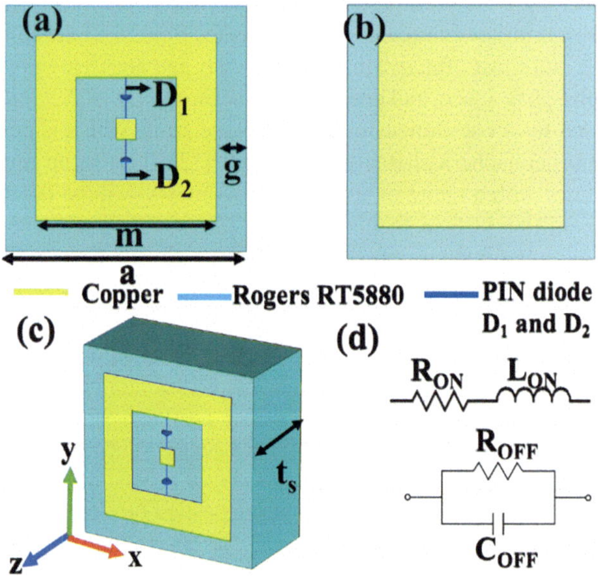

Fig. 1: Schematic of (a) active and (b) passive meta-cell (c) perspective view of meta-cell and (d) equivalent circuits for PIN diodes in ON and OFF states.

The magnitude of the reflection coefficient is shown in Fig. 2(a), it varies from −2.7 to −1.2 dB as frequency increases from 23 to 32 GHz for the OFF state of the diode (see Fig. 2(a)). Similarly, in the ON state, the reflection coefficient magnitude approaches approximately 0 dB, indicating maximum reflection of electromagnetic waves. These findings suggest that the magnitude is high in both the ON and OFF states. Figure 2(b) presents the phase response of the reflection coefficient for the diodes in these states. The phase ranges from −40° to 85°, with a frequency span from 23 to 32 GHz. In the ON state, the phase of the reflection coefficient approaches 60°, whereas in the OFF state, it tends to 72°.

Microstrip patch antennas are widely favored due to their numerous benefits, such as their low profile, lightweight nature, cost-effectiveness, and ease of fabrication. It provides a narrow bandwidth and operates at a specific or fixed frequency. Therefore in our design, we use a microstrip antenna in combination with metasurface for the beam steering applications. Figure 3(a) illustrates the microstrip antenna, with its dimensions detailed in the caption. This microstrip patch antenna is built on a 0.508 mm thick Rogers RT5880, featuring a relative permittivity of 2.2 and a dissipation factor of 0.0009. The metallic patches are etched onto the substrate (as shown in Fig. 3(a)), while copper is utilized for the ground layer on the bottom. Figure 3(b) presents the patch antenna's reflection coefficient response.A resonance is observed at 24 GHz at which the magnitude reduces to −18 dB. This indicates that the patch antenna is radiating at 24 GHz.

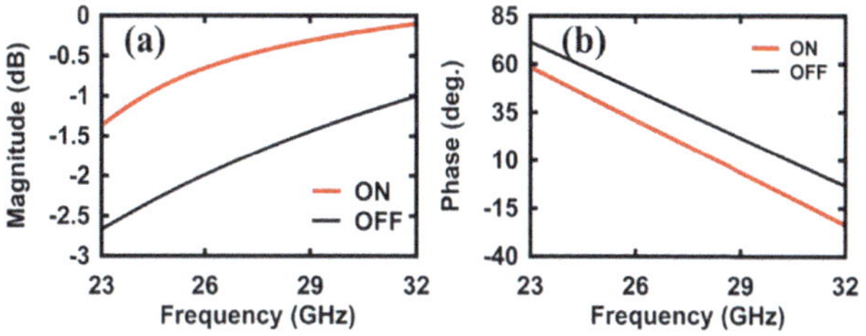

Fig. 2: Simulated reflection coefficient's (a) magnitude and (b) phase when meta-cell is in ON and OFF states.

The full structure of the reflecting metasurface is shown in Fig. 3(c). The complete structure contains both active and passive meta-cells and two PIN diodes are loaded onto the active ones. The meta-cells containing PIN diodes are classified as active, while those without PIN diodes are passive (indicated by the plus-shaped structures). This hybrid combination gives us three phase ranges- diode ON, diode OFF, and passive meta-cell. Thereby, providing us with a phase gradient for the arrangement as shown in Fig. 3(c) for a 5×5 array of the meta-cells. This 5×5 array is divided into four 2×2 subarray of active meta-cells. The PIN diodes in these subarrays are digitally controlled with four different bits. One bit is responsible for one subarray. Figure 3(d) shows the complete system for beam steering, in which the full metasurface is illuminated by a microstrip antenna positioned at a specific height below the metasurface. This height influences the antenna gain and beam direction. This two-layer system features a bottom layer with a microstrip antenna that operates at 24 GHz, and a top layer with a superstrate that includes a partial reflective surface (metasurface). The superstrate is equipped with active and passive meta-cells, which are used for beam steering to produce a boresight beam. This configuration creates a phased gradient from high to low.

3 Result and Discussion

The beam steering for the design structure has been observed for the major states by switching ON or OFF all the PIN diodes of sub-arrays A, B, C, and D through the proper biasing arrangement. In our design, all the PIN diodes in a subarray are simultaneously

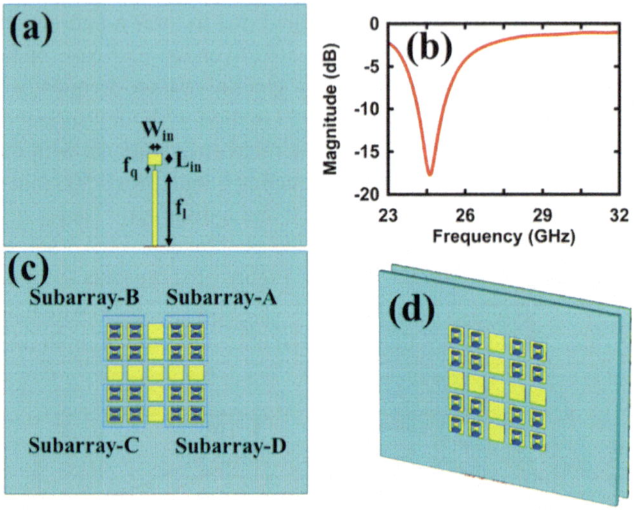

Fig. 3: (a) Microstrip patch antenna with dimensions $W = 90$, $L = 70$, $W_{in} = 4$, $L_{in} = 3.8$, $f_l = 21$, $f_q = 1.5$, , (Units: mm) (b) S-parameter of Microstrip patch antenna (c) Complete structure and arrangement of the active and passive meta-cells (top view), (d) cross-sectional view of the complete structure.

ON or OFF. If all the diodes in a subarray are OFF then that subarray is denoted as state "0". Whereas, state "1" of any subarray indicates that all the diodes in that subarray are in ON. Therefore, with four subarrays we have a total possible sixteen states, i.e., from 0000 to 1111. The state '0000' indicates that diodes in all subarrays are in the OFF state. Similarly, '0110' signifies that the diodes in subarray A and D are in OFF states, and diodes in subarray B and C are in ON states.

The radiation pattern of the proposed structure for some major states is shown in Fig. 4. In the '1111' state, the main lobe of the antenna radiates in a specific direction, achieving a gain of 16.3 dBi and an angular width at 24 GHz (see Fig. 4(a)). Now, when the states of the subarray change from '1111' to '0000' the main lobe direction changes from $\theta = 0°$ to 16° in $\phi = 0°$ plane, as shown in Fig. 4(b). However, a reduction in the gain of 8.41 dBi is observed for this configuration. Similarly, the radiation pattern for states '1100', '0011', '1001', and '0110' are shown in Fig. 4(c)–(f). When the subarray 'A' and 'B' are ON and subarray 'C' and 'D' are OFF (i.e., 1100 state), the main lobe is again directed along $\theta = 0°$ with higher angular width of 24.4^0 and lower gain of 14.5 dBi. From Fig. 4 it can be seen that the main lobe direction can be controlled by changing the states of the subarrays. Hence, a beam steering is obtained by only controlling the states of four subarrays.

The properties of the radiation pattern for all states are tabulated in Table 1. The main lobe of the antenna is steering from -12° at '0001' to $+16^{\circ}$ at '0000'. Therefore, a total of 28° beam steering range is obtained from our proposed structure. Furthermore, a maximum gain of 16.3 dBi is obtained for the '1111' state, whereas a minimum gain of 8.41 dBi is obtained at '0000' state.

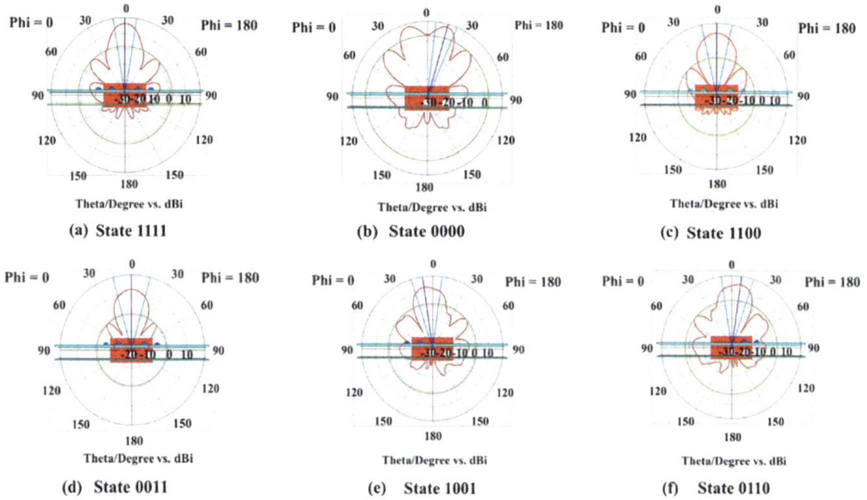

Fig. 4: Radiation pattern of the proposed design for different states of four subarrays (a) 1111, (b) 0000, (c) 1100, (d) 0011, (e) 1001, and (f) 0110.

Table 1: Property of the radiated beam from the design structure for all possible states of the four subarrays at 24 GHz.

State	Θ	φ	Angular width	Side lobe level	Gain
0000	16°	0°	16.7°	−5.8 dB	8.41 dBi
0001	−12°	0°	37.7°	−8.3 dB	10.7 dBi
0010	12°	0°	37.7°	−8.3 dB	10.7 dBi
0011	0°	0°	29.8°	−13.0 dB	12.9 dBi
0100	11°	0°	34.0°	−8.7 dB	11.7 dBi
0101	2°	0°	31.8°	−15.5 dB	12.8 dBi
0110	8°	0°	26.9°	−11.1 dB	14.1 dBi
0111	2°	0°	23.1°	−13.7 dB	15.3 dBi
1000	−11°	0°	33.9°	−8.7 dB	11.2 dBi
1001	−8°	0°	26.7°	−11.1 dB	14.1 dBi
1010	−2°	0°	31.8°	−15.5 dB	12.8 dBi
1011	−2°	0°	23.1°	−13.7 dB	15.3 dBi
1100	0°	0°	24.4°	−16.7 dB	14.5 dBi
1101	−1°	0°	21.4°	−14.7 dB	16.0 dBi
1110	1°	0°	21.4°	−14.7 dB	15.8 dBi
1111	0°	0°	18.2°	−14.8 dB	16.3 dBi

4 Conclusions

A low profile electronically controllable bean steering antenna has been presented in this paper. The direction of the main lobe beam is digitally controlled by adjusting the ON and OFF states of the PIN diodes. Using a microstrip antenna combined with a metasurface, the system can steer the beam at 24 GHz.The designed antenna demonstrated beam steering from $\theta = -12^\circ$ to $\theta = +16^\circ$ in the $\varphi = 0^\circ$ plane at 24 GHz. In state 1111, it achieved a peak gain of 16.3 dBi, a maximum directivity of 17.2 dBi, an angular width of 18.2°, and a side lobe level of −14.8 dB. In contrast, in state 0000, the antenna recorded a minimum gain of 8.41 dBi, a minimum directivity of 8.95 dBi, an angular width of 16.7°, and a side lobe level of −5.8 dB, all in the $\varphi = 0^\circ$ plane. The proposed design can be used for next-generation communication systems, including satellite communications, drone tracking, and radar applications, all of which demand high gain, broad steering range, and minimal cross-polarization.

References

1. Singh, K., Afzal, M.U., Kovaleva, M., Esselle, K.P.: Controlling the most significant grating lobes in two-dimensional beam-steering systems with phase-gradient metasurfaces. IEEE Trans. Antennas Propag. **68**(3), 1389–1401 (2019)
2. Miura, R., Tanaka, T., Horie, A., Karasawa, Y.: A DBF self-beam steering array antenna for mobile satellite applications using beam-space maximal-ratio combination. IEEE Trans. Veh. Technol. **48**(3), 665–675 (1999)
3. Nabeel, M.I., Afzal, M.U., Thalakotuna, D.N., Esselle, K.P.: Beam steering 2D leaky wave resonant cavity antenna for Ka-band satellite communication. In: 2024 18th European Conference Antennas and Propagation (EuCAP), pp. 1–5. IEEE (2024)
4. Zheludev, N.I., Kivshar, Y.S.: From metamaterials to metadevices. Nat. Mater. **11**(11), 917–924 (2012)
5. Kildishev, A.V., Boltasseva, A., Shalaev, V.M.: Planar photonics with metasurfaces. Science **339**(6125), 1232009 (2013)
6. Chen, H.T., Taylor, A.J., Yu, N.: A review of metasurfaces: physics and applications. Rep. Prog. Phys. **79**(7), 076401 (2016)
7. Agrahari, R., Rajbhar, P.K., Mahto, M., Jain, P.K.: Metasurface assisted wideband multifunctional polarizer. J. Appl. Phys. **133**(9), 094902 (2023)
8. Agrahari, R., Lakhtakia, A., Jain, P.K., Bhattacharyya, S.: Pixelated metasurfaces for linear-polarization conversion and absorption. J. Electromagn. Waves Appl. **36**(7), 1008–1019 (2021)
9. Veselago, V.G.: The electrodynamics of media with simultaneously negative values of permittivity and magnetic permeability. Sov. Phys. I-Usp **47**, 509–514 (1968)
10. Naqvi, A.H., & Lim, S.: Low-profile electronic beam-scanning metasurface antenna for Ka-band applications. Waves Random Complex Media, pp. 1–16 (2023). https://doi.org/10.1080/17455030.2023.2166153
11. Matos, S.A., et al.: High gain dual-band beam-steering transmit array for Satcom terminals at Ka-band. IEEE Trans. Antennas Propag. **65**(7), 3528–3539 (2017). https://doi.org/10.1109/TAP.2017.2702658
12. Gheethan, A.A., Jo, M.C., Guldiken, R., Mumcu, G.: Microfluidic based Ka-band beam-scanning focal plane array. IEEE Antennas Wirel. Propag. Lett. **12**, 1638–1641 (2013). https://doi.org/10.1109/LAWP.2013.2294153

13. Di Palma, L., Clemente, A., Dussopt, L., Sauleau, R., Potier, P., Pouliguen, P.: Circularly-polarized reconfigurable transmitarray in Ka-band with beam scanning and polarization switching capabilities. IEEE Trans. Antennas Propag. **65**(2), 529–540 (2016)

Investigation of Holographic Metasurface for Beam-Steering Applications

Aditya Pratap Singh[1], Pranjali Kashyap[1], Rajan Agrahari[2], Dev Kumar Mahato[1], and Manpuran Mahto[2(✉)]

[1] Material Science & Engineering Department, National Institute of Technology Patna, Patna, India

[2] Electronics & Communication Engineering Department, National Institute of Technology Patna, Patna, India

mmahto@nitp.ac.in

Abstract. Beam steering is important for modern communications systems, allowing control of the beam direction without the physical movement of the antenna. For advanced beam steering applications, a holographic metasurface is designed in the paper. The device consists of a rectangular integrated impedance surface that radiates the desired beam, where the beam direction is controlled by a varactor diode by varying the surface impedance. The device with the varactor diode achieved a high gain of 16.8 dBi with a beam steering range of 11°. This capability is important for applications such as autonomous vehicles, payment centers, efficient connections, and satellite communications, where precise and tunable signal orientation is required.

Keywords: Holographic metasurface · Beam steering antenna · Satellite communication

1 Introduction

Beam steering is a technique that is used to change the radiated beam track with some external mechanism without rotating or moving the device in modern wireless communication. The ability to control the radiated beam of the antenna can be useful in various modern devices like satellite communication, autonomous vehicles, 5G networks, etc. that require high directionality and control of the signal. The most common ways of achieving beam steering include—Phase Shifting, Mechanical Steering, Time delay, Frequency Scanning, etc. [1–4]. The performance of the beam steering can be increased by incorporating the metasurface in the device [5].

A metasurface is a 2-D representation of a metamaterial. A metamaterial is an engineered material that can be used for various optical applications that can be rarely observed in naturally occurring materials [6]. It can be used to manipulate waves at the

S. K. Ghosh et al. (Eds.): VEDA 2024, SPPHY 424, pp. 116–122, 2025.
https://doi.org/10.1007/978-981-96-5174-0_17

surface level by altering the phase and amplitude of the wave and can be used for beam steering, sensing, polarization conversion, absorption, etc. [5, 7–9]. A metasurface is a periodic array of unit-cell designs in two dimensions. They are able to manipulate the amplitude, and phase of EM waves. A metasurface that contains the property of holography is called a holographic metasurface.

The explanation of optical holography was first suggested by Gabor and later it was implemented by Chaccacci at microwave frequencies [10, 11]. Holography is best known for creating a unique 3-D image. The other uses can be for data storage, microscopy, etc. The combined properties of the holography and metasurface can be used for manipulating the direction of the wave. One of the key advantages of the holographic metasurface is its ability to change the direction of the radiated beam in the given direction. The verification of hologram's capability of leaky waves was first verified in the field of pattern synthetization by Fong et al. [12]. The holographic metasurface antenna and leaky wave antenna have nearly similar properties. However, the differences lie in the fact that the holographic antenna has the advantage of beamwidth controlling properties, high gain, high directivity, narrow beamwidth, and dynamic beamforming.

The traditional methods of beam steering have various disadvantages. The frequency scanning method in which the beam is steered by changing the frequency of the signal proposes an issue that includes limited beam control, difficulty in narrowband operations, etc. Previously, the sinusoidally modulated graphene-based antenna was used for beam steering at fixed frequencies [13]. However, this type of antenna is difficult to design due to biasing of the graphene ribbon. Later, an electronic beam-steering antenna was proposed, but it employed two varactor diodes for series and shunt connection in each unit cell [14]. Here, the series varactor diode was used for controlling the beam angle, and the shunt varactor diode was used for controlling the beam width. Due to the loading of the series and shunt varactor diode, the device became complex and difficult to design. Furthermore, the device is also limited by high energy loss owing to two varactor diodes in a single unit cell. The mechanical steering method which involves physically rotating the device to steer the antenna includes different problems such as high energy consumption, easy wear and tear of the device, etc. [2]. This paper uses a holographic metasurface for beam steering. The radiated beam track is manipulated by adjusting it's surface impedance using a varactor diode.

2 Unit Cell Design and Analysis

A schematic diagram of the proposed holographic metasurface unit cell is shown in Fig. 1. It is periodic along the x-axis. The ground layer and the top metallic patch of the unit cell are Copper ($t_m = 0.035$ mm). In between the metallic patch and the ground, there is a Rogers RT5880 substrate ($t = 1.575$ mm, and $\epsilon = 2.2$). The metasurface of each unit cell is loaded with a varactor diode on the metallic patch for tuning the phase difference. The above unit cell represents a periodic structure with a period of a. By

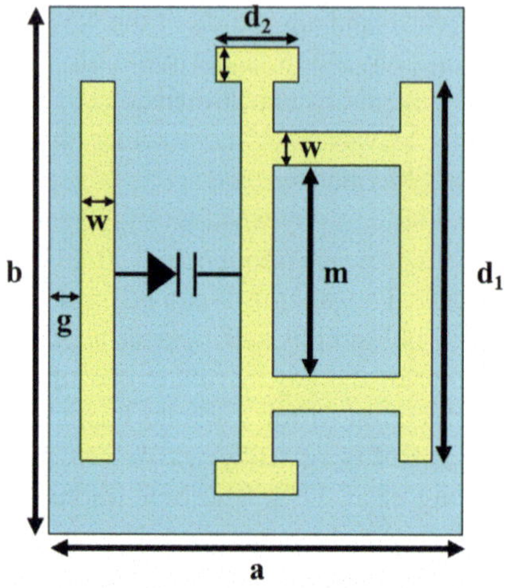

Fig. 1: Top view of the unit cell with the dimensions a = 4 mm, w = 0.4 mm, d1 = 3.6 mm, d2 = 0.8 mm, and m = 2 mm with the thickness of Rogers RT5880 is 1.575 mm. Here, g is the variable gap size to implement TrigMIS.

choosing the varactor diode which has a lower capacitance range, the steering range of the device can be increased. The track of the radiated beam can be manipulated by either tuning the capacitance of the varactor diode (C_v) or varying the frequency when the gap size 'g' is fixed. The unit cell design is simulated on CST Microwave Studio 2021 software.

Figure 2 shows the dispersion characteristics of the unit cell for different values of 'g' and 'C_V'. It shows the relationship of the gap size 'g', varactor diode capacitance 'C_V', and phase. The phase of the unit cell is approximately constant as 'g' increases (See Fig. 2(a)). However, increased in the phase is observed when the frequency is increased. Similarly, the phase-frequency plot for different value of capacitance 'C_V' is plotted in Fig. 2(b). At a fixed frequency, the phase increased as we increased the Varactor diode capacitance 'C_V'. The track of the radiation pattern can be altered by varying the capacitance of the Varactor diode.

The surface of the beam reconfigurable holographic metasurface is designed after designing the unit cell. Figure 3 shows the schematic picture of the proposed beam steering device with the varactor diode SMV1405-079LF. The range of the capacitance for the varactor device used varies from 0.63 to 2.67 pF. The device is divided into three main parts:—microstrip transmission line, matching transition, and triangularly modulated impedance pattern (TrigMIS). The TrigMIS consists of six modulation period with each period consisting of six-unit cells that is placed at different gap sizes for implementing low and high impedance that results in a triangular variation in surface impedance by a single modulating period.

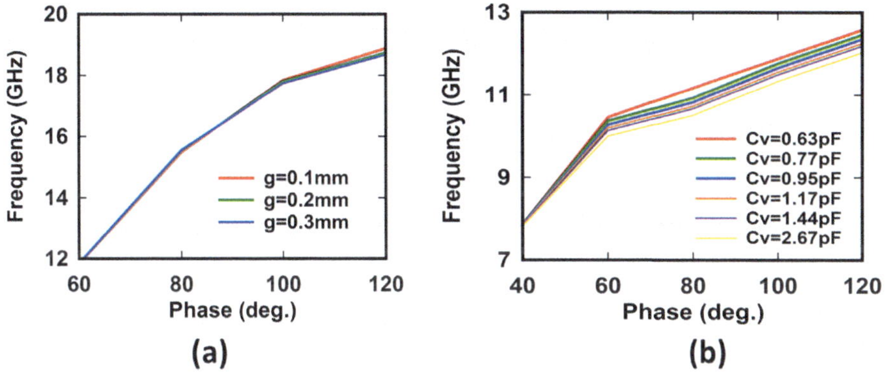

Fig. 2: Dispersion characteristics of the unit cell for; (a) different gap size 'g' (b) different C_V (varactor diode capacitance) with fixed g = 0.1 mm.

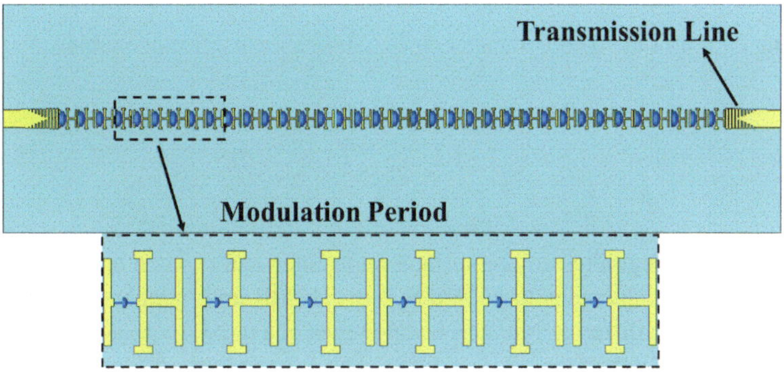

Fig. 3: Full structure of holographic metasurface-based beam steering device.

The length of the one modulation period is chosen to be 30 mm. Beam steering at a fixed frequency requires a dynamic surface impedance which is provided by the varactor diode-loaded unit cell. The capacitance of the varactor diode is controlled through a bias line that results in the tuning of the beam angle.

3 Result and Analysis

For the CST simulation, the varactor diode is replaced with the series RLC circuit given in the datasheet of the varactor diode SMV1405-079LF. At a fixed frequency, the beam steering requires a dynamic surface impedance which is provided by the varactor diode-loaded unit cell. According to the datasheet, the varactor diode capacitance (C_v) values for different biasing voltages are tabulated in Table 1. As we increase the biasing voltage from 0–30 V, the C_V decreases from 2.67 to 0.63 pF.

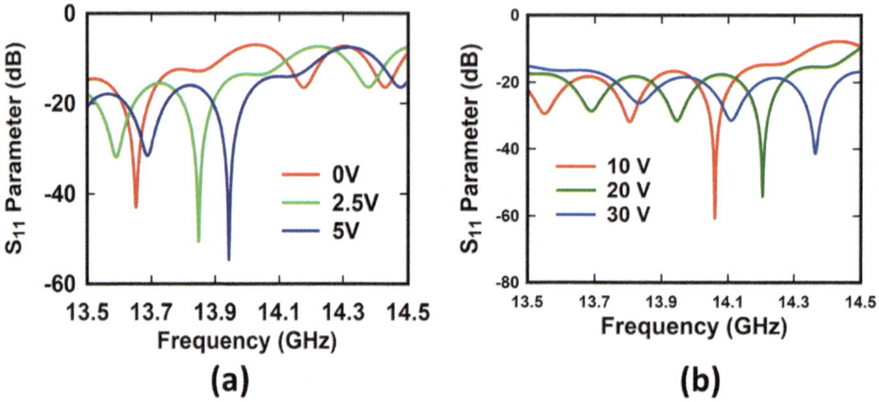

Fig. 4: S_{11} parameter of the beam steering device for different values of biasing voltages: (a) 0, 2.5, and 5 V; (b) 10, 20, and 30 V.

Table 1: SMV1405-079LF Varactor diode Capacitance at different biasing voltage

Biasing voltage in V	0	2.5	5	10	20	30
Capacitance (C_v) in pF	2.67	1.44	1.17	0.95	0.77	0.63

A proper matching between the surface impedance and momentum is required for the wide range of frequencies to ensure $S_{11} < -10$ dB and independence of gain of the radiated beam on different radiated beam angles at a fixed frequency. The proposed design is chosen to be radiated at 13.8 GHz. The S-parameter of the device is shown in Fig. 4 for different values of biasing voltage. The S_{11} for the proposed device is approximately below -10 dB for 13.5–14.5 GHz. It is important to note that the $S_{11} < -10$ dB for all capacitance at different biasing voltages.

Figure 5 shows the radiation patterns of the device when the biasing voltage $V_B \in [0, 2.5, 5, 10, 20, 30]$ V at a fixed frequency of 13.8 GHz. It is noticed that the radiation angle θ increases as the biasing voltage increases. The radiation property of the device is given in Table 2. We get a maximum gain of 16.8 dBi with $\theta = -26°$ for 0 biasing voltage.

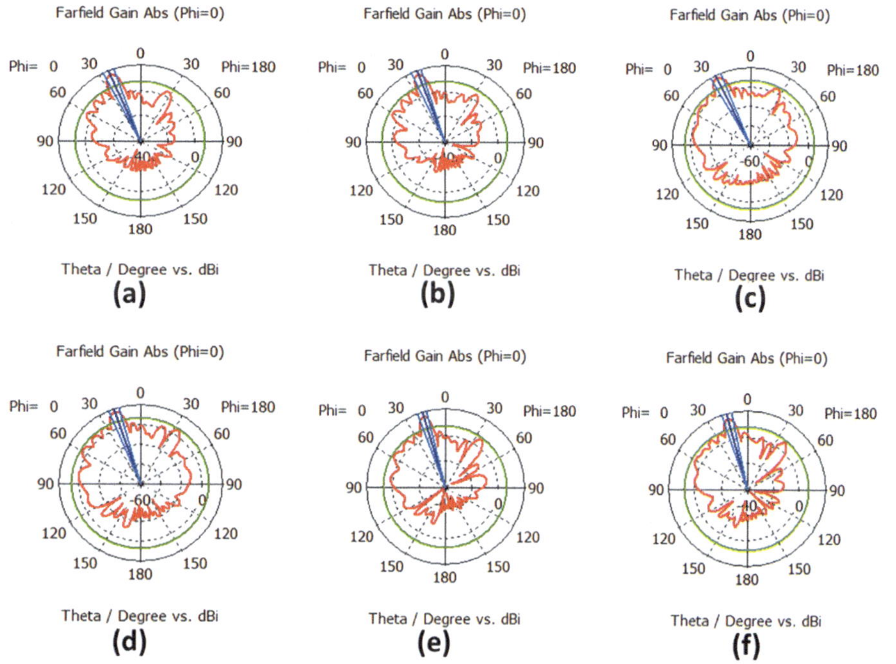

Fig. 5: Radiation pattern of the device with SMV1405-079LF at different biasing voltage (a) 0 V, (b) 2.5 V, (c) 5 V, (d) 10 V, (e) 20 V and (f) 30 V at 13.8 GHz.

Table 2: Radiation parameters for different biasing voltages.

Voltage (V)	Phi (Deg.)	Theta (Deg.)	Angular width (dB)	Gain (dBi)
0	0	−26	7.8	16.8
2.5	0	−23	7.7	16.7
5	0	−21	7.7	16.5
10	0	−19	7.6	16.5
20	0	−17	7.2	16.5
30	0	−15	7.1	16

4 Conclusion

In this paper, the design and simulation of the holographic metasurface has been presented. By changing the surface impedance of the device at a fixed frequency, the track of the radiated beam would be controlled electronically. A triangularly modulated impedance surface has been used for radiating the desired wave when it is excited by the surface wave. The triangularly modulated impedance surface has been found superior in many performances when compared with the sinusoidally modulated impedance surface. From the device with the given varactor diode a gain of 16.8 dBi with a beam steering range of 11° is obtained. The results of the proposed device can be used in various applications like satellite communication etc.

References

1. Ali, Q., et al.: Recent developments and challenges on beam steering characteristics of reconfigurable transmitarray antennas. Electronics **11**(4), 587 (2022)
2. Ahmed, F., Singh, K., Esselle, K.P.: State-of-the-art passive beam-steering antenna technologies: challenges and capabilities. Ieee Access **11**, 69101–69116 (2023)
3. Fu, X., Yang, F., Liu, C., Wu, X., Cui, T.J.: Terahertz beam steering technologies: from phased arrays to field-programmable metasurfaces. Adv. Opt. Mater. **8**(3), 1900628 (2020)
4. Ataloglou, V.G., et al.: Static and reconfigurable Huygens' metasurfaces: Use in antenna beamforming and beam steering. IEEE Antennas Propag. Mag. **64**(4), 73–84 (2022)
5. Babu, R., Agrahari, R., Mevada, P., Jain, P.K., Mahto, M.: Experimental study of holographic metasurface for beam steering applications. J. Appl. Phys. **134**(16) (2023)
6. Ali, A., Mitra, A., Aïssa, B.: Metamaterials and metasurfaces: A review from the perspectives of materials, mechanisms and advanced metadevices. Nanomaterials **12**(6), 1027 (2022)
7. Agrahari, R., Dwivedi, S., Jain, P.K., Mahto, M.: High sensitive metasurface absorber for refractive index sensing. IEEE Trans. Nanotechnol. (2023)
8. Babu, R., Agrahari, R., Jain, P.K., Mahto, M.: Bifunctional graphene-graphite-based THz wave metasurface absorber for sensing and electromagnetic shielding. IEEE Trans. Circuits Syst. I: Regul. Pap. (2024)
9. Agrahari, R., Lakhtakia, A., Kumar Jain, P., Bhattacharyya, S.: Pixelated metasurfaces for linear-polarization conversion and absorption. J. Electromagn. Waves Appl. **36**(7), 1008–1019 (2022)
10. Gabor, D., Stroke, G., Brumm, D., et al.: Reconstruction of phase objects by holography. Nature **208**, 1159–1162 (1965)
11. Checcacci, P., Russo, V., Scheggi, A.: Holographic antennas. IEEE Trans. Antennas Propag. **18**(6), 811–813 (1970)
12. Fong, B.H., Colburn, J.S., Ottusch, J.J., Visher, J.L., Sievenpiper, D.F.: Scalar and tensor holographic artificial impedance surfaces. IEEE Trans. Antennas Propag. **58**(10), 3212–3221 (2010)
13. Esquius-Morote, M., Gómez-Dı, J.S., Perruisseau-Carrier, J.: Sinusoidally modulated graphene leaky-wave antenna for electronic beamscanning at THz. IEEE Trans. Terahertz Sci. Technol. **4**(1), 116–122 (2014)
14. Lim, S., Caloz, C., Itoh, T.: Metamaterial-based electronically controlled transmission-line structure as a novel leaky-wave antenna with tunable radiation angle and beamwidth. IEEE Trans. Microw. Theory Tech. **52**(12), 2678–2690 (2004)

Dual-Band Antenna with AMC Reflector for 5G and L-Band Applications

Smita Kumari and Manpuran Mahto(✉)

National Institute of Technology Patna, Patna, India
{smitak.ph23.ec,mmahto}@nitp.ac.in

Abstract. A dual-band monopole antenna with coplanar waveguide feed, with artificial magnetic conductor as reflector has been designed and simulated using CST Microwave Studio. The antenna operates effectively at 3.5 GHz with the integration of AMC it operates at one additional band 1.5 GHz for L-Band applications. The antenna is designed using Rogers RT5880LZ having a relative permittivity of 2 and a loss tangent of 0.0021 with defected ground plane on the other side of the substrate. AMC unit cell having concentric ring structure is designed on the same substrate as used in the antenna. The antenna's performance parameters reflection coefficient gain and backward radiation show improved results along with minimization of antenna size making it suitable for applications in the ISM band and GPS.

Keywords: Metamaterial · AMC · CSRR · Antenna

1 Introduction

With the advancement in wireless communication systems multiband antennas are in demand as with single antenna devices can operate on multiple frequency bands. This multiband functionality supports multiple applications such as mobile communications, Wi-Fi, and IoT devices, Microstrip antennas are light in weight, compact, and easy to fabricate so are preferred for wireless communication. In recent years, these antennas have been studied for improvement in antenna gain, bandwidth, efficiency, and size reduction. The aim is to achieve performance improvement while reducing antenna size. The miniaturization of an antenna can be achieved in many ways. One way is to design radiating patches of different shapes like E-shaped [1], elliptical [2], triangular [3], and U-shaped antennas [4]. The antenna bandwidth can be increased by adding slots in the radiating patch. As antenna size remains unaffected by adding slots, this one is an effective method of size reduction.

Artificial magnetic material (AMC) is a type of metamaterial not found naturally but whose properties can be altered by the arrangement of materials that already exist in nature [5].

AMC is a two-dimensional metasurface widely used in planar antennas. The AMC is a type of metamaterial that has the property of the PMC at a certain frequency. With the use of AMC, the Antenna's performance is enhanced in terms of Directivity and Gain

S. K. Ghosh et al. (Eds.): VEDA 2024, SPPHY 424, pp. 123–129, 2025.
https://doi.org/10.1007/978-981-96-5174-0_18

[6, 7]. Backward radiation of the patch antenna is also reduced [8]. The AMC layer is integrated with the radiating patch at a small distance while keeping the overall size of the integrated structure minimum.

This paper proposes an antenna with an AMC reflector placed at the back of the antenna at a 15 mm distance. The addition of split rings in the AMC structure achieves one more resonance frequency band at 1.5G Hz. The distance between the reflector and radiating patch can be decreased using AMC in place of PEC (perfect electric conductor). With the use of geometrically symmetric AMC, polarization insensitivity is achieved and the polarization of the radiating antenna will not be disturbed. At the lower band, AMC acts as a perfect electric conductor and hence acts as the ground for the antenna. At a higher band of frequency acts as the perfect magnetic conductor and suppresses parallel plate mode and hence can be used as a reflector to magnetic dipole.

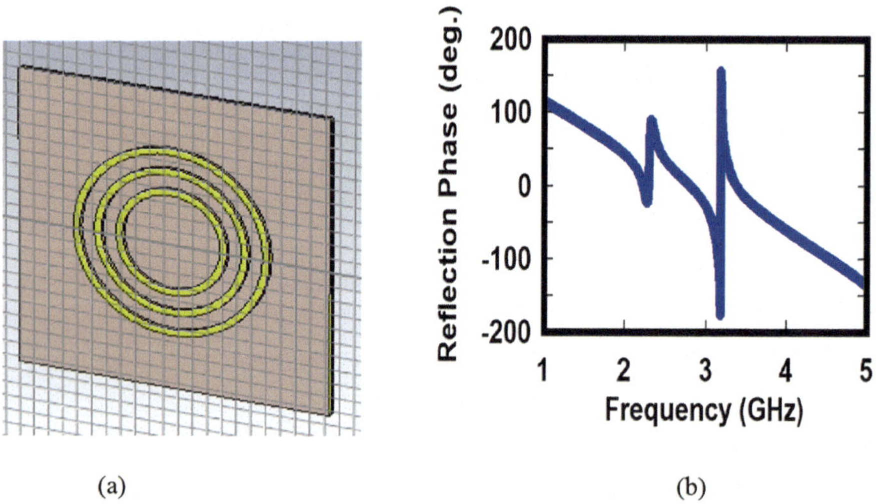

Fig. 1. (a) Unit cell Design (b) Reflection coefficient phase plot

The design described involves a rectangular patch antenna enhanced with two concentric rings, which improves the gain at 3.5 GHz. This configuration concentrates the radiated power and enhances the impedance matching, increasing the overall efficiency and performance of the antenna at that specific frequency.

2 AMC-Backed Antenna Design

2.1 AMC Design

AMC is designed using substrate Rogers RT5880LZ with a thickness of 0.20 mm, dielectric permittivity of 2, and a loss tangent of 0.0021. The unit cell of AMC is square-shaped with concentric rings with a width of 0.5 mm. The radius of concentric rings is 5.1 mm, 8.2 mm, 11.3 mm, and 14.4 mm from the innermost ring to the outermost ring respectively. The complete size of the unit cell is $12 \times 12 \times 0.25$ mm^3. Figure 1(a) shows the structure of the unit cell with rings and Fig. 1(b) shows the phase of the reflection coefficient. It can be seen that the phase of the reflected wave is in phase with that of the incident wave.

2.2 Antenna Design

Patch antenna and concentric rings in the ground plane are designed on Rogers RT5880LZ substrate with relative permittivity 2 and $\tan\partial = 0.0021$. The dimension of the antenna is $30 \times 46 \times 0.20$ mm^3. A microstrip feed line feeds the rectangular radiating patch. A stub is attached to the feed line. The top part of the dual-band antenna is shown in Fig. 2(a) with dimensions in Table 1. In Fig. 2(b) back part of the antenna is shown. The stepwise design and corresponding result are shown in Fig. 3. With ground rings there is gain enhancement at 3.5 GHz as shown in Fig. 4.

(a) (b)

Fig. 2. (a) Top View of Antenna (b) Back View of Antenna

Table 1. Design parameters

Parameters	Values (mm)	Parameters	Values (mm)
L_s	46	L_f	26
W_s	30	W_f	2
L_p	11	L_d	14
W_p	24	G_1	0.3

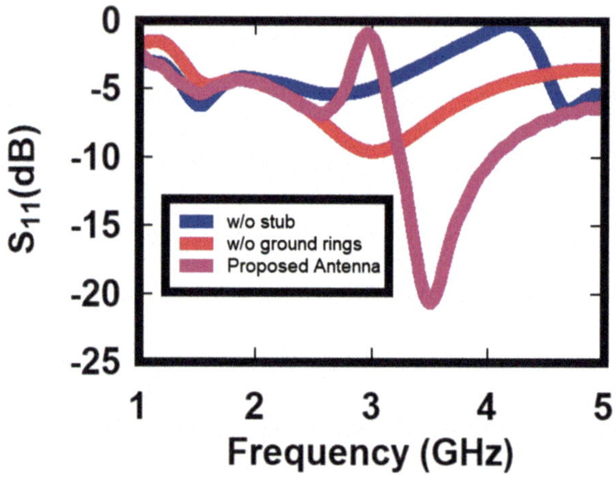

Fig. 3. Reflection coefficient plot stepwisee

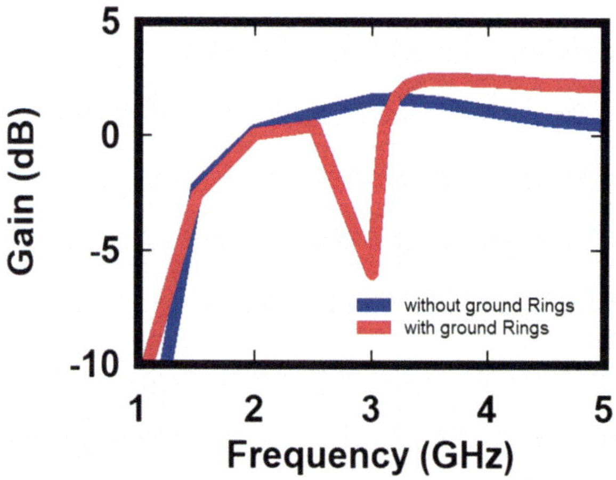

Fig. 4. Gain plot with and w/o Ground Rings

2.3 Design of Antenna with AMC

An array of two AMC unit cells is placed 15 mm below the antenna as shown in Fig. 5. The gap is filled with air. The Antenna with CPW feed and a radiating patch is placed 15 mm above the AMC array, and it acts as an electromagnetic reflector. Different gaps between the Reflector and antenna parameters were calculated. The best results were obtained at 15 mm (Fig. 6).

Figure 7 Compares the antenna's gain with and without AMC.

There is a significant improvement in gain particularly at 3.5 GHz over the frequency range (3.27–4) GHz.

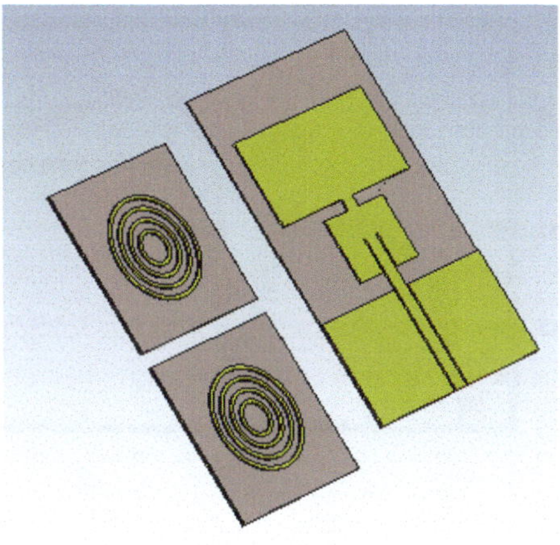

Fig. 5. Antenna with AMC

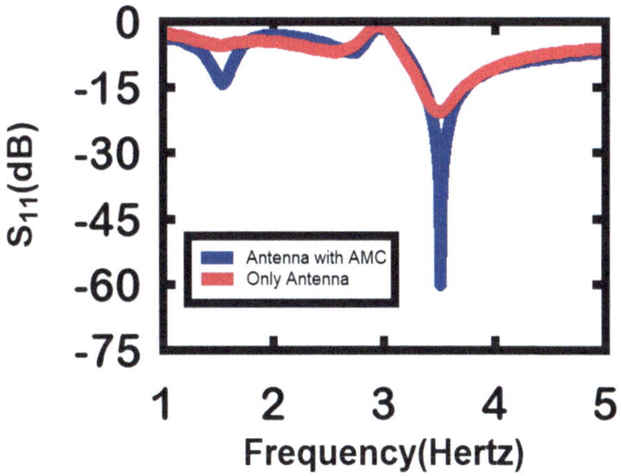

Fig. 6. Reflection coefficient vs Frequency plot

3 Result and Discussion

The proposed antenna is simulated on CST Microwave Studio. The reflection coefficient and gain are measured at every step and compared. The effect of the addition of stubs in feed line rings in the ground is studied. With the addition of the AMC layer, it is noticed that there is an addition of another band of frequency and this way antenna turns to a dual band. The reflection coefficient at both frequencies is below −10 dB at 1.5 and 3.5 GHz. Without AMC, the Antenna resonates only at 3.5 GHz with a reflection coefficient of −20 dB and a gain of 2.5. With the integration of AMC, there is an addition of L band frequency at 1.5 GHz, having a reflection coefficient of −13 dB, making the antenna operate in dual-band mode. It also improves the reflection coefficient to −60 dB from −20 dB without AMC shown in Fig. 6. Figure 7 shows the gain versus frequency variation

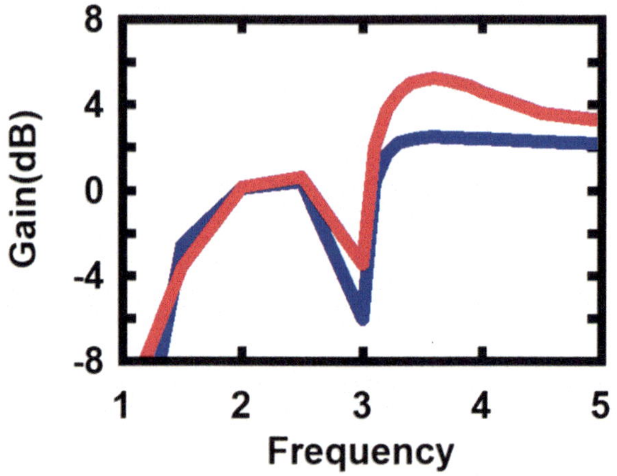

Fig. 7. Gain plot

with and without AMC. With this placement of AMC below the antenna improved the gain.

4 Conclusion

A dual-band antenna has been designed with AMC as a reflector in place of a PEC reflector to enhance its gain while maintaining a low profile. Gain improves to 5.2 dB from 2.5 dB with the integration of AMC. There is a great improvement in the reflection coefficient from −20 to −60 dB at 3.5 GHz. This proposed design finds use for 5G applications. A lower band resonant frequency was introduced with the addition of AMC that finds application in the L band GPS application.

References

1. Alibakhshi-Kenari, M., Naser-Moghadasi, M., Ali Sadegh-zadeh, R., Singh Virdee, B., Limiti, E.: New compact antenna based on simplified CRLH-TL for UWB wireless communication systems. Int. J. RF Microw. Comput.-Aided Eng. **26**(3), 217–225 (2016)
2. Dastranj, A.: Very small planar broadband monopole antenna with hybrid trapezoidal-elliptical radiator. IET Microwaves Antennas Propag. **11**(4), 542–547 (2017)
3. Ju, J.: A novel compact UWB monopole antenna with enhanced bandwidth using triangular defected microstrip structure and stepped cut technique. Microw. Opt. Technol. Lett. **58**(6), 1561–1563 (2016)
4. Hu, H., Chen, F., Chu, Q., Member, S.: A wideband U-shaped slot antenna and its application in MIMO terminals. IEEE Antennas Wirel. Propag. Lett. **15**, 1–4 (2015)
5. Sakli, H., Abdelhamid, C., Essid, C., Sakli, N.: Meta material-based antenna performance enhancement for MIMO system applications. IEEE Access **9**(4), Article ID 38546 (2021)
6. Elwi, T.A.: Remotely controlled reconfigurable antenna for modern 5G networks applications. Microw. Opt. Technol. Lett. **63**(8), 2018–2023 (2021)

7. Bakhtiari, A., Sadeghzadeh, R.A., Moghadasi, M.N.: Gain enhanced miniaturized microstrip patch antenna using metamaterial superstrates. IETE J. Res. **65**(5), 635–640 (2019)
8. Suvarna, K., Ramamurthy, N., Vardhan, D.V.: Minia-turized and gain enhancement of tapered patch antenna using defected ground structure and metamaterial superstrate for gps applications. Prog. Electromagn. Res. C **108**, 187–200 (2021)

Development of Anode Power Supply for Gyrotron

Rasesh Dave[1(✉)], Dishang Upadhyay[1], Kush Mehta[1], Niranjanpuri Goswami[1], Bhavin Raval[1], Amit Patel[1,2], Sandip Gajjar[1], Darshan Parmar[1,2], Hitesh Dhola[1], Vikrant Gupta[1], Jatin Patel[2], Dharmesh Purohit[2], K. G. Parmar[2], Harshida Patel[2], Hardik Mistry[2], Ravindra Solanki[2], Aruna Thakar[1], B. K. Shukla[2], N. P. Singh[1,2], and Ujjwal Baruah[1,2]

[1] ITER-India, Institute for Plasma Research, Bhat, Gandhinagar 382428, India
rasesh.dave@iterindia.in

[2] Institute for Plasma Research, Bhat, Gandhinagar 382428, India

Abstract. The High Power Gyrotrons use mainly two high voltage power supplies cathode voltage power supply (~−50 kV) and anode voltage power supply (~+ 25 kV) to deliver high power microwave. Both the power supplies are critical for the safe and reliable operation of Gyrotron. The Cathode (Main) voltage power supply has been already developed indigenously, the Anode Power Supply (APS) also has been recently developed in-house and integrated successfully with 42GHz Gyrotron.

Pulse Step Modulation (PSM) technology has been extensively used and a popular choice for high-voltage, high-power systems (MW level). While industrial low power high voltage solutions exist, they are often optimized for steady-state applications with limited dynamic response. In order to address the required fast and programmable dynamic response for Anode of MW class Gyrotrons, a 25 kV-100 mA, APS was indigenously developed and tailored for use in Electron Cyclotron Resonance Heating (ECRH) systems at IPR. This advanced power supply operates in both pulsed and continuous wave (CW) modes.

Additionally, it offers programmable rise/fall times (1–5 ms) and voltage modulation capabilities from 0% to 100% of the set voltage. The switched power supply modules, rated for 400V and 100 mA, are housed in PVC enclosures and installed with progressive isolation to enable it to test at high-voltage DC (HVDC) up to 30 kV. The controller, which is CE-certified, operates in an optically isolated manner and reduced channel count is achieved through passive fiber optic splitters.

Keywords: Anode Power Supply · Gyrotron · ECRH

1 Introduction

1.1 IPR ECRH System for SST-1 and Aditya-U

The Electron Cyclotron Resonance Heating (ECRH) system is used for plasma start-up (breakdown), heating and current drive in the Tokamaks. The EC waves transfers energy to the electrons in the plasma at the electron cyclotron resonant frequency. The

S. K. Ghosh et al. (Eds.): VEDA 2024, SPPHY 424, pp. 130–137, 2025.
https://doi.org/10.1007/978-981-96-5174-0_19

42 GHz ECRH system consists of a high power Gyrotron tube installed on a cryo-magnet, a Matching Optic Unit (MOU), corrugated waveguide-based transmission line, dummy load and quasi-optical launcher. This ECRH system is used in SST-1 and Aditya-U Tokamaks to carry out for Electron Cyclotron (EC) experiments i.e. ECRH assisted breakdown, start-up and heating experiments at fundamental (1.5T operation) and second harmonics (0.75T operation) on SST-1 and Aditya-U Tokamaks. The Gyrotron delivers 500kW power at 42 GHz for 500 ms duration.

The Gyrotron power supply system consist of regulated high voltage cathode power supply, APS, cryo-magnet power supply, filament power supply and ion pump power supply. Figure 1 is a schematic of the 42 GHz Gyrotron system at IPR [1].

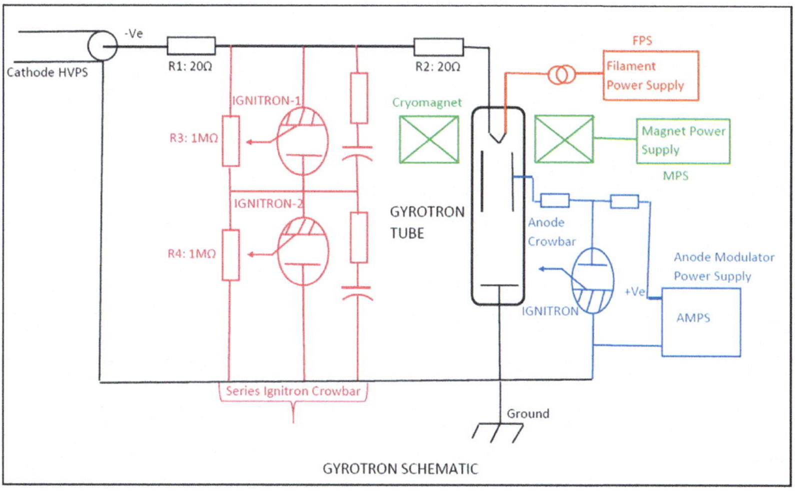

Fig. 1. 42GHz Gyrotron Schematic

2 Anode Power Supply

The 42 GHz Gyrotron is a depressed collector design diode-type tube. With a Magnetron Injection Gun (MIG) that creates a hollow circular electron beam when an acceleration potential is supplied between the cathode and the anode. The cathode high voltage power supply and the anode power supply for the Gyrotrons have strict requirements for safe and reliable operation. Gyrotrons are delicate tubes because their critical fault energy is around 10-Joule. The APS is a 42GHz Gyrotron power supply that gives beams the necessary acceleration to obtain microwave power with fast turn off time.

2.1 APS Specifications and Configuration

Table 1 summarizes the important technical features of the PSM based APS [2–5] for the IPR ECRH system

The APS components are listed in Table 2.These components are housed in an industrial cabinet of size of 0.8 × 1.2 × 2m. The APS schematic is illustrated in Fig. 2.

Table 1. Technical specifications of Anode Power Supply

Sr. no	Parameter	Value
Steady state specification		
1	Operating voltage range	10–25 kV (Nominal 25kV)
2	Current rating max	100 mA
3	Settling time	< 20 ms
4	Voltage accuracy & repeatability	$\leq \pm 1\%$ of nominal value
5	Voltage resolution	500V or better
6	Voltage ripple	$< 2\%$ p-p of nominal value
Dynamic parameter specification		
7	Ramp-up/down time	1–5 ms programmable
8	Overshoot/undershoot during ramp up/down	$\pm$ 500 V
9	Voltage modulation frequency	Up to 100 Hz (ON/OFF)
10	Total energy delivered to the arc in case of fault	Less than 10 J
11	Shutdown time	$\leq 10\ \mu s$

Table 2. Components for Anode Power Supply

Components	Details
Multi-secondary transformer	415V $\pm$ 10%, 50Hz, Three Phase, Delta Connection
SPS modules	390V AC Input, 410V, 2A DC Output, 72#
Controller	Electronic Corporation of India (ECIL), Hyderabad make CE Certified Zynq 7000 SoC Controller with Signal Conditioning Units

3 Description of Anode Power Supply

3.1 Multi-secondary Transformer

A multi-secondary transformer is an essential part of a PSM based HVPS. Here a 10KVA Vacuum Pressure Impregnated (VPI) with 72 secondary winding is utilized. A unique approach of single-phase secondary winding is undertaken here, with higher interwinding capacitance compared to formerly installed large size three phase multi-secondary transformers [6] (Table 3).

3.2 Switched Power Supply Modules

Switched power supply (SPS) modules consists of a single-phase rectifier section followed by a DC link filter and a synchronous buck converter section, for IGBT short

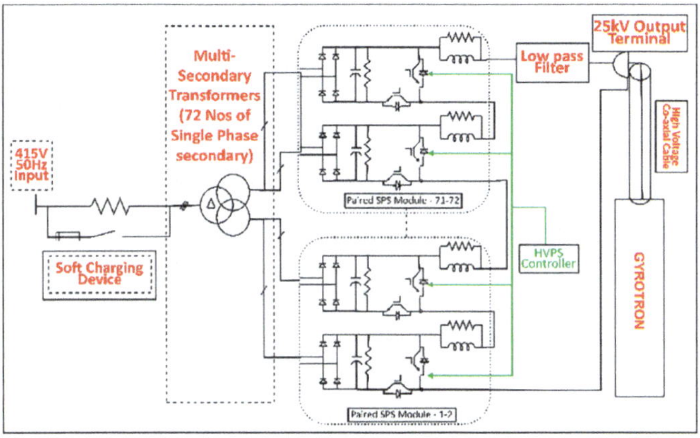

Fig. 2. Schematic of Anode Power Supply

Table 3. Multi-secondary Transformer Specification

Rating	10 KVA, VPI
Primary winding	415V ± 10%, 50 Hz, Three Phase, Delta Connection
Secondary winding	290V AC, Single Phase, 72 Nos
Test voltages for insulation test	Secondary to Primary winding & Core: 37.5 kVdc Primary & Secondary Windings with respect to Core: $5kV_{dc}$ and $3kV_{ac}$
Cooling	Air Natural
Insulation class	Class H
Standard	IEC 60076-11

circuit protection output inductor is used. PWM reception and reproduction with dead band of 2μs among the IGBT switches of SPS modules are managed on board. The SPS modules generate pulsating DC output in accordance with the PWM input generated by the HVPS controller. Each SPS module has a DC link of 410V with 2A peak current capacity and 200mA average current capacity, power schematic is shown in Fig. 2. SPS modules communicates with HVPS controller via fiber optic links, a voltage to frequency based optical link sharing DC link value and an optical input PWM link. A single PCB contains two such SPS modules shown in Fig. 3, encapsulated in polycarbonate enclosure, these will help to maintain isolation among the modules in compact footprint. A resistor based soft charging is placed in 415V three phase input, for controlling capacitor and magnetic inrush current.

3.3 Physical Layout

The physical layout of power supply is divided into three parts as shown in Fig. 4.

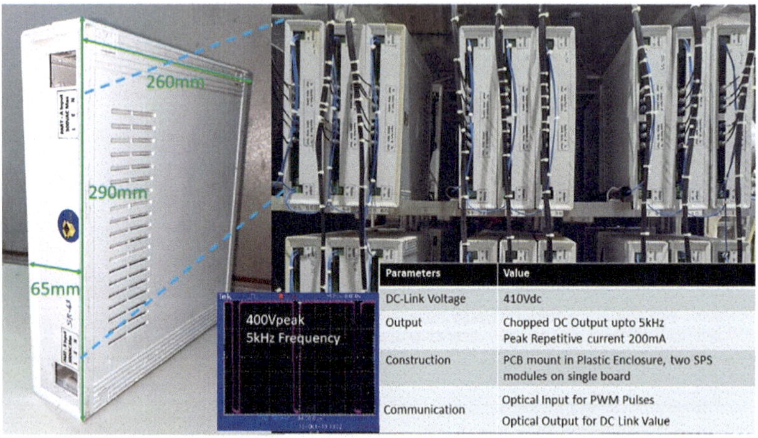

Fig. 3. Enclosed switched power supply modules

Fig. 4. APS Physical Layout in standard industrial cubicle

The multi-secondary transformers heavy weight section at bottom followed by two tier SPS section and a low voltage controller section. High voltage isolation from ground (30kV) is maintained by utilizing acrylic sheets, the cables among the tiers are passed through acrylic pipes to maintain the isolation while the polycarbonate body of SPS modules helps to maintain isolation among the modules. The low voltage controller section is kept at top.

3.4 Controller

Pulse Width Modulation (PWM) pulses are applied to SPS modules in PSM based HVPS with an aim to deliver the desired output voltage. PWM pulses for all SPS modules are phase shifted, with identical switching frequency and PWM duty.

ECIL make CE certified AMD/Xilinx Zynq 7000 SoC Controller [7] based Signal Conditioning Unit (TX/RX) monitors the entire HVPS, selects the operation mode and sets the parameter. The controller performs voltage control loops for output/ feedback control and takes care of fast protection.

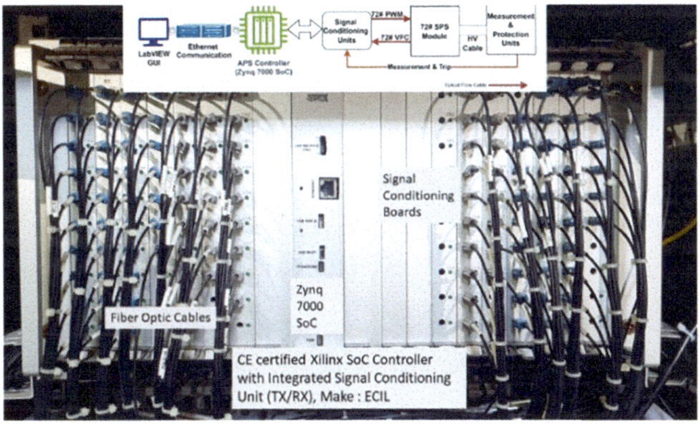

Fig. 5. APS Controller Hardware and Architecture

All signals from SPS modules and Field Devices are transmitted to the controller unit through optical fibers for electrical isolation. Commercial-off-the-shelf (COTS) passive fiber optic splitters (1x2) are used to reduce the controller channel requirements. Controller architecture and hardware are as shown in Fig. 5.

GUI is connected to the controller through the optical fiber media converter to establishes the Ethernet communication. The output voltage can be preset by either LabVIEW based GUI or VFC based optical signal from ECRH Controller.

3.5 Measurement and Protection

Anode power supply demands continuous monitoring for output voltage. To measure high voltage up to 25 kV, a resistive divider with a ratio of 10000:4 is selected, the resulting voltage in a range of 0–10 V is converted in to frequency by a voltage monitoring (signal conditioner) unit (VMU) with 0.25% accuracy. The unit gives optical frequency in a band of 100 kHz to 300 kHz, then measured by optically isolated controller. The equivalent voltage can be displayed on GUI for monitoring. The same VMU has over voltage detection circuit with settable threshold, it gives optical fail-safe pulse in case of over voltage event to controller; to switch off power supply for load protection (Fig. 6).

Additionally, the Gyrotron is to be protected against over current and the anode power supply needs protection against short circuit. A comparator circuit designed with settable threshold along with an LEM current transducer CTSR 0.3-P gives a failsafe active low light pulse to controller when detects over current. The controller then switches off the power supply as fast as possible, output voltage cab be reduced to less than 10% within 10μs.

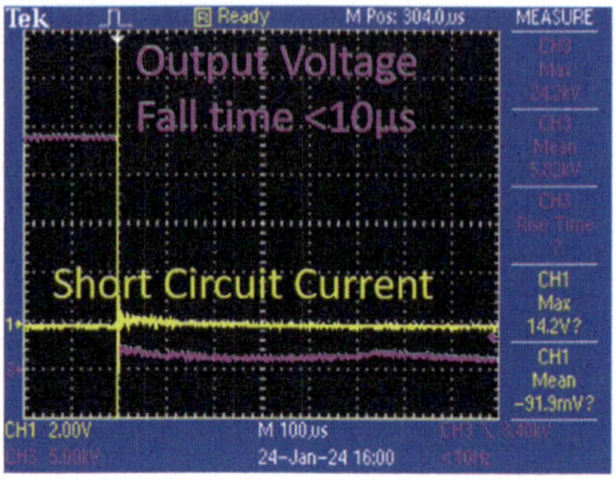

Fig. 6. Fast output voltage turn-off in case of fault

4 Testing on Non-inductive Resistor Load Bank

Validation of APS was performed on non-inductive pulse dummy Load (1s) & simulated fault conditions. Figure 7 shows Dummy Load Test setup with schematic.

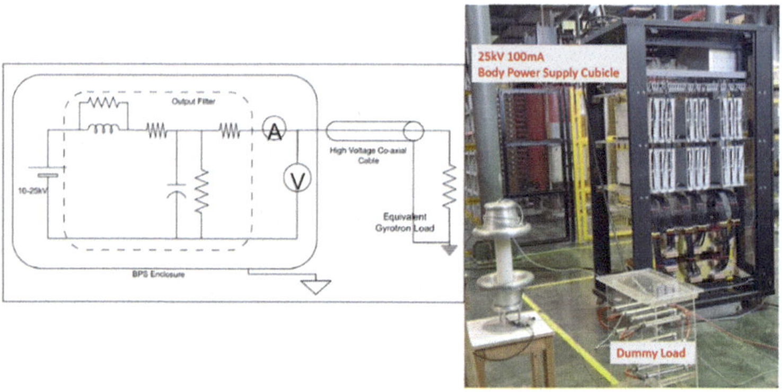

Fig. 7. Dummy Load Test setup with schematic.

Prior to integration with Gyrotron, power supply was successfully verified for programmable rise/fall time, line/load regulation, resolution and accuracy. Mandatory wire burn test was performed to demonstrate energy limiting capabilities. Figure 4 shows power supply test set up with dummy load (10mA). Fig. 8 are the demonstrated test results for performance parameters.

5 Conclusion and Future Work

The Anode power supply was successfully integrated with ECRH system of IPR and operated to support the 200 kW Gyrotron operations. It represents a unique and advanced development in high-voltage low-current power supplies, offering improved reliability, stability, accuracy, modularity, programmable rise and fall time, ability for fast replication and low stored energy. This APS has been successfully tested up to 25 kV for a duration of 1 second. Its modulation capabilities have been tested up to 100 Hz.

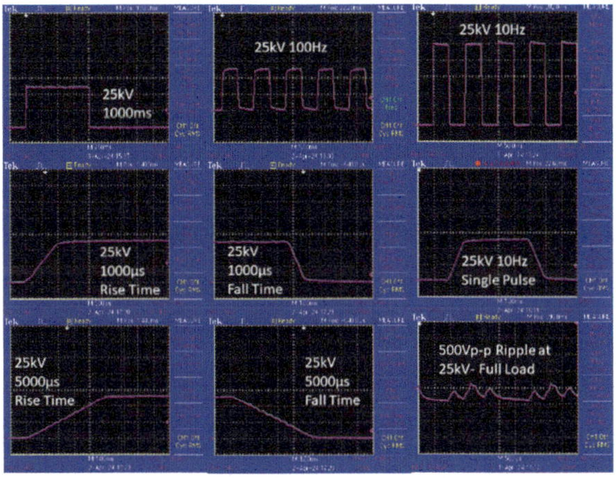

Fig. 8. Test Results with gyrotron equivalent dummy load

Necessary changes in APS SPS Module design for partial depth voltage modulation and integrated operation of power supply with 42 GHz Gyrotron shall be considered as a major part of future work.

References

1. Shukla, B.K., et al.: 42GHz 0.5 MW ECRH system for tokamaks SST-1 and aditya. IEEE Trans. Plasma Sci. **40**, 1234–1238 (2012)
2. Jin, Y.K., et al.: The anode power supply based on the pulse step modulation technology for the ECRH system on J-TEXT tokamak. Fusion Eng. Des. **146**, 2654–2657 (2019)
3. Patel, A. et al.: Development of 3 MW dual output high voltage power supply for ICRH system. In: 2016 IEEE International Power Modulator and High Voltage Conference (IPMHVC), San Francisco, CA, USA, 2016, pp. 95–100. https://doi.org/10.1109/IPMHVC.2016.8012896
4. Patel, A., et al.: Initial operation of 3 MW dual output high voltage power supply with IC RF system. Fusion Eng. Des. **126**, 59–66 (2018)
5. Upadhyay, D.V., et al.: Assessment and optimization of output filter for microsecond transient in PSM-based megawatt HVPS. IEEE Trans. Plasma Sci. **51**(7), 1988–1995 (2023). https://doi.org/10.1109/TPS.2023.3283605
6. ARYA TRANSFORMER "Factory Test Report of 10kVA VPI Multi-secondary Transformer_v1.2", M/S ARYA TRANSFORMER, VADODARA, GUJARAT, INDIA
7. Rasesh, D., et al.: Development of control system for multi-converter High voltage Power supply using programmable SoC. J. Phys: Conf. Ser. **823**, 12035–12043 (2017). https://doi.org/10.1088/1742-6596/823/1/012035

Effects of Slot Lines in the Metamaterial-Inspired Dual-Band Slow Wave Structure

Gowre Soumya, Rajnish Kumar, and Manpuran Mahto(✉)

National Institute of Technology Patna, Patna, India
mmahto@nitp.ac.in

Abstract. The effects of slot lines in a dual-band slow wave structure assisted by metamaterials are investigated in this work. Our results show that when the number of slot lines rises, the concurrently, frequency rises as well. We performed a cold test using the Eigenmode solver in simulation to calculate the dispersion and high impedance of interaction in the C- and X- bands at resonant frequencies i.e. 6.8 GHz and 11.2 GHz. The dual-band DB-CeSRR stands for complementary electric split ring resonator. In a circular waveguide, unit cells are integrated to measure the Dual-band MTM SWS transmission and reflection coefficients.

Keywords: Metamaterial (MTM) · Slow wave structure (SWS) · DB-CeSRR · dual-band

1 Introduction

Veselago introduced the concept of left-handed metamaterials (MTMs) in 1968 [1], but it wasn't experimentally realized until 2000 [2]. MTMs are artificial materials, combining the words 'meta' (meaning beyond) and 'materials' (meaning matter), that do not occur naturally. MTMs exhibits unconventional characteristics, such as a reverse Doppler Effect, reverse Cherenkov radiation and negative R.I (refractive index). These materials are used in variety of fields, including absorptive antennas, cloaking devices, automotive and aerospace industries and super-lens [1, 3].

Radio frequency (RF) interaction structures based on MTM technology are of great interest to researchers, especially Vacuum electron devices: slow wave structure (SWS) (VEDs). SWS are employed to match the RF's velocity analogous to an electron beam signal. Lately, dual-band VEDs have reaped a great deal of interest due to their potential among other things, short-pulse RADAR system applications [4, 5]. Various SWSs have been proposed to generate dual-band signals in conventional dual-band VEDs [6, 7]. There are two primary challenges that need to be addressed: (a) achieving high efficiency and (b) miniaturizing the device. To tackle these challenges, the metamaterial slow wave structures are utilizing as RF modulation section in vacuum electron devices, which aids in miniaturization and also enhances the efficiency of the device [8, 9].

Initially, MTM-based VEDs RF interaction region, split ring resonators (SRRs) were used, which led to dielectric breakdown issues [10, 11]. To address this, [12] designed

S. K. Ghosh et al. (Eds.): VEDA 2024, SPPHY 424, pp. 138–144, 2025.
https://doi.org/10.1007/978-981-96-5174-0_20

CeSRR MTM-based planar all-metal SWS, effectively solving the problem. [13] suggested an interaction region for radio frequency applications, using a circular waveguide integrated with MTM unit cells based on complementary electric split-ring resonator (CeSRR) operating at 2.45 GHz. Additionally, an equivalent circuit method was used by Wang et al. [14] to analyze an all-metal MTM that was introduced. A dual-band backward wave oscillator (BWO) with a coaxial coupler and an MTM design was studied by Tang et al. [15]. Kumari et al. [16] developed a dual-band metamaterial SWS based on complementary electric split ring resonators (CeSRRs), which generates C and X frequency bands. Most of these reported works is limited by their minimum number of slot lines and lower frequency range.

In this paper, we examine the effects of slot lines in a MTM assisted dual-band SWS that can generate two different frequency bands. The frequency increases with the number of slot lines added. MTM SWS is composed of DB CeSRR unit cells arranged in a circle waveguide, which generates the two unique frequency ranges, the X and C bands.

2 Model Representation

Figure 1 illustrates the two-dimensional-schematic of a DB-CeSRR MTM unit cell. The design features six slot lines, which enable the generation of two distinct frequency bands: C-band and the X-band. The C band resonates at 6.8 GHz and while the X band at 11.2 GHz. The unit cell is arranged with specific parameters, as provided in above Table 1. Further, by incorporating the unit cell based on DB-CeSRR into circular waveguide, the MTM SWS is formed, enabling the reflection and transmission characteristics for the DB-CeSRR MTM SWS.

Fig. 1. Schematic of DB-CeSRR MTM unit cell

3 EM Properties and Simulations

This section focuses on simulating the metamaterial-slow wave structure's (MTM SWS) electromagnetic characteristics. We examine the metamaterial-slow wave structure via dual-band complementary electric split ring resonator assistance, in more detail by

Table. 1. Design parameters of CeSRR

Parameters	Values (mm)
Inner radius of inner circumference (r_1)	2
Outer radius of inner circumference (r_2)	4
Inner radius of middle circumference (r_3)	5
Outer radius of middle circumference (r_4)	7.6
Inner radius of outer circumference (r_5)	9.9
Outer radius of outer circumference (r_6)	11.8
Groove height (h_1)	0.5
h_2	1.5
Groove metal width (w)	1
Periodicity (p)	7
Thickness (t)	1

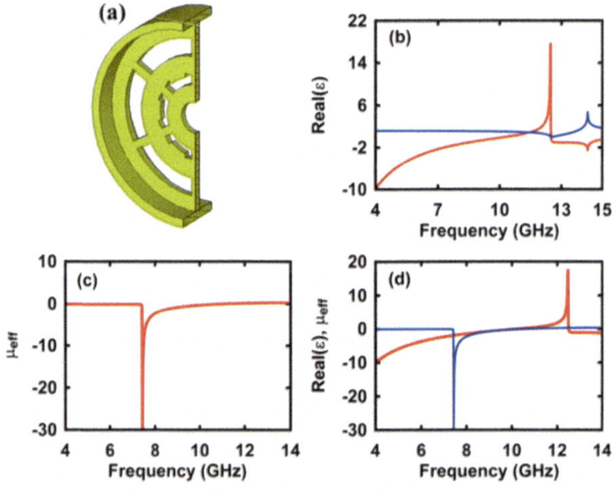

Fig. 2. (a) Schematic of DB-CeSRR unit cell loaded in circular waveguide (b) Effective permittivity (ε_{eff}) (c) Circular waveguide's effective permeability (μ_{eff}), and (d) Combination of ε_{eff} and μ_{eff} of MTM SWS.

assessing the permittivity and effectiveness of permeability. Finding the ε and μ_{eff} of the dual-band MTM SWS is the initial step towards analyzing its electromagnetic properties. These parameters are essential for understanding the material's behaviour across multiple frequency ranges. The ratio of λ to p—where the periodicity of free space is p = 7 mm, and its wavelength is λ—is 6.3 and 3.8 for the frequencies of resonance 6.8 GHz and 11.2 GHz, respectively, showing that there is no electromagnetic difference between the dual-band MTM SWS. The permittivity and permeability values are derived using a 'parametric extraction' method' [17], the S-Parameters acquired from the "CST Frequency Domain" module are analysed by this.

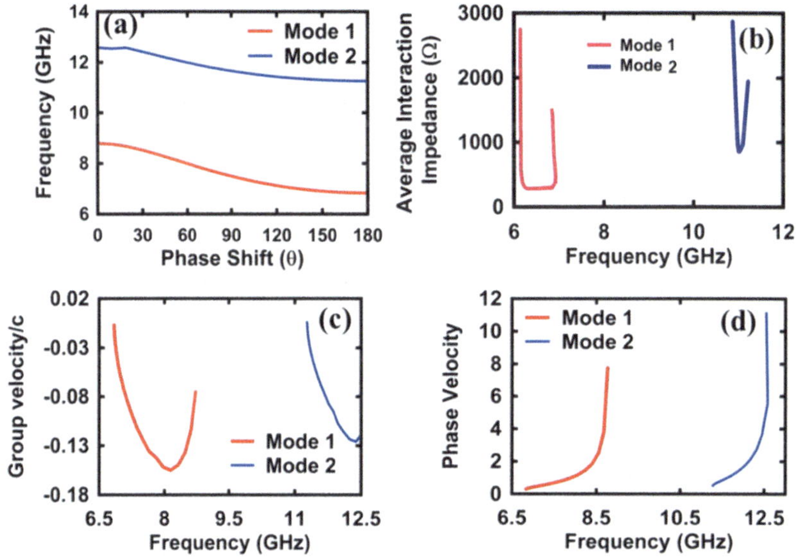

Fig. 3. (a) Dispersion Curve for mode 1 and mode 2 (b) Average interaction impedance (c) Group velocity and (d) Phase velocity for mode 1 and mode 2.

The permeability and effective permittivity values for the DB-CeSRR metamaterial are shown in Fig. 2. Specifically, Fig. 2a displays a diagram of DB-CeSRR unit cell placed inside the circular waveguide. In Fig. 2b, we can see that permittivity (ε) values become negative within certain frequency ranges, indicating a strong electric response from DB-CeSRR within these bands. Figure 2c shows that the circular waveguide has effective permeability (μ_{eff}) values lower than the cut-off frequency for fundamental transverse magnetic (TM) Mode. As a result, when the waveguide is loaded with DB-CeSRR, both permittivity and permeability are negative within specific frequency ranges, as illustrated in Fig. 2d. This confirms that the DB-CeSRR possesses double-negative metamaterial characteristics.

The high frequency properties of a DB-CeSRR-based metamaterial slow wave structure were investigated using electromagnetic (EM) simulations with Eigen mode solver in CST Studio Suite. The dispersion curve, average interaction impedance and additional dispersion characteristics of MTM SWS are investigated in this work. The dispersion curve for the DB-CeSRR MTM SWS is obtained through EM simulations, as indicated in Fig. 3a. The curve displays negative dispersion in the ranges of frequencies that for mode 1 at 8.5–9 GHz and for mode 2 at 12.5–12.9 GHz. This implies that the spatial harmonic with n = 0 functions as a "backward wave" in these ranges of frequencies.

The average interaction impedance of the DB-CeSRR-based metamaterial SWS is shown in Fig. 3b, which is important information to evaluate the device's efficiency. To achieve the enhanced efficiency and better beam-wave interaction, high interaction

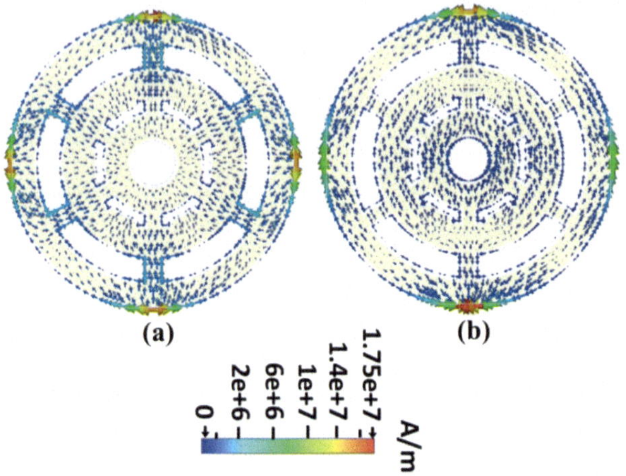

Fig. 4. Surface Current Distribution of DB-CeSRR MTM SWS.

impedance is required. This impedance is calculated using a post-processing method. The impedance can be calculated using the given equation [18, 19]:

$$\frac{K_0 = |E_z|^2}{2\beta_z^2 W v_g}$$

Here, E_z stands for the component of the localized electric field along the x-axis, β_z is the propagation constant, W signifies the energy stored and v_g represents the group velocity. Figure 3b shows that mode-1 and mode-2 have average interaction impedance greater than 250 Ω and 800 Ω. Figure 3c and 3d illustrate the MTM slow-wave structure's group and phase velocities. According to Fig. 3c, the phase velocity is positive within the frequency bands of 6.5–8.5 GHz and 10.8–12.5 GHz. However, the group velocity is negative, ranging from −0.01c to −0.16c, which suggests that the electromagnetic (EM) waves are propagating in a backward direction. Figure 4 displays the surface current distribution of the dual-band MTM SWS for mode 1 and 2.

4 Conclusion

We have explored the influence of slot-lines in a dual-band CeSRR metamaterial slow-wave structure (SWS) that can produce two distinct frequency bands: the C-band and X-band, which have respective resonant frequencies of 6.8 GHz and 11.2 GHz, respectively. For mode-1 and mode-2, the MTM SWS exhibits an average interaction impedance of more than 250 Ω and 800 Ω, respectively. Consequently, the DB-CeSRR-based MTM SWS exhibits promise for application in vacuum electron devices as a modulation region for electronic observation, target detection, and other related fields.

References

1. Veselago, V.G.: The electrodynamics of media with simultaneously negative values of permittivity and magnetic permeability. Sov. Phys. I-Uspekhi **47**, 509–514 (1968). https://doi.org/10.1007/978-94-007-1067-2_5
2. Smith, D.R., Padilla, W.J., Vier, D.C., Nemat-Nasser, S.C., Schultz, S.: Phys. Rev. Lett. **84**, 4184 (2000). https://doi.org/10.1103/PhysRevLett.84.4184
3. Wu, B.-I., Lu, J., Kong, J.A., Chen, M.: Left-handed metamaterial design for Cerenkov radiation. J. Appl. Phys. **102**(11), 114907 (2007). https://doi.org/10.1063/1.2818066
4. Ginzburg, N.S., Rozental, R.M., Sergeev, A.S.: Dual band operation of the relativistic BWO. In: 4th IEEE International Conference on Vacuum Electronics, 2003, Seoul, Korea (South), 2003, pp. 181–182. https://doi.org/10.1109/IVEC.2003.1286225
5. Wang, T., Qian, B.-L., Zhang, J.-D., Zhang, X.-P., Yi-Bing, C., Zhang, Q.: Preliminary experimental investigation of a dual-band relativistic backward wave oscillator with dual beams. Phys. Plasmas **18**(1), 013107 (2011) https://doi.org/10.1063/1.3537818
6. Yue, L. et al.: Design of helix slow-wave structure for Ka/Q dual-band TWT. In: 2017 Eighteenth International Vacuum Electronics Conference (IVEC), London, UK, 2017, pp. 1–2. https://doi.org/10.1109/IVEC.2017.8289597
7. Tang, Y., Luo, Y., Xu, Y., Yan, R., Jiang, W., Zheng, Y.: Design of a novel dual-band gyro-TWT. IEEE Trans. Electron Devices **61**(11), 3858–3863 (2014). https://doi.org/10.1109/TED.2014.2353851
8. Lyu, Z., Luo, H., Wang, X., Jiang, S., Jin, D., Gong, Y., Chen, M., Duan, Z.: Compact reversed Cherenkov radiation oscillator with high efficiency. Appl. Phys. Lett. **120**(5), 053501 (2022). https://doi.org/10.1063/5.0078211
9. Tang, X., et al.: Miniature metamaterial backward wave oscillator with a coaxial coupler. IEEE Trans. Electron Devices **69**(3), 1389–1395 (2022). https://doi.org/10.1109/TED.2022.3141330
10. Smith, D.R., Padilla, W.J., Vier, D.C., Nemat-Nasser, S.C., Schultz, S.: Composite medium with simultaneously negative permeability and permittivity. Phys. Rev. Lett. **84**(18), 4184–4187 (2000). https://doi.org/10.1103/physrevlett.84.4184
11. Shiffler, D., Seviour, R., Luchinskaya, E., Stranford, E., Tang, W., French, D.: Study of split-ring resonators as a metamaterial for high power microwave power transmission and the role of defects. IEEE Trans. Plasma Sci. **41**(6), 1679–1685 (2013). https://doi.org/10.1109/TPS.2013.2251669
12. Duan, Z., Hummelt, J.S., Shapiro, M.A., Temkin, R.J.: Subwavelength waveguide loaded by a complementary electric metamaterial for vacuum electron devices. Phys. Plasmas **21**(10), 103301 (2014). https://doi.org/10.1063/1.4897392
13. Wang, Y., Duan, Z., Tang, X., Wang, Z., Zhang, Y., Feng, J., Gong, Y.: All-metal metamaterial slow-wave structure for high-power sources with high efficiency. Appl. Phys. Lett. **107**(15), 153502 (2015). https://doi.org/10.1063/1.4933106
14. Wang, X., et al.: Characterization of metamaterial slow-wave structure loaded with complementary electric split-ring resonators. IEEE Trans. Microw. Theory Techn. **67**(6), 2238–2246 (2019). https://doi.org/10.1109/TMTT.2019.2908360
15. Tang, X., Zhang, Y., Jiang, S., Zhang, J., Zhang, L., Wang, Q., Li, X., Duan, Z.: Metamaterial-inspired backward wave oscillator using a dual band coaxial coupler. IEEE Electron Dev. Lett. early access (2023). https://doi.org/10.1109/LED.2023.3294759
16. Kumari, S., Jain, P.K., Mahto, M.: Design and cold-test measurement of a novel metamaterial assisted dual-band slow wave structure. IEEE Electron Device Lett. **44**(10), 1752–1755 (2023). https://doi.org/10.1109/LED.2023.3303939

17. Numan, A.B., Sharawi, M.S.: Extraction of material parameters for metamaterials using a full-wave simulator [education column]. IEE Antennas . Mag. **55**(5), 202–211 (2013). https://doi.org/10.1109/MAP.2013.6735515
18. Wang, X., et al.: Experimental demonstration of compact S-band MW-level metamaterial-inspired klystron. IEEE Electron Device Lett. **44**(1), 152–155 (2023). https://doi.org/10.1109/LED.2022.3223909
19. Wang, X., et al.: Novel S-band metamaterial extended interaction klystron. IEEE Electron Device Lett. **41**(10), 1580–1583 (2020). https://doi.org/10.1109/LED.2020.3015036

Investigation on Micro-Fabrication Methods for the Development of W-band Interaction Structure

Sahil Jain[1,2](✉), Vishant[2], Prerna Unadkat[1,2], Rajesh Meena[2], and Niraj Kumar[2]

[1] Academy of Scientific and Innovative Research (AcSIR) Ghaziabad, Ghaziabad 201002, India
sahil.ceeri22a@acsir.res.in

[2] (CSIR)–Central Electronics Engineering Research Institute (CEERI), Pilani 333031, India

Abstract. This paper demonstrates the investigation of micro-fabrication methods for the development of W-band interaction structure. Vacuum Electron Devices (VED) based high-frequency and high-power sources consists of three major components namely, electron gun, interaction structure, and spent beam collector. The interaction structure is an important component for high-power generation and its dimensions reduces with increase in the frequency. Therefore, it becomes very challenging to develop this sub-mm interaction structure with stringent fabrication tolerances (<10 microns) and good surface finish (<100 nm). Hence, this paper presents a comprehensive investigation into different micro-fabrication methods aimed at optimizing the performance and development of W-band interaction structure within the limits of desired fabrication tolerances and surface finish. The results indicate the hybrid approaches, combining multiple fabrication techniques, offering significant advantages in achieving the desired precision and performance. The work done will be useful in the development of vacuum electronics based micro structures for high-power THz generation.

Keywords: W-band · THz · interaction structure

1 Introduction

Terahertz has been a wide area of research in recent years and many studies have been performed to develop high power sources. This serves as a new area of interest for the development of more powerful, and coherent sources of terahertz (THz) radiation. The availability of sources in the submillimeter wavelength frequency range has wide potential for high data rate communications, and spectroscopy, which could provide new information for chemistry and biochemistry, sensing and imaging, which can be used for quality control and inspection in manufacturing sector [1–3]. One of the major bottlenecks in realization of such devices is the fabrication of sub-mm beam-wave interaction structure as with increase in the frequency of the device, the wavelength of the electromagnetic wave reduces and hence the dimensions of the structure decreases. The devices operating at high frequency requires structure with sub-mm dimensions and wavelength becomes very small which can be of dimensions comparable to surface

S. K. Ghosh et al. (Eds.): VEDA 2024, SPPHY 424, pp. 145–153, 2025.
https://doi.org/10.1007/978-981-96-5174-0_21

roughness. The surface roughness of metal walls can easily exceed skin depth, leading to an unpredictable increase in losses [4, 5]. These losses need to minimized with selection of proper vacuum grade material during fabrication, as surface roughness highly depends on type of material and on process development of interaction structure. Traditional fabrication procedures become challenging when the fabricated parts are only visible under a microscope [6–8]. Thus, the introduction of microfabrication techniques to replace the conventional methods of fabrication is not just a fair practice but a significant advancement. It allows vacuum electron devices (VEDs) to be used in a variety of previously inaccessible applications [9–11]. Over the past few decades, the world has focused its research efforts on developing a portable source with high frequency and high power that fits into a small package, meets strict fabrication tolerances, and has a better surface finish. Therefore, in this paper efforts have been made to develop the W-band interaction structure using different micro fabrication techniques and results for the same has been presented.

2 Experimental Setup

2.1 Wire-EDM (Electric Discharge Machining)

The initial method used for the development of W-band interaction structure is Wire-EDM (Electric Discharge Machining). This machining operation was performed on CHMER G32FD wire-EDM machine. The workpiece was mounted on the base plate with the aid of clamping screws and the starting coordinates are marked with the machining codes. Further de-ionized water is turned on continuously to serve as a dielectric medium during discharge. After that, the entire system is sealed off inside a chamber, and the machining process begins. During the machining process, a 0.25 mm diameter brass wire was employed, and additional parameters like feed rate and voltage were added to get the desired structure. The developed structure has been shown in the Fig. 1. In the Wire-Cutting method, brass wire is used as the continuous cutting tool electrode, which enabled us to eliminate the non-uniformity issue during the fabrication.

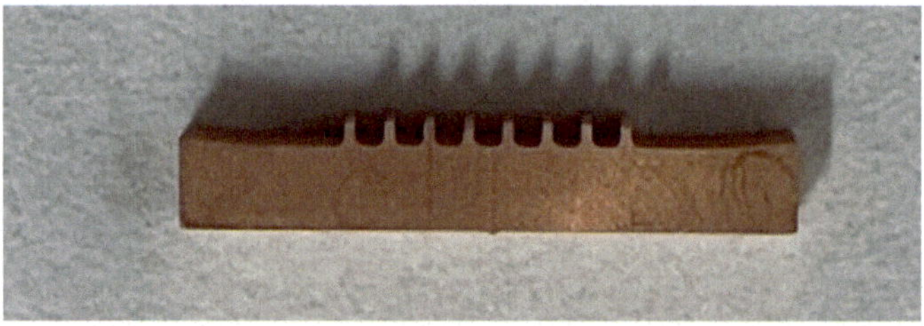

Fig. 1. W-band interaction structure developed using Wire-EDM machining method

2.2 Micro-Milling

The second method used for the development of W-band interaction structure is Micro-milling. It is similar to conventional milling but operates on a much finer scale, with tools that have diameters as small as a few micrometers. The manual tool clamping and precision control system of the hybrid μ-EDM machine expands the machine's usage. The workpiece has been clamped with the clamping screws on the monolithic granite table to perform the machining operations. The spindle is brought to its origin coordinates prior to beginning the machining processes, and the reference value is then set at the top of the workpiece. Further, the coolant is sprayed all over the workpiece to avoid excess heating and wash away the etched debris. The feed rate was held constant at 0.05 mm/min while the spindle speed was adjusted between 10,000 and 20,000 RPM. For the purpose of fabricating the intended planar interaction structure, machining operations were carried out after generating machining codes based on the design. Figure 2 shows the interaction structure developed using the micro-milling method.

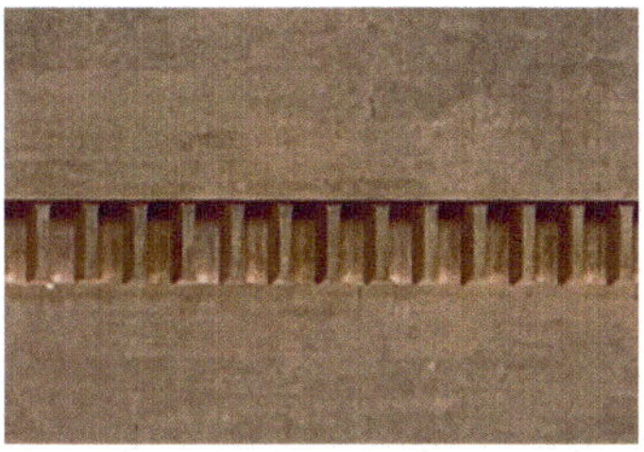

Fig. 2. W-band interaction structure developed using Micro-milling machining method

3 Simulation

The eigenfrequency analysis for the developed interaction structure was conducted using the RF module in COMSOL Multiphysics. This analysis is critical for identifying the natural resonant frequencies of the structure. The eigenfrequency modes give valuable information about the structure's inherent electromagnetic properties, including resonance behavior, which plays a key role in determining performance characteristics. To carry out this analysis, a detailed 3D model of the unit cell was created, as shown in Fig. 3. This unit cell represents the fundamental building block of the periodic structure. The dimensions utilized for the calculations, including those used for generating the dispersion plot, were sourced directly from Table 1. The dispersion plot is a graphical representation that displays the relationship between frequency and phase, providing insights into how electromagnetic waves propagate through the structure. Additionally, a comparative analysis of the dispersion characteristics was performed to evaluate the impact of different micro-fabrication techniques as illustrated in Fig. 4. Using the same dimensions as the unit cell previously designed, an array consisting of multiple such unit cells was developed in the COMSOL Multiphysics software. The RF module within COMSOL was utilized to perform a frequency domain analysis, which allows for the accurate simulation of the system's response to varying frequencies across the specified range. For this simulation, a total of 9-unit cells were modeled to assess the scattering parameters (S-parameters) of the interaction structure, with a particular focus on the S11 parameter. By analyzing S11, we were able to evaluate how much of the incident signal is reflected, thus providing insights into the effectiveness of the impedance matching and overall energy transmission. The results from these simulations were then used to conduct a comparative analysis between the reflection coefficients (S11) of various interaction structures. The outcomes of this comparison, including the reflection behaviors at different frequencies, are visually represented in Fig. 5, which displays the performance characteristics of the interaction structures across the simulated frequency range.

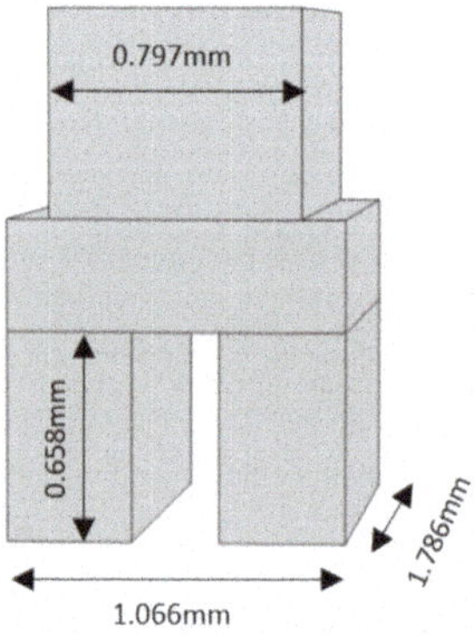

Fig. 3. 3D model of the unit cell of the interaction structure

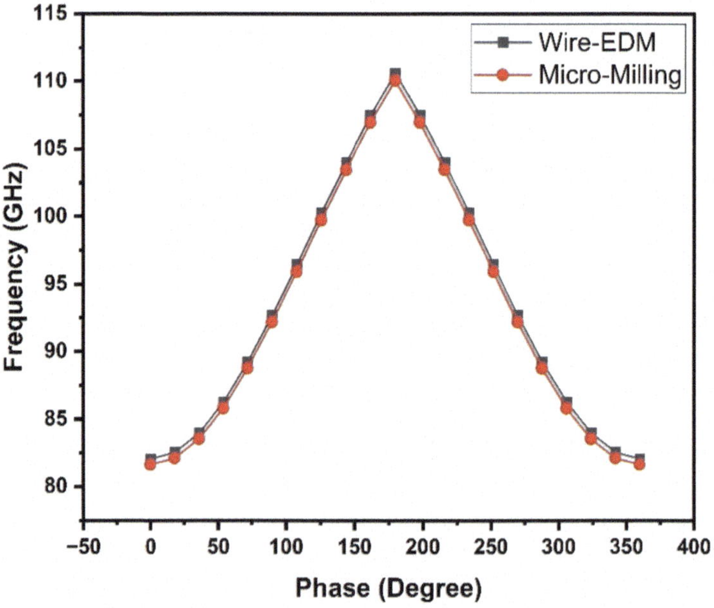

Fig. 4. Dispersion analysis of the interaction structure developed using two different micro fabrication methods

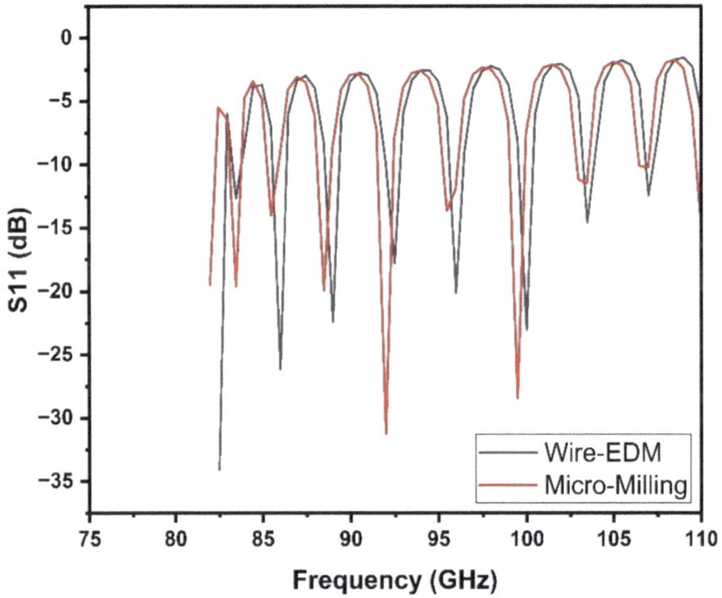

Fig. 5. Comparison of the simulated S11 measurements of the developed structure

Table 1 Analysis of the differences in dimensions deviations

S.no	Parameters	Desired Values (~0.1 THz)	Wire-EDM		Micro-Milling
1	Width (mm)	0.797	0.806	0.792	
2	Wall thickness (mm)	0.269	0.277	0.297	
3	Height (mm)	1.786	1.820	1.830	

4 Results and Discussions

We conducted a comprehensive analysis of the variations in the dimensional properties along both the length and width of the W-band interaction structure fabricated using two microfabrication methods. To ensure precise and detailed measurements, we used the scanning electron microscopy (SEM) technique. This allowed us to examine and compare the structural differences at the microscopic level, highlighting the distinct effects that each fabrication method had on the overall geometry and surface features of the interaction structure. Through SEM imaging, we were able to visualize and quantify these dimensional variations with high resolution and accuracy. The SEM images of the interaction structure developed by Wire-EDM and micro-milling techniques are presented in Fig. 6a and b respectively. The structures produced through both microfabrication processes appear to be uniform in their overall design and dimensional characteristics.

We observed minimal variations between the two structures, indicating a high degree of consistency in the fabrication outcomes. The overall average dimensional variation for the complete length of interaction structure in both the structures is less than ~20 microns. However, there are variations in the surface roughness values as shown in Table 2, highlights a significant difference between the surface finish obtained using the two fabrication methods. The structure produced using the micro-milling technique exhibited a much smoother surface finish with average roughness value of ~150 nm, whereas the structure developed through the wire-EDM method has 6 times higher average roughness value as compared to micro-milling. This indicates that the micro-milling process offers superior surface quality as compared to wire-EDM. To verify the simulation results with the experimental data, we conducted a cold test analysis on the fabricated structure using the VNA (model E8364C). This test provided us a direct comparison between the measured and simulated outcomes. The results of this analysis show a good correlation between the simulated and experimental values of the developed structure. As illustrated in Fig. 7, the experimental data closely matched the simulated results.

5 Conclusion

In this article an investigation on two different micro-fabrication methods for the development of W-band interaction structure has been performed. The structure developed with both the techniques is uniform and is consistent along the length of the interaction structure with dimensional variation less than 20 microns. Also, the micro-milling technique provided a smoother surface with average roughness value of 150 nm as compared to Wire-EDM technique. Hence, micro-milling proves to be better technique. This comparative analysis done will be useful in the development of vacuum electronics based micro structures for high-power THz generation.

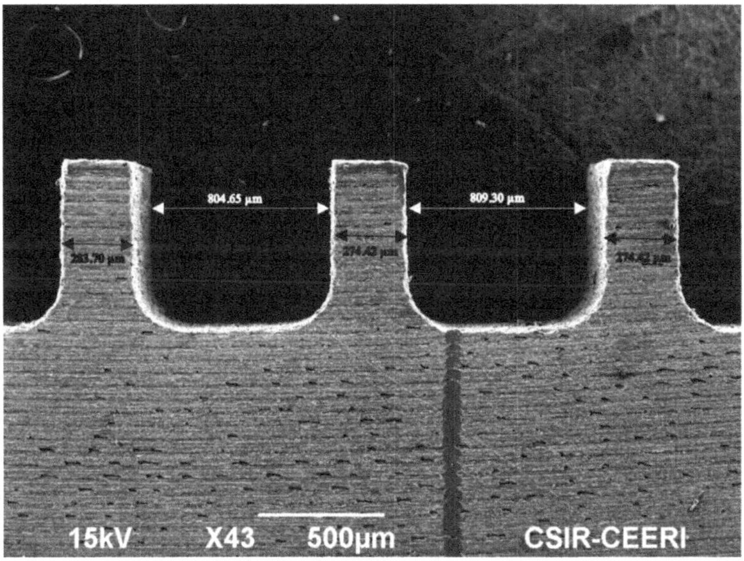

(a)

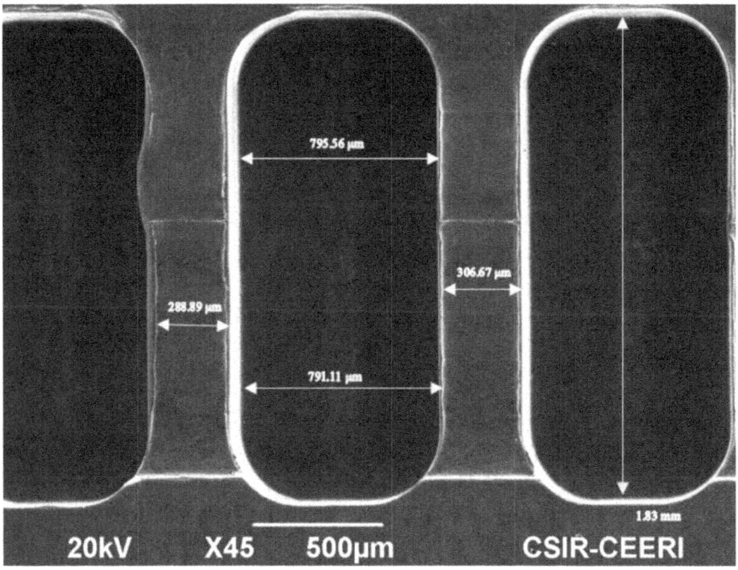

(b)

Fig. 6: SEM image of the developed interaction structure using, **a** Wire-EDM, **b** Micro-milling

Table 2 Analysis of surface roughness

S.no	Parameters	Wire-EDM	Micro-Milling
1	Ra (μm)	0.9636	0.1544
2	R_{pv} (μm)	7.288	2.716
3	R_p (μm)	3.402	1.513
4	R_v (μm)	3.887	1.203
5	R_z (μm)	4.578	2.090

(Ra), arithmetic average of deviations from mean
(Rpv), maximum peak to valley difference
(Rp), maximum peak height
(Rv), maximum valley depth
(Rz), average of (Rp - Rv) from five sections

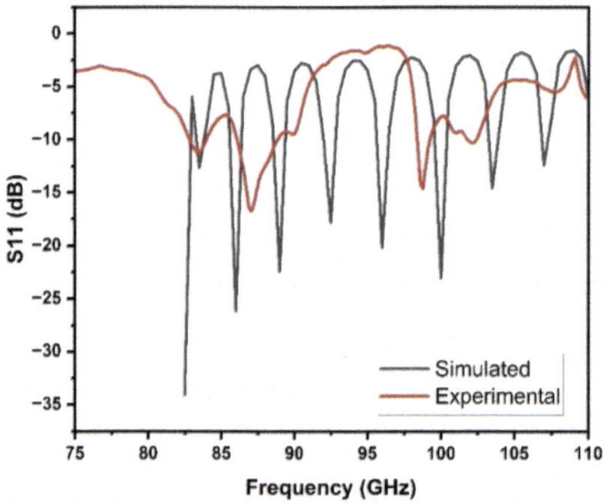

Fig. 7: Comparison between the simulated and experimental measurements of the developed structure

Acknowledgment. The authors are thankful to the Director, CSIR-CEERI, and all the colleagues of the plasma devices technology laboratory, CSIR-CEERI.

References

1. Paoloni, C., et al.: THz backward-wave oscillators for plasma diagnostic in nuclear fusion. IEEE Trans. Plasma Sci. **44**(4), 369–376 (2016)
2. Jain, S., Vishant, K.M., Kumar, N.: Parameterization of micro-milling process for fabrication of 0.22 THz interaction structure with improved surface roughness and dimensional tolerance. In: 2023 IEEE Microwaves, Antennas, and Propagation Conference (MAPCON), pp. 1–5 (2023)

3. Ryskin, N.M., Ozhogin, I.S., Starodubov, A.V., Serdobintsev, A.A., Kozhevnikov, I.O., Bahteev, I., Molchanov, S., Vishant, Abhishek A., Kumar, N.: Microfabrication of prototypes of millimeter-band electromagnetic components by using DLP 3D printing and magnetron sputtering. In: 2024 8th IEEE Electron Devices Technology & Manufacturing Conference (EDTM), pp. 1–3 (2024)
4. Kumar, N., Vishant, Ranjan, O., Abhishek, A., Starodubov, A.V., Titov, V.N., Ryskin, N.M.: A technique for planar integration of components of W-band backward-wave oscillator to eliminate misalignment. IEEE Electron Device Lett. **44**(2), 313–316 (2023)
5. Starodubov, A., Torgashov, R., Galushka, V., Pavlov, A., Rozhnev, A., Emelyanov, V., Ryskin, N.M., Abhishek, A., Vishant, Kumar, N.: Study on a microfabricated slow-wave structure for the millimeter-band backward-wave oscillator based on the pseudospark-source electron gun. In: 2022 23rd International Vacuum Electronics Conference (IVEC), pp. 43–44 (2022)
6. Mineo, M., Paoloni, C.: Corrugated rectangular waveguide tunable backward wave oscillator for terahertz applications. IEEE Trans. Electron Devices **57**(6), 1481–1484 (2010)
7. Gamzina, D., Himes, L.G., Barchfeld, R., Zheng, Y., Popovic, B.K., Paoloni, C., Choi, E., Luhmann, N.: Nano-CNC machining of sub-THz vacuum electron devices. IEEE Trans. Electron Devices **63**(10), 4067–4073 (2016)
8. Narasimhan, P., Jain, S., Gurjar, N., Kumar, N., Ghosh, S.: Design of thin wire metamaterial-based interaction structure for backward wave generation. IEEE Trans. Electron Devices **67**(3), 1227–1233 (2020)
9. Kumar, N., Abhishek, A., Vishant, Singhal, K., Gurjar, N., Jain, S., Starodubov A.V., Ryskin, N.M.: Pseudospark-driven high-current miniaturized voltage-tunable sheet-electron-beam source. IEEE Trans. Electron Devices **68**(12), 6482–6486 (2021)
10. Unadkat, P., Jain, S., Kumar, N.: A novel multi-aperture r-θ configuration-based pseudospark-driven electron beam source. IEEE Electron Device Lett. **45**(8), 1520–1523 (2024)
11. Unadkat, P., Kumar, N.: Study of space charge potential for propagation of plasma-assisted electron beam. IEEE Trans. Plasma Sci. **52**(5), 1669–1675 (2024)

Design Approach Towards Unit Cell of Folded Waveguide Slow-Wave Structure for (W-Band) TWT

Suvojit Bhattacharjee(✉), Subhradeep Chakraborty, and Amitavo Roy Choudhury

VED Group, CSIR-Central Electronics Engineering Research Institute, Pilani 333031, India
suvojitbhattacharjee4@gmail.com

Abstract. W-band Folded Waveguide Travelling Wave Tubes (FW-TWT) works as an amplifier in the frequency region of 100–110 GHz. This paper presents the electrical design approach of a FW-TWT in the range 100–110 GHz. The dispersion curve of the structure was theoretically calculated and further designed in the simulation software CST where modification was made in order to attain optimised interaction bandwidth between the dispersion curve and the beam line i.e., the beam wave interaction in the above-mentioned region.

Keywords: Folded waveguide travelling wave tubes · beam wave interaction · dispersion curve · beam line · serpentine structure

1 Introduction

Hutter in 1960 first perform analysis on Folded Waveguide Travelling Wave Tube (FW-TWT), later in 1990 Dohler also worked on FW-TWT. FW-TWT has wide varies of applications in radar, deep space exploration, communication, military and medical fields, especially at millimetre-wave or terahertz frequencies. The FW-TWT have advantages such as they are easy to fabricate, due to presence of waveguide the losses due to radiations are minimum with respect to other TWT and fairly good bandwidth. In FW-TWT structure electron gun emits electron beam that is accelerated by the applied voltage into the interaction region where the exchange of energy from the electron to the RF field happens to ensure that the dimensions of the folded waveguide structure is so adjusted that on every interaction point the electron beam interacts with the RF field in retarding phase, so that maximum energy is transferred from the electron to the RF field before electron gets dumped into the collector. In this paper serpentine structure is been incorporated as a slow wave structure, the transverse height of the serpentine structure is such that electron travels in the gap in less than half the time period of the EM wave, due to the periodicity of the structure a phase reversal is observed by the beam, the electron find itself in the same phase in the consecutive gaps. In the initial gaps due to incidence

S. K. Ghosh et al. (Eds.): VEDA 2024, SPPHY 424, pp. 154–158, 2025.
https://doi.org/10.1007/978-981-96-5174-0_22

of electron to different phases of the RF field, some gets accelerated some gets deaccelerated, the accelerated electron catchup with the deaccelerated electron and produce bunches voltage supply is then modified to ensure more energy transfer from electron to the RF field as a result EM wave can be greatly amplified (Fig. 1).

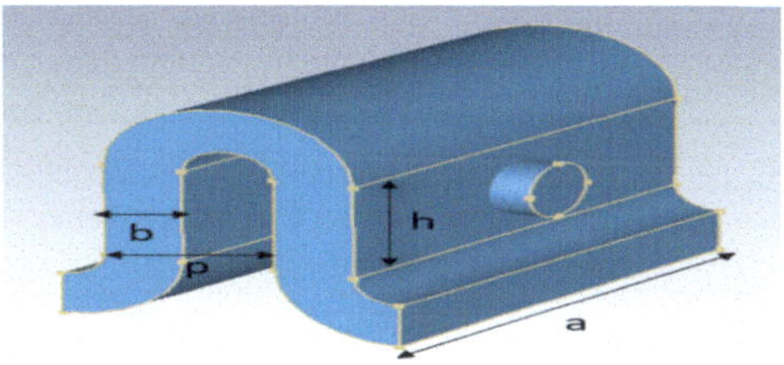

Fig. 1: Unit cell of a serpentine waveguide TWT.

2 Design Approach

In this paper a mathematical analysis was made to estimate the parameters value then those values are used to design the unit cell of a FW-TWT, after which using the values of the obtained parameters the CST design had been carried out, followed by different parameters were optimised is order to achieve a perfect synchronisation between the dispersion curve and the beam line, here in this paper the author has used the relations mentioned below to analytically obtain the dispersion curve and the beam line.

Following formula was employed to obtain the dispersive curve,'

$$F = \frac{c}{360} * \sqrt{\left(\frac{\phi - 180}{L}\right)^2 + \left(\frac{180}{a}\right)^2} \; where, L = h + \frac{\pi P}{2}$$

The alternate form of the above dispersive curve equation,

$$F = f_c * \sqrt{1 + \left(\frac{a}{L}\right)^2 \left(\frac{\Phi}{\pi} - 1\right)^2}$$

And the beam line equation,

$$F = \frac{v \times \phi}{P \times 360} where, v = 0.593 \times 10^6 \sqrt{V_0}$$

From the theoretical analysis we obtained the values of different parameters as given below (Table 1).

Following equations were plotted using analytical tool and following graph had been obtained in Fig. 2.

The above parameter values obtained are then used to design a unit cell of the FW-TWT in CST software, the obtained result from CST differ from that obtained

Table 1: Summary of different variables and their corresponding values

Parameters	Values	Description
a	2.000 mm	Broad dimension of FW
b	0.230 mm	Narrow dimension of FW
p	0.443 mm	Pitch length
h	0.398 mm	Transverse height
r_tunnel	0.149 mm	Radius of the beam tunnel
fc	75 GHz	Cut off frequency
V_0	9.5 kV	Beam voltage

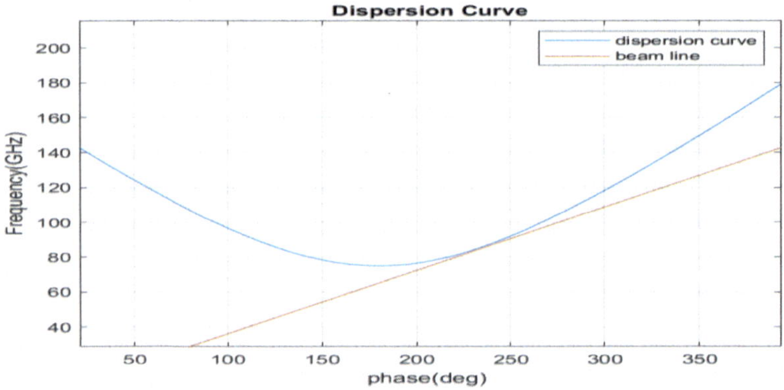

Fig. 2: Theoretical plot obtained for dispersion curve and beam line

theoretically for obvious reasons such as we haven't taken the beam tunnel dimension, beam scattering, RF losses, etc., into account. Then the values of the parameters are optimised to obtain synchronisation region of the dispersive curve and the beam line, at first the value of *p* was optimised followed by the optimisation of *h*, after that for different values of a dispersive curve had been plotted and finally for different values of the beam voltage beam line equation was plotted accordingly to obtain a reasonable interaction bandwidth.

Following optimised values are used to design the unit cell of a FW-TWT in the CST software for different vales of *a* and their corresponding value of V_o for proper synchronisation in Tables 2, 3 and 4.

Table 2: Summary of different variables and their corresponding values

Parameters	Value	Description
a	2.000 mm	Broad dimension of FW
b	0.230 mm	Narrow dimension of FW
p	0.465 mm	Pitch length
h	0.349 mm	Transverse height
r_tunnel	0.149 mm	Radius of the beam tunnel
f_c	75 GHz	Cut off frequency
V_0	6.7 kV	Beam voltage

Table 3: Summary of different variables and their corresponding values

Parameters	Value	Description
a	1.85 mm	Broad dimension of FW
b	0.230 mm	Narrow dimension of FW
p	0.443 mm	Pitch length
h	0.398 mm	Transverse height
r_tunnel	0.149 mm	Radius of the beam tunnel
f_c	82 GHz	Cut off frequency
V_0	7.3 kV	Beam voltage

Table 4: Summary of different variables and their corresponding values

Parameters	Value	Description
a	1.75 mm	Broad dimension of FW
b	0.230 mm	Narrow dimension of FW
p	0.443 mm	Pitch length
h	0.398 mm	Transverse height
r_tunnel	0.149 mm	Radius of the beam tunnel
f_c	86 GHz	Cut off frequency
V_0	7.9 kV	Beam voltage

From the above obtained sets of value, the graph the graph that had obtained is given below.

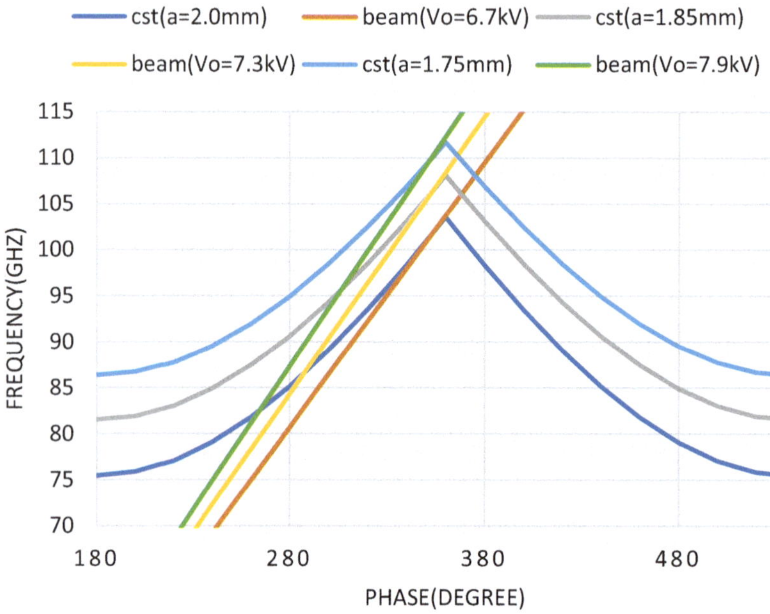

3 Conclusion

Above graph concludes that for a maximum interaction bandwidth we may opt for a = 2.0 mm, V_0 = 6.7kV or a = 1.85 mm, V_0 = 7.3kV for the following values of the parameters p = 0.465 mm, b = 0.23mm, h = 0.349mm and r_tunnel = 0.149 mm, form these parametric values the interaction bandwidth that had been obtained lies around 3–4 GHz. In this region maximum electron bunching occurs that ensures maximum energy transfer from electron to the RF field.

Acknowledgements. The authors would like to thank Director, CSIR-CEERI for their valuable guidance and encouragement to carry out this research work.

References

1. Gilmour, A.S.: Principles of Traveling Wave Tube. Artech House, Boston, London (1994)
2. Srivastava, A., Christie, V.L.: Design of a high gain and high efficiency W-band folded waveguide TWT using phase-velocity taper. J Electromagn Waves Appl (2018)
3. Srivastava, A., Penmetsa, S., Christie, L., Bhat, K.N.: Electromagnetic design of a 220 GHz BWO with experimental study of micro-fabricated folded waveguide structure. J Electromagn Waves Appl (2019)
4. Malek, F.: The analytical design of a folded waveguide travelling wave tube and small signal gain analysis using Mandey's Theorem", progress in electromagnetics research
5. Zhang, C., Gong, Y.: Analytical exploration of folded waveguide circuit design for high-power travelling wave tube amplifier
6. CST Design Suite 2019

Design and Analysis of S-band Waveguide Components for an Electron Accelerator

B. V. Ramarao(✉), Manjiri Pande, and Gopal Joshi

Bhabha Atomic Research Center, Mumbai 400094, India
bvram@barc.gov.in

Abstract. Generally, the electron linear accelerator (e-LINAC) is used for various Industrial, Scientific, and Medical applications. It is powered by a high power S-band klystron at 2856 MHz for acceleration of its electron beam through it. The typical RF power output expected from klystron is close to 6.2 MW and an average power of around 25 kW. The WR 284 based waveguide components like straight sections, bends, directional coupler, RF load etc. are required to test the klystron and to transmit the RF power to accelerating cavities. The electromagnetic simulation software has been used to design these waveguide components. Design is based on EIA WR284 standard waveguide. The RF performance parameters of the devices have been optimized while changing the physical dimensions of the devices. The designed directional coupler has directivity greater than 30 dB at 2856 MHz. The E & H-plane bends are of sweep bends and RTN losses are better than 40 dB. An RF water load is designed using the tapered water wedge. To isolate the SF-6 and water column a ceramic barrier is used. The RTN loss of the designed load is greater than 30 dB at 2856 MHz.

Keywords: Waveguide · Coupler · S-band

1 Introduction

In view of various applications of electron linac for industrial scientific, and medical applications and which requires an S-band klystron at 2856 MHz [1]. The BARC has taken up a challenge to develop an indigenous S-band klystron. To test RF performance parameters of such a development, it requires various S-band waveguide components for its test set-up. The indigenous design of these waveguide components developments is taken up. The paper presents the design details of various S-band waveguide components like bends, load, directional coupler etc.

2 Design and Simulations

The various waveguide components have been designed using a standard rectangular waveguide of WR-284. The dominant mode in this rectangular waveguide is TE_{10} mode and the operating frequency is 2856 MHz. The calculated characteristics of this waveguide at 2856 MHz are shown in Table 1.

Where c is the velocity of light in vacuum, f is the operating frequency; 2856 MHz, and η is the free space impedance.

S. K. Ghosh et al. (Eds.): VEDA 2024, SPPHY 424, pp. 159–162, 2025.
https://doi.org/10.1007/978-981-96-5174-0_23

Table 1. Calculated parameters of the WR-284 waveguide for TE10 mode at 2856 MHz.

Characteristic	Formula	Calculated value
Width (mm)	a	72.14
Height (mm)	b	34.04
Cut-off frequency (MHz), f_c	$c/(2a)$	2079
Guide wavelength (mm), λ_g	$\lambda_0/\{1-(f_c/f)^2\}$	153.2
Guide impedance (Ω), Zg	$\eta/\{1-(f_c/f)^2\}$	549.8

2.1 Design of Waveguide Bends

Generally, in an application of S-band klystron to accommodate the location of applicator and klystron in different plane waveguide bend are required [2]. Both E-plane and H-plane bends in meter and sweep types have been designed. The metered bends have mean length of $\lambda_g/4$, and whereas sweep bends have radius of 2.5 times a and b respectively for H and E-plane bends. The designed waveguide metered bends are shown in Fig. 1. The RF performance of these metered bends have been simulated. The simulated insertion loss and return loss of these bends are shown in Fig. 2. The metered bends have input match of less than 37 dB. The sweep bends along with its' E-fields are shown in Fig. 3. The sweep bends have RTN loss better than 41 dB as shown in Fig. 4.

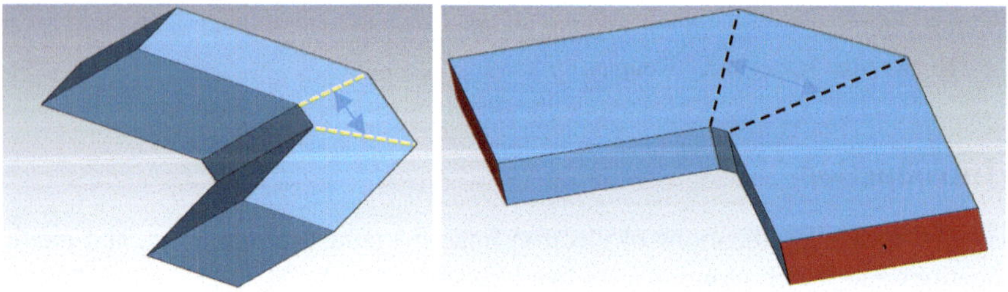

(a) (b)

Fig. 1. The designed waveguide (a) E-plane, and (b) H-plane metered bends.

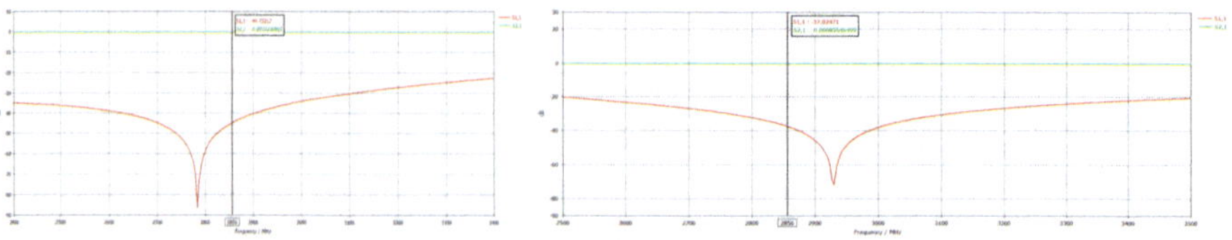

Fig. 2. Simulated RF performance; $|S_{11}|$ and $|S_{21}|$ (a) E-plane and (b) H-plane metered bends

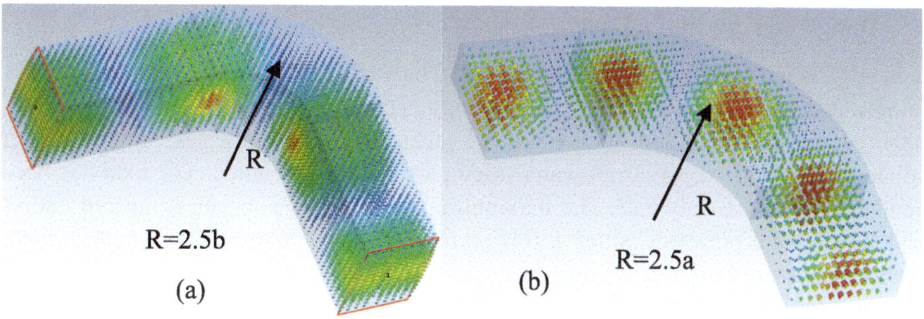

Fig. 3. Simulated E-filed of (a) E-plane and (b) H-plane sweep bends.

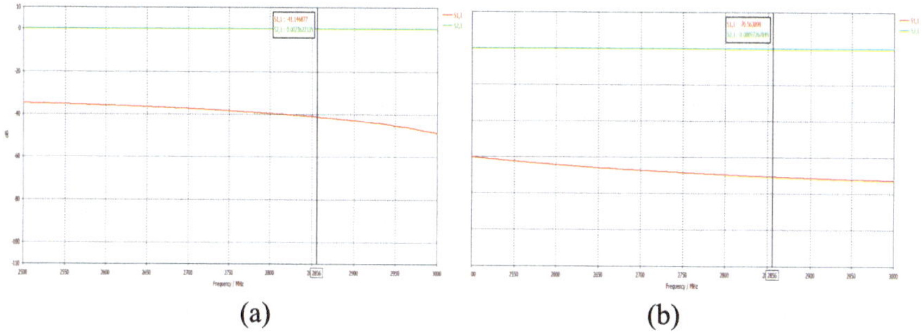

Fig. 4. Simulated RF performance; $|S_{11}|$ and $|S_{21}|$ of (a) E-plane and (b) H-plane sweep bends.

2.2 Design of Directional Coupler

There are various types of directional coupler designs available in literature like single, double, and multi-hole couplers [3]. All the directional couplers are designed using principle of small aperture radiations, which have directional properties in the forward and backward directions [4]. A small aperture on the broad wall of the waveguide radiates the fields into an auxiliary coaxial line. The designed directional coupler is shown in Fig. 5a. The simulated and optimized result of the coupler is show in in Fig. 5b. The coupler has directivity better than 30 dB.

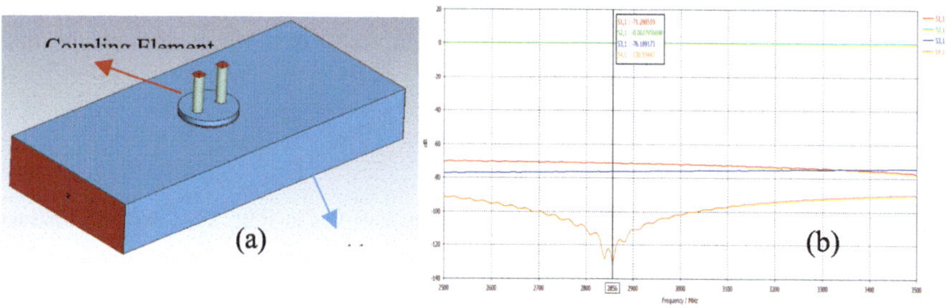

Fig. 5. The simulation results of directional coupler (a) model, (b) S-parameters at 2856

2.3 Design of Waveguide Load

Generally, waveguide loads are of two types. These are (1) wet, and (2) dry type loads. To have compact design a wet type load design is chosen. Figure 6 shows the sketch of the simulated wet load with its E field and water as main absorber. A ceramic barrier isolate the SF-6 filled part from water filled part of the waveguide. The load has been simulated using RF software. The maximum power densities occur at tapered end of load. The typical power densities at this portion are one kW/sq in. The length of taper has been optimized to have low power densities while matching the input. The RTN loss of the load is <30 dB at 2856 MHz. The simulated input match $|S_{11}|$ of the water load is shown in Fig. 6c.

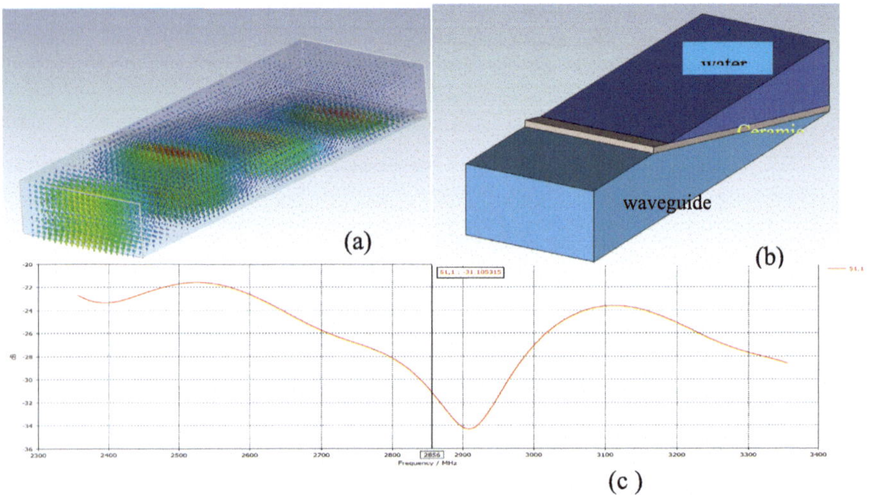

Fig. 6. The simulation results of waveguide water load (a) E-field, (b), model, and (c) $|S_{11}|$ at 2856 MHz.

3 Conclusions

The waveguide components like meter and sweep bends, directional coupler, and water load in a WR-284 waveguide have been designed. These have better RF performance parameters like $|S_{11}|$ and $|S_{xy}|$ at 2856 MHz. The fabrication of these devices are under progress.

References

1. Dwivedi, J.: Electron linacs and E-beam irradiation facilities at RRCAT, Book of Abstracts, InPAC-2022, pp. 5–6
2. Venkataramarao, B., et al.: Status of commissioning of the high-power waveguide systems for LEHIP, Book of Abstracts, InPAC-2022, pp. 28
3. Pozar, D.M.: Microwave Engineering, 4th edn. Wiley, New York (2018)

The First TWT Developed in India

S. N. Joshi[1], Amitavo Roy Choudhury[1(✉)], S. K. Ghosh[1], and B. N. Basu[2]

[1] CSIR-Central Electronics Engineering Research Institute, Pilani 333031, India
arc@ceeri.res.in

[2] Supreme Knowledge Foundation Group of Institutions, Mankundu, West Bengal 712139, India

Abstract. An account of the first Indian traveling-wave tube (FITWT), designed and developed in 1977, is presented in this paper. The tube was developed at CSIR-Central Electronics Engineering Research Institute (CEERI), Pilani under a competence development project executed by the FITWT team at CEERI under the supervision of S.S.S. Agarwala. S.N. Joshi, one of the authors of this paper, was the leader of the FITWT team, and B.N. Basu, yet another author of this paper, was one of the members of the team. The FITW, as per design, delivered 2W (CW) RF output power in the octave band of 2–4 GHz with 30 dB small-signal gain, <30 dB noise figure and 3°/dB AM-to-PM conversion coefficient.

Keywords: Traveling-wave tube · slow-wave structure · dispersion characteristics · helical coupler · electron gun

1 Introduction

The traveling-wave tube (TWT) is a microwave vacuum electron device (VED) amplifier [1–4] (Fig. 1).The TWT is categorized as a slow-wave device in which an electron beam interacts with an RF wave of phase velocity less that the speed of light supported by an interaction structure (slow-wave structure). In this device, the electron beam is constrained to move linearly under the influence of a DC magnetic field created by a magnetic focusing structure placed outside the device (Fig. 1). The magnetic field of the focusing structure, parallel with the linear electron beam, confines the electron beam in the device but does not take part in the beam-wave interaction in the device [3].

The electron gun of the device is of the Pierce type comprised of a cathode, a beam focusing electrode (BFE), and an anode (Fig. 1) [2, 3]. The slow-wave structure of the device is typically a helix (Fig. 1) [3]. The device is provided with RF input and output couplers (shown in Fig. 1) and also with an attenuator (not shown in Fig. 1), the latter for preventing the device from oscillation due to the reflection of RF waves from the input and output ends of the device caused by improper matching of the slow-wave structure with the couplers [3].

The account of the first Indian traveling-wave tube (FITWT), designed and developed in 1977, is presented in this paper. The tube was developed at CSIR-Central Electronics Engineering Research Institute (CEERI), Pilani. Some of the members of the team of

S. K. Ghosh et al. (Eds.): VEDA 2024, SPPHY 424, pp. 163–167, 2025.
https://doi.org/10.1007/978-981-96-5174-0_24

CEERI to whom the credit of the successful development of the FITWT goes are: C. Dattareyan, K.C. Chhabra, Dipti Das, R. Bendale, S.L Bawalia, Srinivasan, R.S Raju, B.N. Basu (one of the authors of this paper), S.N. Joshi (team leader) (one of the authors of this paper), and S.S.S. Agarwala (team advisor and supervisor).

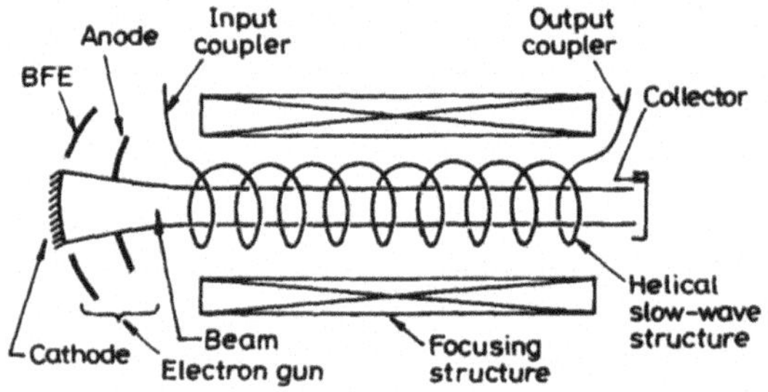

Fig. 1. Schematic of a TWT (*Source*: Fig. 8.2.1 of [2])

The FITWT was an all-glass cylindrical tube blown in two different diameters—one of an increased diameter to house the electron gun and another of a decreased diameter into which the wire helix closely fits [1]. The design specifications of the FITWT encompass the operating frequency and bandwidth, RF output power, noise figure, and AM-to-PM conversion coefficient [1] (Sect. 2). The specifications also include the parameters of the Pierce electron gun for the formation of electron beam [2, 3] and those of the external solenoid Brillouin magnetic field for the confinement of the electron beam [3] (Sect. 3). The parameters included are also those of the helix, coupler and attenuator of the FITWT (Sect. 4).

2 Design Performance Specifications

The design performance specifications of the FITWT are given in Table 1.

Table 1: Design specifications of FITWT [1]

Operating frequency	2–4 GHz
Bandwidth	1 octave
RF output power	2W (CW)
Small-signal gain [2]	30 dB
Noise figure	<30 dB
AM-to-PM conversion coefficient	3°/dB

The AM-to-PM conversion coefficient in Table 1 is defined as the change in the output phase for a 1-dB change in the power sweep applied to the input of the amplifier, here FITWT [1].

3 Electron Gun and Magnetic Focusing Parameters

The parameters of the Pierce electron gun and those of the solenoidal magnetic field of the FITWT are depicted in Table 2. An indirectly heated oxide-coated cathode [2] was used in the electron gun. The shapes of the gun electrodes were found by electrolytic tank method [1, 4]. The required Brillouin magnetic field [1, 3] is provided by a solenoid external to the device. The position of the solenoid with respect to the device is adjusted for minimum beam scalloping [1].

Table 2: Electron gun and solenoidal magnetic field parameters of the FITWT [1]

Cathode diameter	3.5 mm
Beam voltage	1.2 kV
Beam current	20 mA
Area convergence	6
Beam filling factor	0.55
Brillouin magnetic field (calculated)	0.02T
Solenoidal magnetic field (applied)	0.04T

4 Slow-Wave Structure, Coupler, and Attenuator Parameters

The contra-wound helices were used for the coupling of RF power in and out of the main helix slow-wave structure of the FITWT (Fig. 1) [1].The basic form of a helical coupler is depicted in Fig. 2.

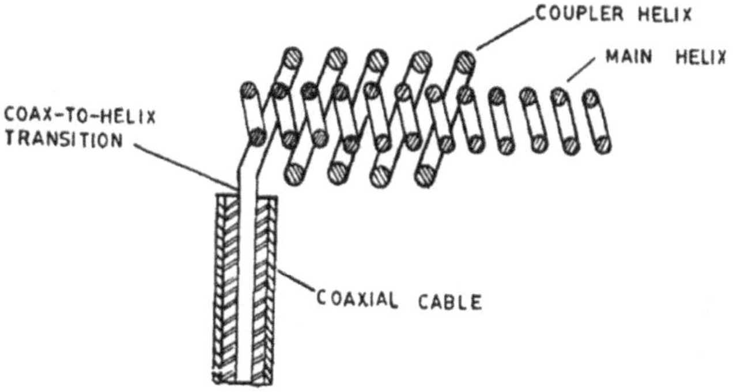

Fig. 2. Basic form of helical coupler [1].

The coupler helices were placed outside the device, and their positions relative to the device are adjusted for maximum power coupling (Sect. 1). Similarly, from outside of the device a coupled-helix attenuator is provided to prevent the device from oscillating, as mentioned in Sect. 1. The internal diameters of the coupled-helix couplers and the attenuator are so chosen that the couplers could loosely sleeve around and move freely along the length of the glass tube envelope of the FITWT. This enabled one to adjust the positions of the couplers for maximum power coupling and also adjust the position of the attenuator for the avoidance of oscillations in the device. A lossy paper was wrapped around the helix to provide the required attenuation. The coupler helix and attenuator parameters are depicted in Table 3 with respect to the main helix winding diameter, main helix pitch angle ψ_1, coupler helix pitch angleψ_2, outer diameter of the glass tube into which the main helix fits, material of the main helix wire, number of main helix turns, material of the coupler helix, material of the lossy paper of the attenuator, surface resistivity of the lossy paper of the attenuator, and attenuation of the attenuator.

Table 3: Coupler helix and attenuator parameters [1]

$\cot\psi 1$	17.9
$\cot\psi_2$	15.8
Main helix winding diameter	0.319 cm
Outer diameter of the Corning 7052 glass tube	0.549 cm
Material of main helix wire	Tungsten-Rhenium alloy
Number of main helix turns	7.5
Material of coupler helix	Molybdenum/Nichrome
Material of lossy paper of the attenuator	Teledeltos
Surface resistivity of lossy paper of the attenuator	10kΩ/Sq
Attenuation of the attenuator	40 dB

The bandwidth of the coupler is the frequency over which the optimum length of the coupler corresponding to the maximum power coupling remains constant. The following results of the analysis of a system of coupled helical transmission lines are of relevance [5]. The optimum length of the coupler decreases with the decrease of the ratio of the inner-to-outer helix radii and the increase of the dielectric constant of the dielectric sleeve of the coupler helix. The bandwidth of the couplerincreases with the increase of $\cot\psi_2/\cot\psi_1$, that is, with the increase of the pitch of the outer helix for a given pitch of the main helix [5].

5 Conclusion

The FITWT, designed and developed in 1977, is a Corning 7052 glass tube into which a wire helix typically made of tungsten-rhenium alloy closely fits. The coupled-helix couplers, attenuator, and solenoid for magnetic focusing are all positioned outside the tube. The FITWT delivered the RF output power of 2W (CW) in the octave band of 2–4 GHz with the small-signal gain of 30 dB, noise figure less than 30 dB, and AM-to-PM conversion coefficient of 3°/dB. The competence of developing the FITWT became the landmark and subsequently triggered the development of several helix and coupled-cavity TWTs in India.

Acknowledgment. The paper is based on a lecture delivered by SN Joshi, the first author of this paper, in a webinar programme held on 7th November 2020, arranged by BN Basu, the third author of this paper, on the WhatsApp group: Thinkers in VED. We wish to dedicate this paper to Andrei Haeff, who invented the TWT in 1933; to SSS Agarwala, who guided the team to which goes the credit of designing and developing the FITWT; and to Beena Joshi, wife of the team leader SN Joshi.

References

1. Joshi, S.N.: Design and development of the first TWT in India. In: Kesari, V., Singh, R., Basu, B.N. (Eds.) Proceedings of the the Online Webinar of WhatsApp Group: Thinkers in VEDA group (2020)
2. Pierce, J.R.: Traveling-Wave Tubes. D. Van Nostrands, New York (1950)
3. Basu, B.N.: Electromagnetic Theory and Applications in Beam-Wave Electronics. World Scientific Publishing Co., Inc., Singapore, New Jersey, London, Hong Kong (1996)
4. Gittins, J.F.: Power Traveling-Wave Tube. American Elsevier, New York (1965)
5. Singh, V.N., Basu, B.N., Paul, B.B., Vaidya, N.C.: Equivalent circuit analysis of a system of coupled helical transmission lines in a complex environment. J. Appl. Phys. **54**, 4141–4146 (1983)

Systematic Performance Investigation of a Ka-Band, Three-Stage Clustered-Cavity Gyroklystron Amplifier

Ankhi Debnath, M. V. Swati(✉), and Gaurav Singh Baghel

National Institute of Technology Silchar, Silchar, Assam 788010, India
swati@ece.nits.ac.in

Abstract. The bandwidth advancement of a Ka-band, conventional three-cavity gyroklystron amplifier using clustered-cavity approach has been explained in detail. The strong consequence of varying the location of clustered buncher cavity on the RF out power, gain, efficiency, and bandwidth has also been studied in detail. It has been observed that the maximum efficiency has been achieved when the clustered buncher cavity is located nearer to the output cavity whereas highest gain is obtained when the clustered buncher cavity is located in between the input and output cavities.

Keywords: Gyroklystyron amplifier · Clustered-cavity · Second harmonic

1 Introduction

Nowadays, there is an increase demand in the design of wideband, high-power, and high-gain gyoklystron amplifiers due to their applications in radar systems, communication systems, linear accelerators etc. [1]. As per the state-of-the art, one of the favored methods nowadays for the bandwidth improvement with minimal degradation in efficiency and gain is clustered cavity approach. The physics behind the approach is that the specific cavity has been substituted by a pair of two or three cavities such that the Q-factors are diminished to half or on-third accordingly which in turn leads to bandwidth enhancement [1, 2]. Figure 1 depicts the geometrical representation of a three-stage, clustered-buncher cavity gyroklystron amplifier.

The generalized mathematical non-linear analysis of the clustered-cavity gyroklystrons has already been explained in detail as per previous literatures [2, 3]. The influence of the buncher cavity location on the performance of the conventional gyroklystron i.e. gain, bandwidth, and efficiency has been discussed in detail by Zasykin et al. in 1995 [4]. The similar study has been performed here for the three-stage, clustered-cavity gyroklystron amplifier to analyze the effect of varying the position of clustered-cavity buncher on the gyroklystron performance. This study will be helpful for the designers of the device.

In the present paper, in Sect. 2, the nonlinear analysis of the three-stage, clustered-buncher cavity gyroklystron amplifier has been explained and is then applied for the

S. K. Ghosh et al. (Eds.): VEDA 2024, SPPHY 424, pp. 168–173, 2025.
https://doi.org/10.1007/978-981-96-5174-0_25

bandwidth improvement of a three-cavity, conventional, second harmonic, gyroklystron amplifier. In Sect. 3, the consequence of varying the location of the clustered-cavity on the device performance has been discussed in detail. In Sect. 4 the significant conclusions are summarized.

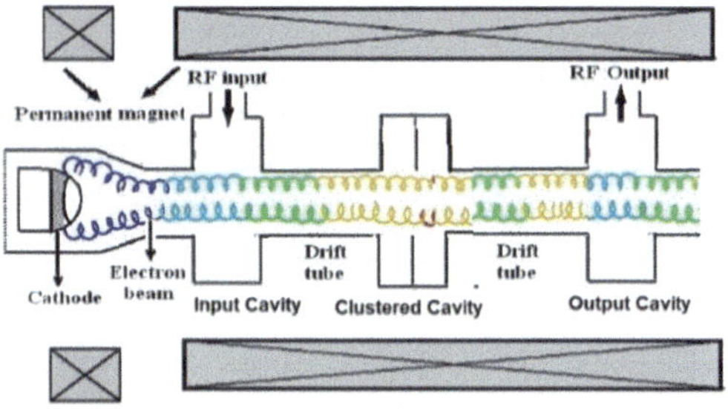

Fig. 1. Geometrical configuration of a clustered buncher-cavity gyroklystron amplifier [3].

2 Non-linear Analysis

The geometrical representation of a typical 3-stage clustered buncher-cavity gyroklystron amplifier is shown in Fig. 1. The interaction of electron beam with RF wave in a clustered-cavity gyroklystron amplifier has been investigated step-by step for each stage using generalized non-linear equations [2]. The electron velocity spread and the space-charge forces are assumed to be ignored. The coupling between the cluster-cavities, stagger-tuning and beam loading are supposed to be ignored.

The equation for the estimation of transverse efficiency is given by [3]:

$$\eta_{\perp} = -2J_1\{q_1\}(|F_3|) - |F_3|^2 \tag{1}$$

The device gain (G) is expressed as [3]:

$$G = 10\log\left\{\frac{1}{A^2}|F_3|^2\right\} \tag{2}$$

The non-linear analysis for the clustered cavity considered above has been applied to an experimental, conventional 32.3 GHz, three-cavity gyroklystron amplifier [4]. The buncher cavity has been replaced by the two cavities of clustered-cavity buncher by reducing the quality factors to half. Using the non-linear analysis discussed above, the RF power and efficiency has been evaluated for the clustered-cavity gyroklystron and the conventional-cavity gyroklystron amplifier and the comparison has been done.

It can be concluded from the Figs. 2 and 3 that the bandwidth ~50 MHz is obtained in the cluster-cavity case, which is relatively much better that that in the conventional cavity, i.e., ~11 MHz. The RF output power of ~310 kW is achieved in case of clustered buncher cavity gyroklystron amplifier, however in conventional case, a peak RF output power of ~300 kW is obtained. Therefore, it is concluded that the inclusion of clustered cavity widens the device bandwidth with the slight increase in the gain and efficiency.

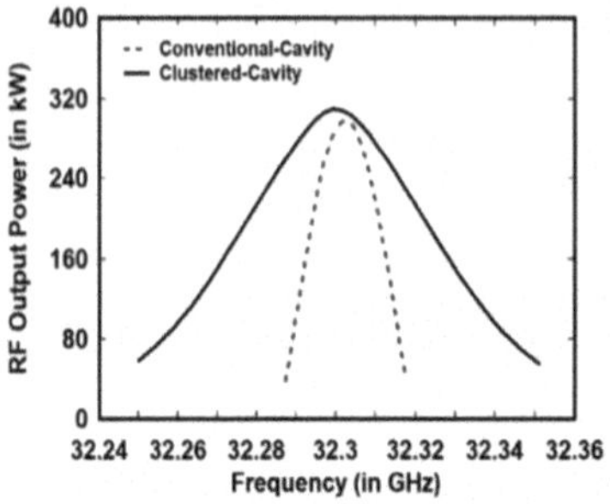

Fig. 2. Bandwidth comparison plots for conventional cavity and clustered-cavity.

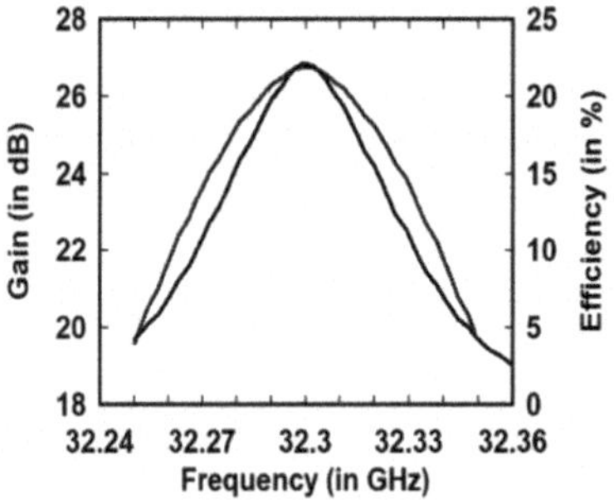

Fig. 3. Gain and efficiency plots for clustered-cavity.

3 Effect of Clustered-Cavity Position on the Device Performance

In this section, the influence of changing the position of clustered buncher cavity w.r.t central position on the gyroklystron performance i.e. gain, efficiency, and bandwidth has been studied in detail. This study is very useful for the designers of the device to achieve effective device performance. Let the parameter r is the relative position of the cluster. Let $L_{\sum}$ is the total interaction length of the device. The length of the drift Sect. 1 (L_{dr1}) and drift Sect. 2 (L_{dr2}) is calculated as:

$$L_{dr1} = rL_{\sum}; L_{dr2} = L_{\sum} - L_{dr1} \tag{3}$$

(a) ***Dependency of bunching efficiency and RF output power on the relative cluster position***

The generalized expression for the bunching efficiency (η_b) of the device is given by [6]:

$$\eta_b = \left| \frac{1}{2\pi} \int_0^{2\pi} \exp(i\theta_{dr}) d\theta \right|^2 \tag{4}$$

where, θ_{dr} is the gyrophase at the entrance of the output cavity and is given by the expression:

$$\theta_{dr} = \theta_0 - q\sin\theta_0 - q(\Delta\psi_1 + \theta_0 - qr\sin\theta_0) - q_1\sin(\Delta\psi_2 + \theta_0 - qr\sin\theta_0) \tag{5}$$

The device output power (P_{out}) can be calculated as in terms of bunching efficiency (η_b) as:

$$P_{out} = \eta_b \times (V_b \times I_b) \tag{6}$$

Corresponding to the different values of the relative cluster position (r), the efficiency and output power is calculated using expression (4) and (5). Figure 4 shows the alteration of bunching efficiency and RF output power versus relative position of the cluster. It has been observed from the figure that, the maximum efficiency of around 23% and maximum power output of around 324 kW is obtained at $r = 0.85$. Hence, it is observed that the clustered-cavity gyroklystrons achieve the maximum efficiency and power values as the intermediate cavities are positioned comparatively closer to the output cavity *w.r.t* to the central location.

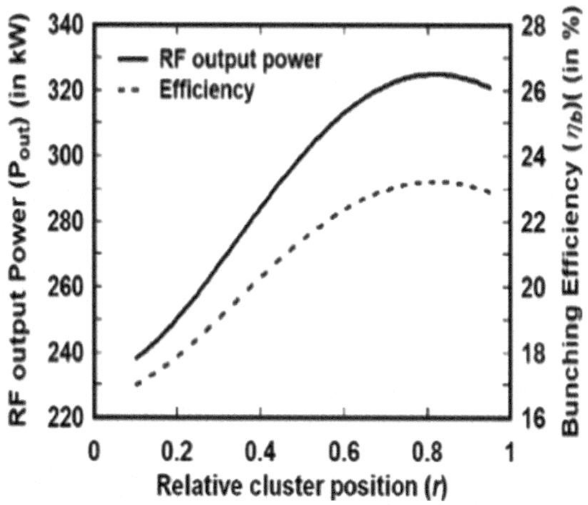

Fig. 4. Dependency of the RF output power and efficiency on the relative cluster position.

(b) ***Dependence of Gain on the relative cluster position***

The generalized expression of the gain (G) for the clustered-cavity gyroklystron is given by [6]:

$$G = 20log\left\{8I_{03}I_{02}\mu_{\sum}^{2}r(1-r)\right\} \tag{7}$$

where, I_{0i} are normalized current parameters, $\mu_{\sum}$ is the total normalized interaction length of the device. Corresponding to the different values of r, the gain G is calculated

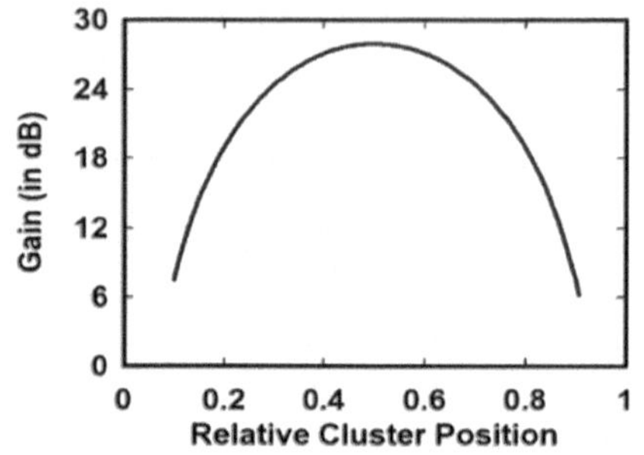

Fig. 5. Dependence of the gain on the relative cluster position.

using expression (7). Figure shows the variation of gain versus relative position of the cluster.

For different values of r, the gain G is calculated using expression (23). Figure 5 shows the variation of gain versus relative position of the cluster. From the fig., it is clear that in the clustered-cavity gryoklystrons, the maximum gain of ~ 27 dB is achieved when the clustered-buncher is positioned at the center of the device i.e. $r = 0.5$. However, the above statement is not reliable with that obtained in case of the efficiency that when buncher clustered cavities are located comparatively closer to the output cavity, efficiency and power is maximum. Therefore, there should be a trade-off between the gain and efficiency, when the choice of location of clustered cavity has to be determined for the design of clustered-cavity gyroklystron. The concept described in this section can also be validated from the selection of the design parameters of an experimental 32.3 GHz gyroklystron amplifier as shown in Table 1. The electronic efficiency of ~ 23%, RF output power of ~ 305 kW, and ~ 27 dB is obtained at $r = 0.5$ i.e. when the buncher cavity is located at the center of the device ($L_{dr1} = L_{dr2}$).

4 Conclusion

The proposed work is crucial to analyze the effect of altering the position of the clustered-buncher cavity on the device performance. The experimental three-cavity Ka band gyroklsytron amplifier has been considered for the investigation. In the current analysis, relative cluster position has been varied. It has been observed from the results that, the maximum efficiency of around 23% and maximum power output of around 324 kW is obtained at $r = 0.85$, however the maximum gain of ~ 27 dB is achieved when the clustered buncher is positioned at the center of the device i.e. $r = 0.5$. Thus, there should be a trade-off between the efficiency and gain with the variation in relative cluster position. The study will be useful for the performance analysis and practical implementation of the experimental gyroklystron amplifiers.

References

1. Nusinovich, G.S., Antonsen, T.M., Guo, H., Granatstein, V.L.: Theory of clustered-cavity gyroklystron. Phys. Plasmas **9**(9), 4032–4039 (2002)
2. Swati, M.V., Chauhan, M.S., Jain, P.K.: Clustered-cavity approach for the performance improvement of a Ka-band second-harmonic gyroklystron amplifier. IEEE Trans. Electron Dev. **67**(3), 1240–1247 (2020). https://doi.org/10.1109/TED.2020.2966650
3. Swati, M.V., Baghel, G.S.: Design, analysis, and simulation of a three-stage stagger-tuned, clustered-buncher cavity gyroklystron amplifier. IEEE Trans. Plasma Sci. **51**(3), 660–667 (2023). https://doi.org/10.1109/TPS.2023.3244628
4. Zasypkin, E.V., Moiseev, M.A., Sokolov, E.V., Yulpatov, V.K.: Effect of penultimate cavity position and tuning on three-cavity gyroklystron amplifier performance. Int. J. Electron. **78**(2), 423–433 (1995). https://doi.org/10.1080/00207219508926174
5. Antakov, I.I., Gachev, I.G., Zasypkin, E.V.: Experimental studies of a gyroklystron operating in the field of a permanent magnet, Radiophys. Q. Electron **54**(3), 166–173 (2011); Siddiqui, N.F., et al.: Effect of off-axis electron beam on the performance of a gyroklystron amplifier. Phys. Plasmas **31**(7) (2024)
6. Oleksandr, S.: Analyses of advanced concepts in multi-stage gyroamplifiers and startup in high-power gyro-oscillators. Ph.D. thesis, Dept. Elect. Eng., Univ. Maryland, College Park, MD, USA (2004)

Investigation of Electron Beam Misalignment on a 94 GHz Gyroklystron Amplifier

Ankhi Debnath, M. V. Swati, and Gaurav Singh Baghel(✉)

National Institute of Technology Silchar, Silchar, Assam 788010, India
gaurav@ece.nits.ac.in

Abstract. The proposed work provides an insight about the practically possible beam misalignment in the gyro-devices. The misaligned beam interacts differently with the induced wave in the cavity. The effect of the shifted guiding center of the misaligned beam on the device performance has been analyzed and then compared with an experimental three cavity gyroklystron. This study shows that increment in the misalignment in terms of radial shift, leads to the decrement of the power, gain and efficiency of the device. The sharp decline in the efficiency and gain curve highlights that it is highly sensitive to radial displacement, where even minor shifts lead to a considerable reduction in performance. The minor shift in the resonant frequency and bandwidth has been noticed while changing the misalignment parameter.

Keywords: Beam misalignment · Radial shift · Gyroklystron · Radar application

1 Introduction

High-frequency gyroklystron amplifiers play an important role as microwave sources for millimeter wave radar applications, communication system, plasma heating, spectroscopy and particle accelerators. One appealing aspect of a gyroklystron is its capacity to function with a smooth-walled cylindrical resonator and a concentric annular electron beam [1]. Researchers have extensively explored the beam-wave interaction technique in gyroklystrons through both theoretical and simulation-based approaches. Various studies have presented mathematical models to aid in the design and analysis of gyroklystrons [2–4].

A key challenge in developing a gyroklystron amplifier is the unavoidable misalignments that occur between the cavity and the electron beam, disrupting the structural symmetry. These misalignments can lead to reduced power, efficiency or, in the worst case, cause the gyroklystron to malfunction [5, 6]. In a gyroklystron amplifier, because of these misalignments the radius of the beam's guiding center changes depending on the angular position relative to the cavity axis. Consequently, the electron beam loses its symmetrical alignment along the axis, causing its characteristics to shift based on its angular location. Nazish Fatima Siddiqui et al. reported an experimental result for a 34.95 GHz gyroklystron amplifier, in which for an ideal beam without any misalignment, the efficiency is measured at 35.2% with output power of 202 kW and 23.41 dB gain.

S. K. Ghosh et al. (Eds.): VEDA 2024, SPPHY 424, pp. 174–179, 2025.
https://doi.org/10.1007/978-981-96-5174-0_26

However, when a radial shift of 1.2 mm occurs, the efficiency significantly decreases to approximately 14% along withoutput power of 82 kW and 19.5 dB gain [6].

In this paper, a new analytical approach, rooted in nonlinear theory, has been explored to investigate the above-mentioned misalignments on an experimental gyroklystron amplifier. Section 2 focuses on the interaction of the misaligned beam with the rf wave induced in the cavity. The complete calculation procedure is depicted using an elaborated flowchart. In Sect. 3, the analytical result for ideal electron beam has been demonstrated. Following this, the paper examines how electron beam misalignment influences the device's performance. In Sect. 4, the conclusions are drawn.

2 Analysis

A gyroklystron amplifier functions within a transverse electric field near the cutoff frequency. The electric field within the cavity (E_0) can be determined using the input RF power and the quality factor [4].

For a three-cavity gyroklystron amplifier, the normalized field amplitude (F_2) in the buncher cavity can be calculated as follows,

$$F_2 = \sqrt{\pi} I_2 \mu_2 e^{-x_2^2} J_1(q) \sin\left(\psi + \frac{\sqrt{3}}{2}\mu_2 \Delta_2\right) \tag{1}$$

In the equation, "ψ" is the cavity phase and "q" represents the bunching parameter of the respective cavity. The normalized beam current I_2 can be expressed as [3],

$$I_2 = \frac{\sqrt{2}}{\pi^{\frac{5}{2}}} \frac{I_0 Q_2 e}{c^3 m_e \varepsilon_0} \frac{\beta_{\perp 0}^{-4}}{\gamma_0} \frac{\lambda}{L_2} \frac{J_{m\pm n}^2(k_\perp R_{el})}{\left(\upsilon_{mp}^2 - m^2\right) J_m^2(\upsilon_{mp})} \tag{2}$$

where,, $k_\perp = \frac{2\pi}{\lambda}$, I_0 is the initial beam current, Q_2 and L_2 is the quality factor and length of buncher cavity respectively, γ_0 represents the relativistic mass factor, R_{el} is the beam radius, υ_{mp} is eigenvalue of operating TE_{mp} mode.

Electron Beam Misalignment Effect

Beam misalignment causes the initial positions of the guiding centers to shift. Figure 1 illustrates a cross-sectional view, where the ideal beam is shown as a solid circle, and the misaligned beam is represented by a dashed circle.

As shown in Fig. 1, the electron beam guiding center has been shifted by a distance d, resulting in a change in position of beam radius from R_{el} to $R\{\emptyset\}$ at an angle $\emptyset$ relative to the cavity axis. The relationship between these two guiding centers can be described geometrically using the cosine rule for triangles [5, 6],

$$R_{el}^2 = R\{\emptyset\}^2 + d^2 - 2R\{\emptyset\}\mathrm{d}\cos\emptyset \tag{3}$$

As noted in the case of the gyrotron oscillator, the equations of motion employed to analyze its performance can also be applied to gyroklystron amplifiers [1]. This is done by substituting the normalized current parameter I_2 for an ideal beam with an

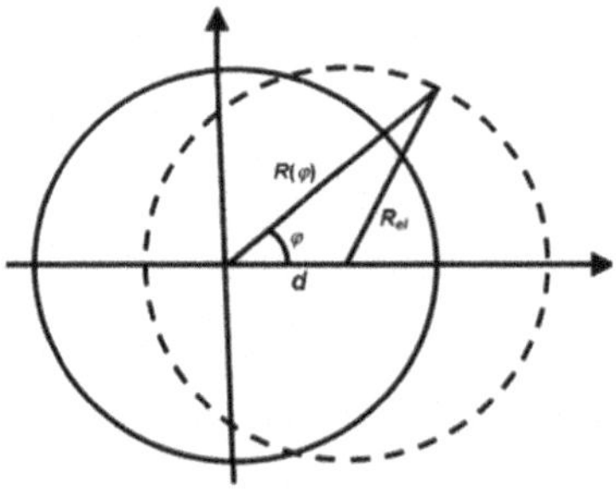

Fig. 1. Schematic illustration of a transversally shifted electron beam (dashed circle) by distance d. The beam's initial position is indicated by the solid circle [5].

azimuth-dependent normalized current parameter $I\{\emptyset\}$. These two parameters can be mathematically related by the following expression [5],

$$I\{\phi\} = \frac{J_{m\pm1}^{2}\left(\frac{2\pi}{\lambda}R\{\emptyset\}\right)}{J_{m\pm1}^{2}\left(\frac{2\pi}{\lambda}R_b\right)} * I \tag{4}$$

Equations (3) and (4) are utilized in the aforementioned analysis to examine electron beam misalignment impacts the performance of the gyroklystron. The procedural flowchart has been presented in Fig. 2 which explains the algorithm and stepwise calculation of azimuth-dependent normalized current parameter (due to the effect of misalignment).

3 Result and Discussion

The previous section details the nonlinear analysis developed to assess the influence of a shifted electron beam on the device's performance metrics. For numerical evaluation of the analysis, a practical three-cavity gyroklystron amplifier operating at 93.2 GHz has been examined, with the relevant design parameters provided in [6].

Figure 3 illustrates the relationship between RF output power and gain as a function of input power. The plot shows that the RF output power reaches its saturation level of 316 kW with an input power of 1.5 kW, while the corresponding gain is measured at 23.23 dB. Figure 4 illustrates the relationship between RF output power and frequency for different values of d. From the figure, it's clear that when there is no misalignment, i.e., $d = 0$, the gyroklystron's center frequency is 93.2 GHz, with the max. RF output power of 316 kW. Additionally, the figure reveals that as the shift rises, the output power declines. When $d = 0.9$ mm, the output power falls to 75 kW.

Figure 5a illustrates the changes in resonant frequency in response to variations in the electron beam's radial position. With the increase in radial shift, from 0 mm to 0.6 mm, frequency shift rises from 0 to 0.01%. Beyond this range, the frequency shift shows negligible sensitivity.

Figure 5b depicts the effect of radial shift on efficiency and gain. It is clear from the figure that the gain and efficiency both steadily decreases with the increase in the shift. As the shift increases from 0 to 0.9 mm, the gain drops from 23.23 to 16.99 dB, and

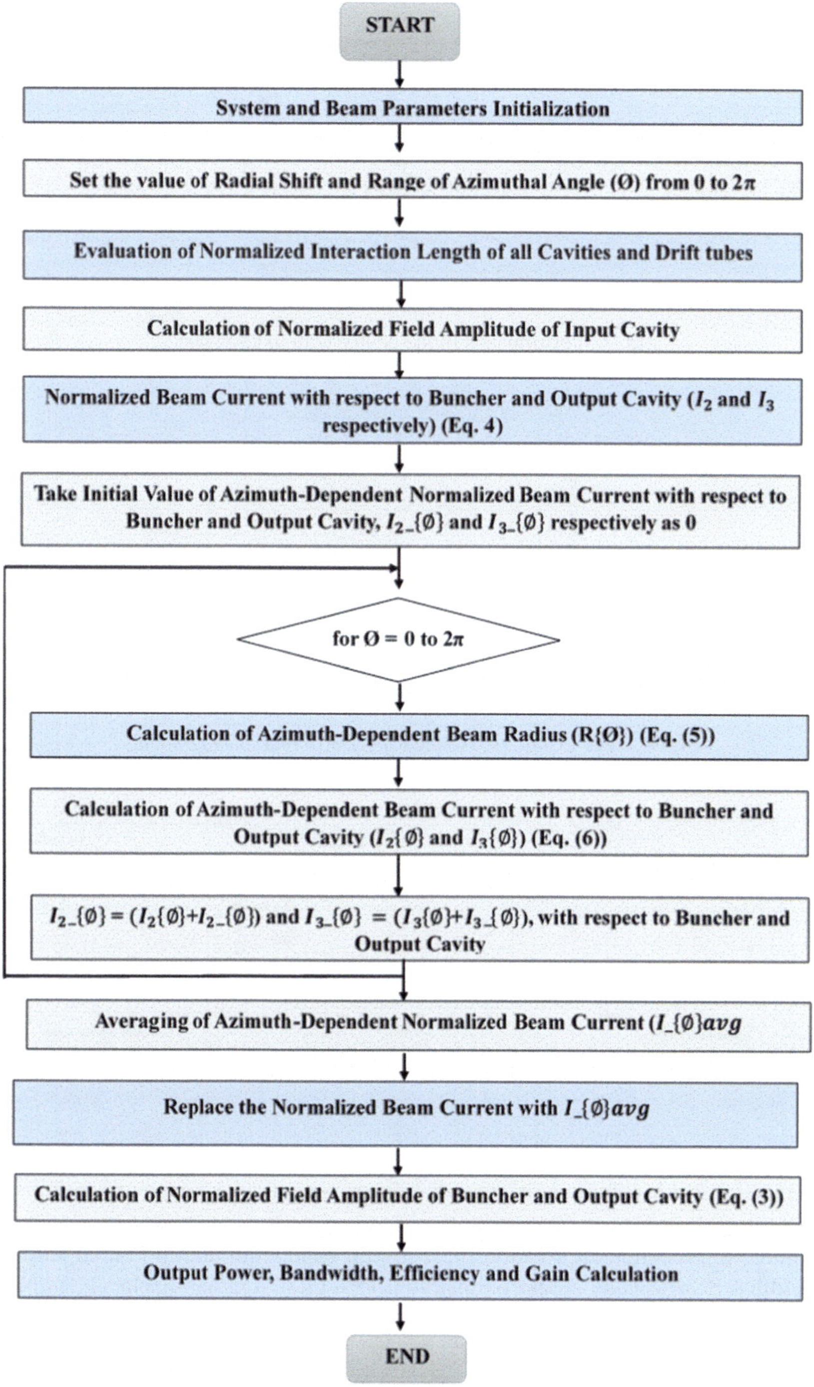

Fig. 2. Flowchart illustration of the non-linear analysis of electron beam misalignment effect on gyroklystron amplifier performance

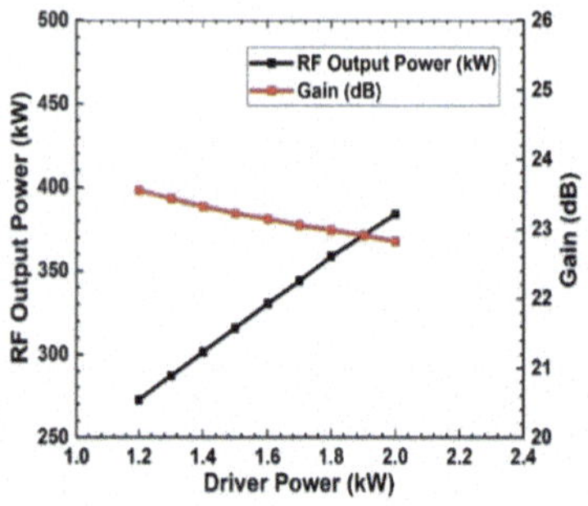

Fig. 3. Performance of the device with input power.

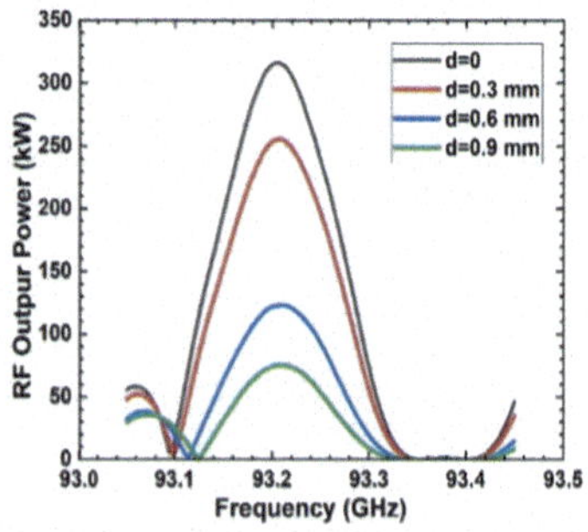

Fig. 4. Peak Power vs. Frequency for different misalignment.

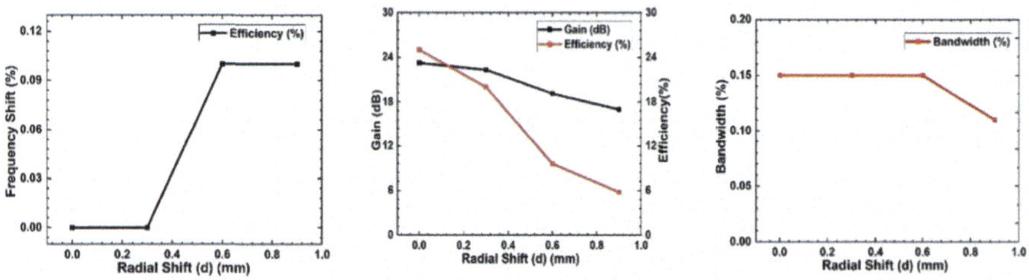

Fig. 5. Effect of radial shift on (a) Frequency Shift, (b) Gain and Efficiency, (c) Bandwidth

efficiency reduces from 25 to 5.8%. The sharp decline in the efficiency and gain curve highlights that it is highly sensitive to radial displacement, where even minor shifts lead to a considerable reduction in performance.

Figure 5c illustrates the relationship between bandwidth and the radial shift. The bandwidth remains steady at approximately 0.15% up to a radial displacement of d = 0.6 mm. Beyond this point, it decreases to approximately 0.11%.

4 Conclusion

The proposed work is crucial to analyze the effect of misaligned electron beam on the gyro-devices. The paper focuses mainly on the analysis of performance characteristics of gyroklystron amplifier in detail. In the current analysis, 0.3–0.9 mm shifts have been employed. As the shift rises from 0 to 0.9 mm, the output power reduces from 316 to 75 kW, the gain drops from 23.23–16.99 dB, and efficiency falls from 25 to 5.8%. The analysis shows that the output power, gain and efficiency is highly sensitive to the misalignment however the device resonant frequency and bandwidth is minimum affected. The study will be useful for the performance analysis and practical implementation of the experimental gyroklystron amplifiers.

References

1. Chauhan, M.S., et al.: Nonlinear analysis of a gyroklystron amplifier with misaligned electron beam. J. Fusion Energy **35**, 289–298 (2016)
2. Xu, S.-X., et al.: Numerical simulation of a W-band four-cavity gyroklystron amplifier. Phys. Plasmas **19**(3) (2012)
3. Choi, J.J.: A high-gain, 28 GHz, 200 kW gyroklystron amplifier. Int. J. Inf. Millimeter Waves **19**, 1681–1691 (1998)
4. Tran, T.M., et al.: Optimization of gyroklystron efficiency. Phys. Fluids **29**(4), 1274–1281 (1985)
5. Airila, M.I.: Spatio-temporal chaos in a gyrotron resonator with an off-axis electron beam. In: Twenty Seventh International Conference on Infrared and Millimeter Waves. IEEE (2002)
6. Siddiqui, N.F., et al.: Effect of off-axis electron beam on the performance of a gyroklystron amplifier. Phys. Plasmas **31**(7) (2024)
7. Zasypkin, E.V., et al.: Experimental study of a W-band Gyroklystron amplifier operated in the high-order TE_{021} cavity mode. Radiophys Q. El **55**, 309–317 (2012). https://doi.org/10.1007/s11141-012-9370-4

PIC Simulation of Echelon Type Strapped Magnetron Using Realistic Magnetic Field

Sandeep Kumar Vyas[1(✉)], Narugopal Nayek[1], Sandeep Kakatkar[1,2], and Sharad Chavan[1,2]

[1] SAMEER Center for High Power Microwave Tube and Component Technology, Guwahati, India
{sandeepkumar.vyas,sharad}@sameer.gov.in

[2] LINAC Technology Division, Society for Applied Microwave Electronics Engineering and Research, IIT Campus, Mumbai 400076, India

Abstract. The manuscript presents the design of pulsed power tunable S-band magnetron using electromagnetic and PIC (particle in cell) simulation. The PIC simulation has been carried out including the all part of magnetron as coupling loops, output glass window, output launcher, tuner, and anode block with echelon type strapping. For the realistic analysis the simulation has been carried out using a realistic 3D magnetic field profile. An electromagnet has been designed using magneto statics simulation in CST studio. The 1565 Gauss is achieved in center of electromagnet for the 21 Amp coil current. The output power of magnetron with this 3D realistic magnetic field is found 2.78 MW at 45.5 kV anode voltage. The anode current and overall efficiency are 95.2 Amp and 62% respectively.

Keywords: Pulsed magnetron · Hole and slot type · Echelon type strapping · Electromagnet

1 Introduction

High power microwave sources have various uses in scientific, industrial, medical, space, and defense field [1–3]. Magnetron is a compact size, low cost and high efficiency microwave oscillator. Please note that the first paragraph of a section or subsection is not indented. The first paragraphs that follows a table, figure, equation etc. does not have an indent, either. A linear accelerator (LINAC) system uses high-power microwave source to accelerate electron beam for production of X-ray. The magnetron is most suitable source for LINAC due to its compact size, high efficiency and low cost. The S-band 6 meV LINAC developed by SAMEER needs 2.6 MW pulse power magnetron at 2.998 $\pm$ 0.005 GHz.

This paper presents the design of twelve hole and slot type magnetron with echelon type strapping using electromagnetic and PIC simulation. Many research paper based on PIC simulation of CW, pulse magnetron and relativistic magnetron are published in last two decade [2–4]. In these papers the magnetron has either no strapping or ring type strapping. The simulation is done with uniform magnetic field. There is no paper available

S. K. Ghosh et al. (Eds.): VEDA 2024, SPPHY 424, pp. 180–183, 2025.
https://doi.org/10.1007/978-981-96-5174-0_27

on PIC simulation of echelon type strapped magnetron as per our best knowledge. The only electromagnetic simulation of magnetron with echelon type strapping is reported in one paper [4]. The accurate modeling and meshing of this kind of strapping is making difficult to do accurate simulation. Now these days the some available power full GPU based hardware and supporting PIC simulation code made possible to us to visualize a virtual prototype design on computer. This paper presents the PIC simulation of complete magnetron including all parts and using a realistic non uniform 3D magnetic field. The 3D magnetic field profile is generated by the magneto static simulation of magnet circuit of magnetron. The magnetic field plays a crucial role in the electron beam wave interaction for magnetron compare to O-type tube. The 3D electromagnet system has been designed to generate required magnetic field for the magnetron. The electromagnet allows us to operate magnetron in broad operating region.

The paper has been divided into four sections including the introduction Sect. 1. Section 2 presents the electromagnetic and magneto static simulation of magnetron and electromagnet. The PIC simulation of magnetron with non uniform magnetic field is given in section. Finally, the conclusion is given in Sect. 4.

2 Design of Magnetron and Electromagnet

2.1 Electromagnetic Simulation of Magnetron

The electromagnetic simulation of the magnetron was carried out using CST Microwave Studio to determine the resonance frequency and Q value. The eigenmode solver and frequency domain solver were utilized for this purpose. The simulation models and results are shown in Figs. 1 and 2(a–b). The eleven cavities are alternatively connected by using 22 echelon-type straps. The strap-breaking technique is used to suppress the space harmonic of π-1 mode. Output power from the cavity is coupled using loop coupling and launched through the output launcher with the help of loop support. The output launcher section contains the impedance matching network, loop support, and RF window. The Q (quality factor) value and circuit efficiency are calculated with the help of Fig. 2 (b). The results are summarized in Table 1.

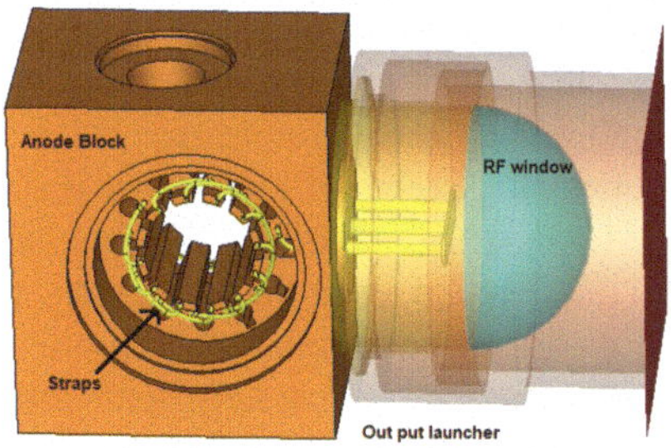

Fig. 1. Electromagnetic simulation model of magnetron

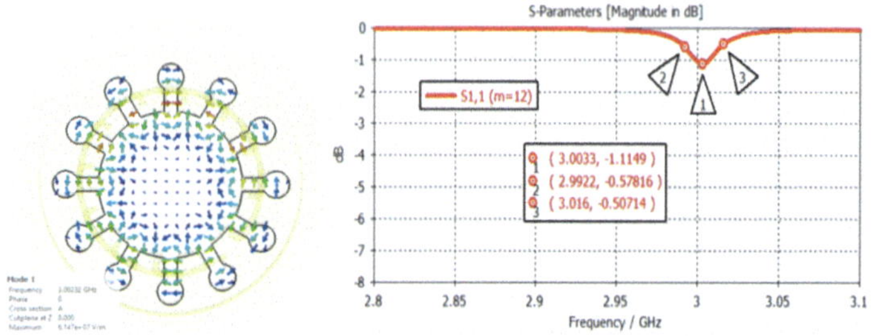

Fig. 2. (a) Vector plot of RF electric field of π-mode and (b) resonance spectrum (S11).

Table 1. Cold test simulation results.

Parameter	Value
Frequency	3.003 (GHz)
R_L	1.1 (dB)
VSWR	15.81
Δf	24 (MHz)
Q_0	2103
Q_L	125
Q_{ect}	133
Circuit efficiency (η_{ckt})	93%

2.2 Design of Electromagnet Simulation of Magnetron

The electromagnet offers the ability to smoothly control and adjust the strength of the magnetic field, allowing for operation of the magnetron over a wide range. This means that the same magnetron can deliver different output power and operate at different voltages. To achieve a compact design, two solenoid coils are used at the pole piece. A yoke is utilized to provide a return path for the magnetic field and to shield other equipment from the magnetic field. Figure 3(a) shows the simulation model of electromagnet. The Magneto static solver of CST studio is used for design. The equal number 255 turns is kept in both coils. A magnetic field strength of 1565 Gauss is achieved at the center with a solenoid coil current of 21 Amp. A graph of the magnetic field versus coil current is plotted and compared with the standard electromagnet graph, showing good agreement in the results (as seen in Fig. 3(b)). The magnetic field profile obtained from this simulation is then used for PIC simulation to accurately predict the output performance of the magnetron.

3 PIC Simulation of Magnetron

The CST Particle Studio's PIC solver was utilized to analyze the performance of the magnetron. The simulation involved a non-uniform realistic magnetic field, with an axial variation of around 6% as depicted in Fig. 3 (c). An operating voltage of 45.5 kV was applied between the cathode and anode. Figure 4 shows the PIC simulation results. The output power and overall efficiency are found to be 2.789 MW and 64%.

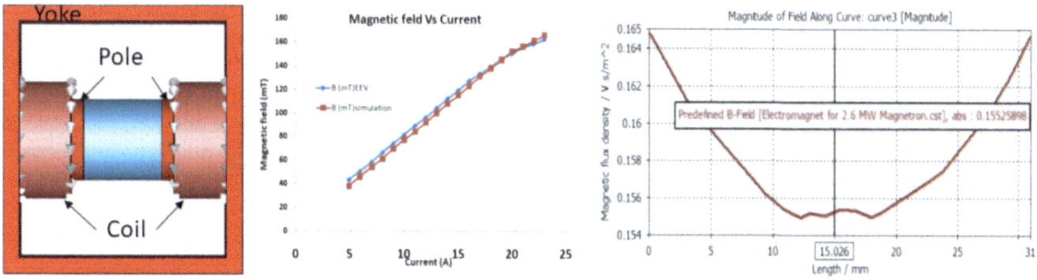

Fig. 3. (a) Simulation model of electromagnet, (b) current vs magnetic field, and (c) variation of magnetic field in the axial direction.

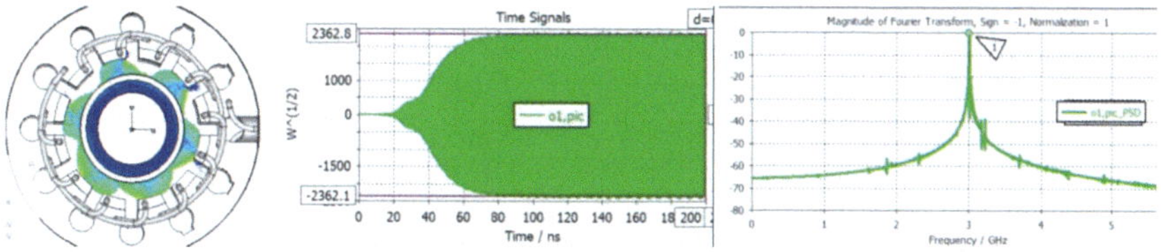

Fig. 4. (a) Electron bunch formation, (b) Output power vs time, and (c) Power spectrum density curve of output signal.

4 Conclusion

The design electromagnet for pulse power 2.6 MW magnetron is completed. The PIC simulation of echelon type strapped magnetron is completed. We have achieved stable output power with a non-uniform realistic magnetic field profile, obtaining 2.789 MW power with a 45.5 kV applied voltage.

References

1. Andreev, D., Kuskov, A., Schamiloglu, E.: Review of the relativistic magnetron. Matter Radiation at Extremes **4**(6) (2019). https://doi.org/10.1063/1.5100028
2. Anilkumar, P., Pamu, D., Tiwari: Historical developments and recent advances in high-power magnetron: A review. IETE Tech. Rev. **39**, 1309–1323 (2022)
3. Vyas, S.K., Verma, R.K., Maurya, S., Singh, V.V.P.: Review of magnetron developments. Freuenz **70**(9–10), 455–462 (2016). https://doi.org/10.1515/freq-2015-0196
4. Maurya, S., et al.: Electromagnetic design and analysis of a high power tunable pulsed magnetron using MAGIC-3D. In: 2011 IEEE International Vacuum Electronics Conference (IVEC). IEEE (2011)

A Novel Interaction Structure for a Wideband $TE_{1,1}$-Mode Gyro-Travelling-Wave Tube

H. R. Shobha[1], Ogar Ram[1], S. Anuradha[1], and Vishal Kesari[2(✉)]

[1] Department of Electronic Science, Bangalore University, Bengaluru 560034, India
[2] Microwave Tube Research and Development Centre, DRDO, Bengaluru 560013, India
vishal_kesari@rediffmail.com

Abstract. A novel interaction structure for a $TE_{1,1}$-mode gyro-travelling-wave tube (gyro-TWT) was proposed in the paper. The interaction structure possesses both the azimuthal and axial periodicities. The interaction structure was designed, modelled and analyzed in simulation for its dispersion characteristics. The sensitivity analyses were carried out for dimensional control on the lower-cutoff-frequency, where the gyro-TWT is being operated. While comparing to the known interaction structures, a dispersion shaping is observed which in turn would be suitable for the wideband performance of a gyro-TWT. The lower-cutoff-frequency of the considered interaction structure increases with the increase in the radius of the projected cylinders, the radial height of the projected cylinders and the number of the projected cylinders in the azimuth, and the decrease of axial periodicity. The radius of projected cylinders and number of projected cylinders in the azimuth control the lower-cutoff-frequency, respectively, the most and least. Thus, the considered structure is a possible interaction structure for a wideband gyro-TWT.

Keywords: Interaction Structure · Gyro-TWT · Azimuthal and Axial Periodic Structure · Dispersion Characteristics · Cutoff-Frequency Study · Simulation

1 Introduction

A wide variety of interaction structures have been proposed in past for their potential application in gyro-TWTs. Some of them have been proven for wide device bandwidth, some for higher device gain and some for both [1–7]. Most commonly, the axial periodic structures have been predicted for wider bandwidth [1–4] and the azimuthal periodic structures for higher device gain [1, 4, 5]. All metal interaction structures [1, 2, 4, 5, 7] are preferred over dielectric-based interaction structures avoiding the issues of dielectric charging and associated heating [6]. These problems are minimized up to some extent in combined metal-dielectric loaded structures [3]. In the recent past, an azimuthal and axial periodic interaction structure in which the azimuthal periodicity appears because of metal vanes and axial periodicity because of the finite length of metal vanes, is proposed for a wideband gyro-TWT [7]. It has been noted that dispersion shaping and thus broadbanding is possible in azimuthal and axial periodic interaction structures [7].

S. K. Ghosh et al. (Eds.): VEDA 2024, SPPHY 424, pp. 184–190, 2025.
https://doi.org/10.1007/978-981-96-5174-0_28

In the present paper, another novel interaction structure is proposed, which also possesses both azimuthal and axial periodicity. It is a circular waveguide, in which number of cylinders are radially projected from the wall of the waveguide at the same axial position forms a unit cell and the axial periodicity of such unit cells defines the infinitely long interaction structure (Fig. 1). The present paper is planned in four sections: Sect. 2 discusses the design, modelling and methodology for the presented simulation based study of the proposed structure; Sect. 3 presents the results and associated discussion including the comparison of proposed interaction structure with known structures; and Sect. 4 concludes the paper.

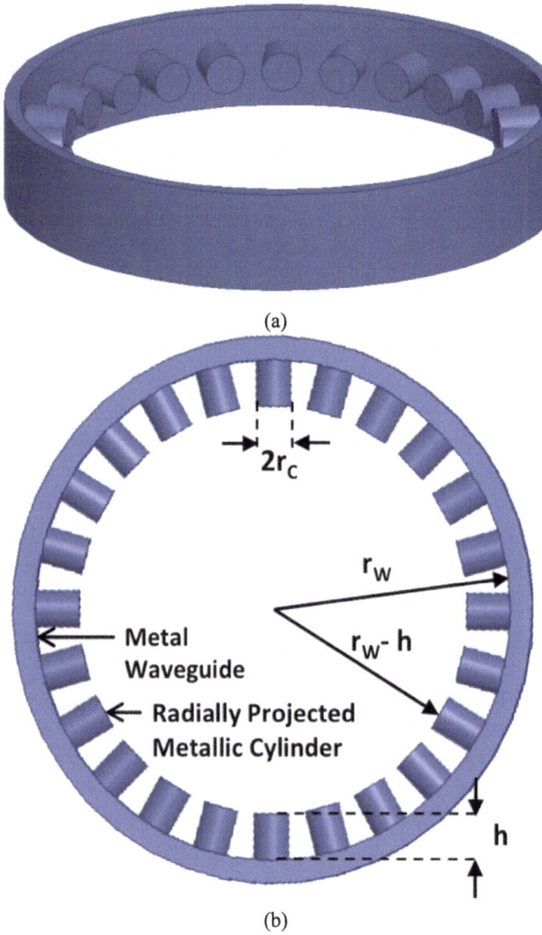

Fig. 1. The unit-cell model of circular waveguide with the radially projected cylinders from the waveguide-wall (an azimuthal and axial periodic interaction structure) in (a) isometric-view and (b) top-view showing the dimensional parameters.

2 Methodology Towards Design and Modelling for Simulation

A theoretical model is postulated to simultaneously achieve both azimuthal and axial periodicity (Fig. 1(a)). The proposed structure comprises a circular waveguide with a smooth inner wall, characterized by an inner radius (r_W). Additionally, the configuration incorporates N cylindrical elements projected radially inward from the waveguide wall, each possessing specified dimensions, including the radius (r_C) and radial height (h) of the projected cylinder. The axial periodicity is defined as p. The azimuthal arrangement of the projected cylinders is orchestrated to ensure equidistant spacing, with each cylinder positioned at angular intervals of 360/N, where N is the designated number of projected cylinders. 360/N also represents the azimuthal periodicity of the proposed interaction structure with the projected cylinders.

In the process of modelling a circular interaction structure with radially projected cylinders using a FEM-based simulation tool: high-frequency structure simulator (HFSS) [8], a smooth-wall circular waveguide is meticulously constructed. The waveguide, characterized by a radius (r_W) of 5.36 mm, is aligned along the z-axis. The fundamental unit-cell of the structure commences at z = -p/2, with the height of the unit-cell matching the axial periodicity (p) set at 2.00 mm. Within this framework, a cylindrical entity is introduced radially to the smooth-wall circular waveguide, positioned at a distance (r_W-h) from the waveguide axis. The central coordinates of the projected cylinder are specified as (r_W-h, 0 mm, 0 mm). For the purpose of achieving a representative simulation, the projected cylinder undergoes duplication around the axis of the waveguide. This replication occurs with a rotation angle of 360/N, N is set as 24 in this particular instance.

3 Results and Discussion

The interaction structure containing a circular waveguide with radially projected cylinders from the waveguide wall is electromagnetically designed. Based on the geometrical parameters obtained in electromagnetic design, the interaction structure is modelled, and the boundary conditions and materials are assigned and the structure is analyzed in a commercially available simulation tool: HFSS [8]. Firstly, one may check the excited mode. The desired mode in the present case is the dominant $TE_{1,1}$-mode. It can clearly be seen that due to the presence of radially projected cylinders from the waveguide-wall in a circular waveguide, the dominant $TE_{1,1}$-mode does not get distorted from that in a smooth-wall circular waveguide (Fig. 2).

Secondly, it is required to check the possibility of dispersion shaping in the proposed interaction structure for its possibility to be used in a wideband gyro-TWT. For this purpose, the dispersion characteristics of the dominant $TE_{1,1}$-mode in the smooth-wall, disc-loaded [1, 2, 4], infinite-length vane-loaded [5], and finite-length vane-loaded [7] circular waveguides along with that of proposed waveguide can be plotted together and can be compared for the possible tailoring of dispersion characteristics (Fig. 3). Here, one may observe that the lower-cutoff-frequencies in various considered structures do not deviate much, however, the upper-cutoff-frequencies show deviations from one another, which bring out the possibility of dispersion shaping and thus widening of operating bandwidth in a gyro-TWT (Fig. 3).

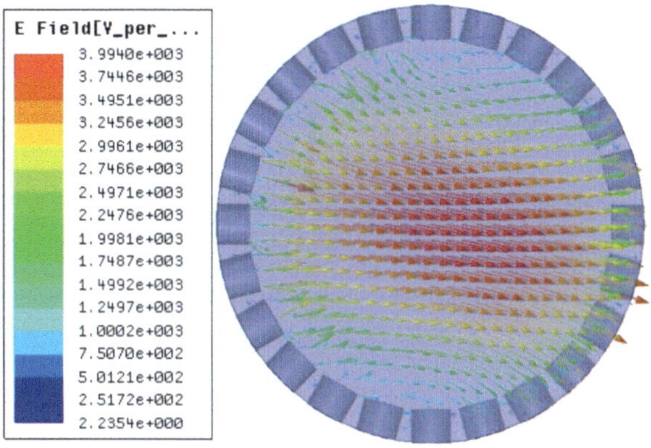

Fig. 2. Vector electric field pattern of the dominant circular $TE_{1,1}$-mode in the proposed waveguide.

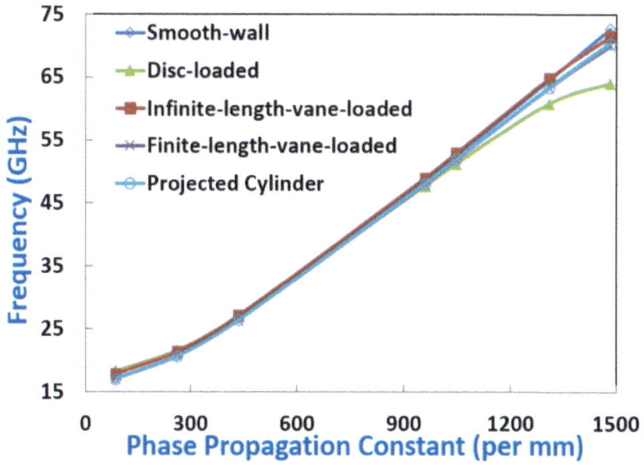

Fig. 3. Comparison of dispersion characteristics of the smooth-wall, disc-loaded [1, 2, 4], infinite-length vane-loaded [5], and finite-length vane-loaded [7] circular waveguides along with that of the proposed waveguide.

Increasing the radial extent of the radially projected cylinder within the proposed structure corresponds to an elevation in the lower-cutoff-frequency. In the limit where the radius of the radially projected cylinder approaches zero ($r_C \to 0$), the structure converges to a smooth-wall circular waveguide, resulting in the cutoff frequency approaching that of a conventional smooth-wall circular waveguide (Fig. 4). The lower-cutoff-frequency of the dominant $TE_{1,1}$-mode increases with increasing the radius of radially projected cylinders in the proposed structure, while all other geometrical parameters are kept constant. Further, in the special case of zero radius of radially projected cylinders in the proposed structure, the lower-cutoff-frequency of the dominant $TE_{1,1}$-mode in the proposed waveguide passes on to that in a smooth-wall circular waveguide (Fig. 5).

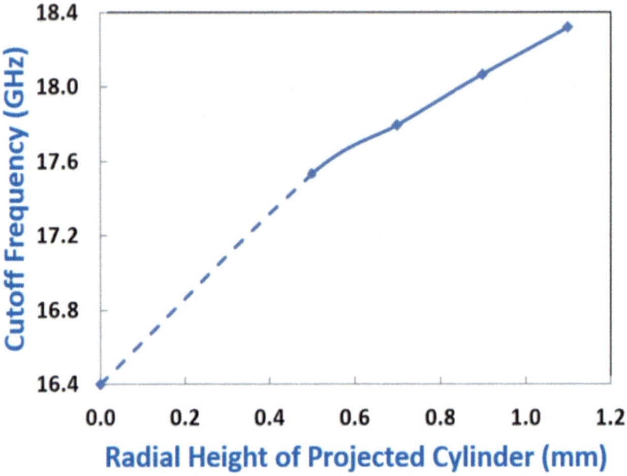

Fig. 4. Variation in lower-cutoff-frequency of the $TE_{1,1}$-mode of the proposed waveguide with change in the radial height of radially projected cylinders (h) taking $r_W = 5.36$ mm, p = 2.00 mm, N = 24 and $r_C = 0.55$ mm. In special case, the proposed waveguide passes on to a smooth-wall circular waveguide for zero value of the height of the radially projected cylinder ($h \rightarrow 0$) (The data point corresponding to the height of the radially projected cylinder equal to 0).

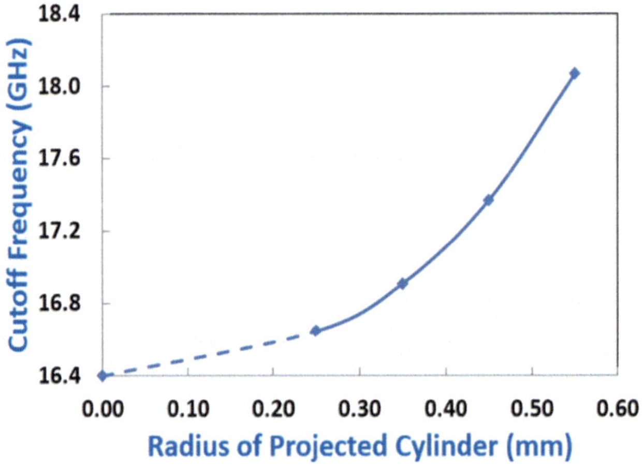

Fig. 5. Variation in lower-cutoff-frequency of the $TE_{1,1}$-mode of the proposed waveguide with change in the radius (r_C) of the radially projected cylinders taking $r_W = 5.36$ mm, p = 2.00 mm, h = 0.9 mm and N = 24. In special case, the proposed waveguide passes on to a smooth-wall circular waveguide for zero value of the radius of the radially projected cylinder ($r_C \rightarrow 0$) (The data point corresponding to the radius of the radially projected cylinder equal to 0).

The augmentation in the count of radially projected cylinders (N) is associated with a proportional escalation in the lower-cutoff-frequency. In the absence of any radially projected cylinders (N = 0), a distinctive scenario arises, resembling a smooth-wall cylindrical waveguide, and consequently, the cutoff frequency aligns with that of a smooth-wall circular waveguide (Fig. 6). Conversely, an augmentation in axial periodicity is associated with a decline in the lower-cutoff-frequency of the proposed structure (Fig. 7).

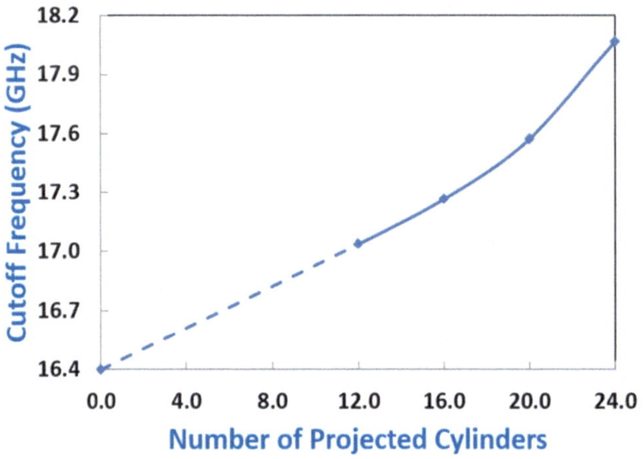

Fig. 6. Variation in lower-cutoff-frequency of the $TE_{1,1}$-mode of the proposed waveguide with change in the number of radially projected cylinders (N) in azimuth of the structure taking r_W = 5.36 mm, r_C = 0.55 mm, p = 2.00 mm, and h = 0.9 mm. In special case, the proposed waveguide passes on to a smooth-wall circular waveguide for zero number of the radially projected cylinders (N = 0) (The data point corresponding to zero number of radially projected cylinder).

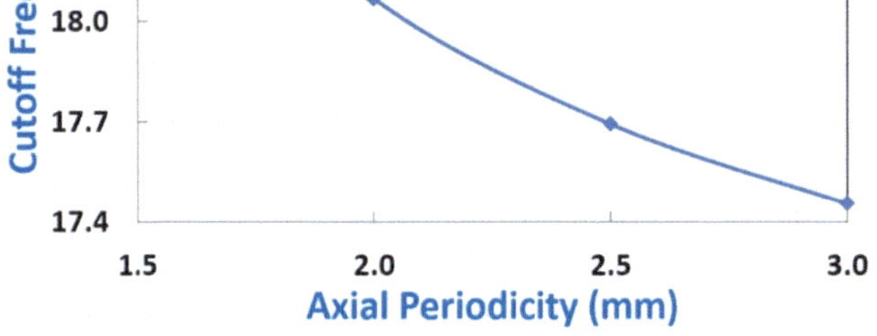

Fig. 7. Variation in lower-cutoff-frequency of the $TE_{1,1}$-mode of the proposed waveguide with change in the axial periodicity (p) of the structure taking r_W = 5.36 mm, r_C = 0.55 mm, h = 0.9 mm and N = 24.

4 Conclusion

A novel projected cylinders based azimuthal and axial periodic interaction structure is proposed, designed, modelled, simulated and analyzed for its application in a wideband gyro-TWT. There is a possibility of tailoring the dispersion characteristics, which in turn may provide a wideband coalescence with beam-mode dispersion-line and hence the wideband device performance. The dispersion characteristics of the proposed waveguide is compared with those of conventional smooth-wall, annular disc-loaded, infinite-length vane-loaded, and finite-length vane-loaded circular waveguides. Further, a sensitivity analysis is carried out with reference to the variations in the radius and the height of projected cylinders, the number of projected cylinders in azimuth of the structure and the axial periodicity of the structure to examine their effect on lower-cutoff-frequency of the $TE_{1,1}$-mode of the proposed waveguide. The variation in lower-cutoff-frequency of the proposed waveguide is important because a gyro-TWT operates in vicinity to the lower-cutoff-frequency of its interaction structure. It may clearly be noted that the geometrical dimensions have a significant control on the lower-cutoff-frequency of the proposed structure.

References

1. Kesari, V., Basu, B.N.: High Power Microwave Tubes: Basics and Trends, vol. 1. Morgan and Claypool Publishers, California for IOP Concise Physics, London (2018)
2. Kesari, V., Jain, P.K., Basu, B.N.: Analysis of a circular waveguide loaded with annular metal discs for wideband gyro-TWTs. IEEE Trans. Plasma Sci. **33**(4), 1358–1365 (2005)
3. Kesari, V., Jain, P.K., Basu, B.N.: Modelling of axially periodic circular waveguide with combined dielectric and metal loading. J. Phys. D: Applied Phys. **38**, 3523–3529 (2005)
4. Kesari, V.: Innovative interaction structures for wideband gyro-TWTs. In: Proc. VEDA-Kolleg-2013, pp. 1–5 (2013)
5. Singh, G., Ravi Chandra, S.M.S., Bhaskar, P.V., Jain, P.K., Basu, B.N.: Analysis of an azimuthally periodic vane-loaded cylindrical waveguide for a gyro-traveling-wave tube. Int. J. Electronics **86**(12), 1463–1479 (1999)
6. Rao, S.J., Jain, P.K., Basu, B.N.: Broadbanding of a gyro-TWT by dielectric-loading through dispersion shaping. IEEE Trans. Electron Dev. **43**(12), 2290–2299 (1996)
7. Ram, O., Shobha, H.R., Kesari, V.: Simulation based study of azimuthal and axial periodic interaction structure for a gyro-travelling-wave tube. In: IEEE Int. Conf. on Recent Innovation in Smart and Sustainable Technology, Presidency University, Bengaluru, India, 08–09 Mar. 2024 (2024)
8. User's Guide—High Frequency Structure Simulator. ANSYS, Canonsburg, PA (2016)

Exploration of Dispersion Shaping in Azimuthal and Axial Periodic Interaction Structure for a Wideband Gyro-Travelling-Wave Tube

K. A. Rahul[1], N. Chaithra[1], S. K. Datta[2], and Vishal Kesari[2(✉)]

[1] Department of Electronics and Communication Engineering, Presidency University, Bengaluru 560064, India

[2] Microwave Tube Research and Development Centre, DRDO, Bengaluru 560013, India

vishal_kesari@rediffmail.com

Abstract. Exploration of a novel interaction structure for a $TE_{0,1}$-mode gyro-travelling-wave-tube (gyro-TWT) was presented in the paper. This interaction structure features both the azimuthal and axial periodicities. The interaction structure was presented, modeled, simulated and analyzed to study it's dispersion characteristics. The dimensional parameters were varied to perform the sensitivity analysis. The dispersion shaping advantageous for augmenting the wideband capabilities for a gyro-TWT was noted while varying the geometrical parameters. The width of radially projected metallic blocks and the structure periodicity are respectively the least and most effective geometrical parameters for controlling the property of dispersion shaping. Hence, the study predicts that the presented structure is the feasible interaction structure for a wideband gyro-TWT.

Keywords: Interaction Structure · Gyro-TWT · Azimuthal and Axial Periodic Structure · Dispersion Characteristics · Simulation

1 Introduction

A diverse array of interaction structures has been recommended previously for their potential implementation in gyro-traveling-wave tubes (gyro-TWTs). Some of the structures have demonstrated a wide device bandwidth, while others have illustrated for higher device gain and some achieve both [1–8]. Typically, the axial periodic structures are anticipated to provide improved bandwidth [1–5] while azimuthal periodic structures are associated with elevated device gain [1, 5, 6]. Metal interaction structures [1–3, 5, 6, 8] are chosen over dielectric-based counterparts due to their ability to evade the problems associated with dielectric charging and resultant dielectric heating [7]. These issues are partially mitigated in unified metal-dielectric configurations [4]. Recently, a structure incorporating both azimuthal and axial periodic interactions has been developed for a wideband gyro-TWT. The azimuthal periodicity arises from metal vanes and axial periodicity arises from finite axial length of those vanes and it has been remarked that such structures facilitate dispersion shaping and thus broadbanding of a gyro-TWT.

S. K. Ghosh et al. (Eds.): VEDA 2024, SPPHY 424, pp. 191–198, 2025.
https://doi.org/10.1007/978-981-96-5174-0_29

In this work, a new interaction structure is introduced, that also features azimuthal and axial periodicities. It is a cylindrical waveguide, where rectangular metallic blocks are placed inside the wall of waveguide symmetrically creating azimuthal and axial periodicities (Fig. 1). The aim of this paper is to explore the possibility of dispersion shaping in the proposed interaction structure for its potential use in a wideband gyro-TWT. The present paper is structured in four sections. Section 2 deliberates on the design, modelling and methodology for the conducted simulation study of the proposed interaction structure. Section 3 unveils the outcomes and the discussion associated which includes the comparison of proposed structure with established structures, and Sect. 4 provides conclusion of the study.

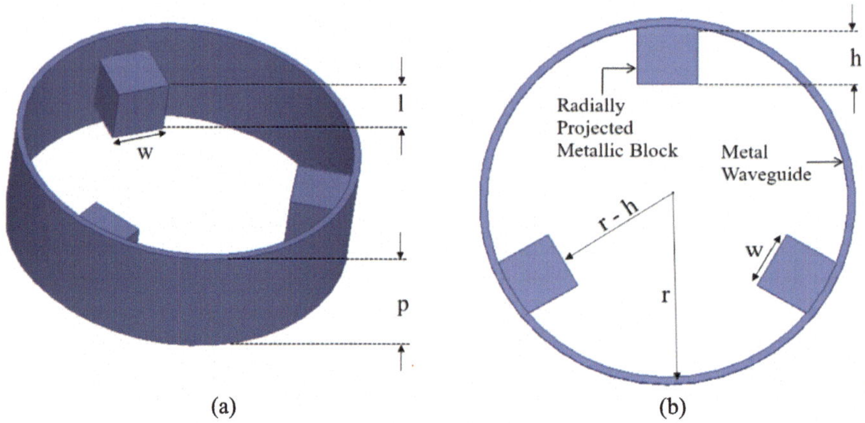

Fig. 1. The unit-cell structure of a cylindrical waveguide loaded with the rectangular blocks from the waveguide-wall (an azimuthal and axial periodic interaction structure) in (a) isometric-view and (b) top-view showing the dimensional parameters.

2 Methodology Towards Design and Modelling for Simulation

A theoretical model is hypothesized to accomplish both azimuthal and axial periodicity simultaneously (Fig. 1). The presented structure encompasses a cylindrical waveguide, which is described by an inner radius (r). Furthermore, the design comprises a number of rectangular blocks placed inward the wall of the cylindrical waveguide, each having defined dimensions, which includes radial height (h), axial height (l), width (w) and periodicity (p). The azimuthal orientation of the rectangular blocks inside the waveguide is coordinated to ensure equal distant spacing, with each rectangular block placed at angular segments of 360°/N where N is the number of rectangular blocks in azimuth inside the wall of cylindrical waveguide.

In the course of modelling a cylindrical interaction structure with rectangular blocks projected from inward cylindrical wall with the help of a Finite Element Method (FEM) established simulation tool: HFSS (High-Frequency Structure Simulator) [9], a smooth-wall cylindrical waveguide is diligently constructed. The cylindrical waveguide is described by a radius (r) of 6.0 mm, is projected along the z-axis. The fundamental unit cell of the structure which is the cylindrical waveguide, embarks at z = -p/2, with the height of the block corresponding the periodicity (p) established at 4.0 mm. Within this cylindrical waveguide, a rectangular block is introduced inward to the wall of waveguide, established at a distance (r-h), from the axis of waveguide. The median coordinates of projected rectangular blocks are enumerated as (r-h, -w/2, (p-l)/2). With the intention of attaining an illustrative simulation, the inserted rectangular block is subjected to duplication around the axis of the cylindrical waveguide. This occurs with an angle of rotation of 360°/N where N is assigned to three in the present model.

3 Results and Discussion

The model encompassing a cylindrical waveguide with inserted blocks inward the wall of the waveguide is designed electromagnetically. The structure is centered around the geometrical parameters which is obtained in the electromagnetic design, and with the help of HFSS [9] the materials and boundary conditions are delegated and the analysis of interaction structure is carried out. Primarily, one may assess the excited mode. In the current case, the preferred mode is the first azimuthally symmetric $TE_{0,1}$-mode. It is evident that because of the existence of rectangular blocks inward the wall of cylindrical waveguide, the azimuthally symmetric $TE_{0,1}$-mode remains undistorted in comparison with that of a smooth-wall cylindrical waveguide (Fig. 2).

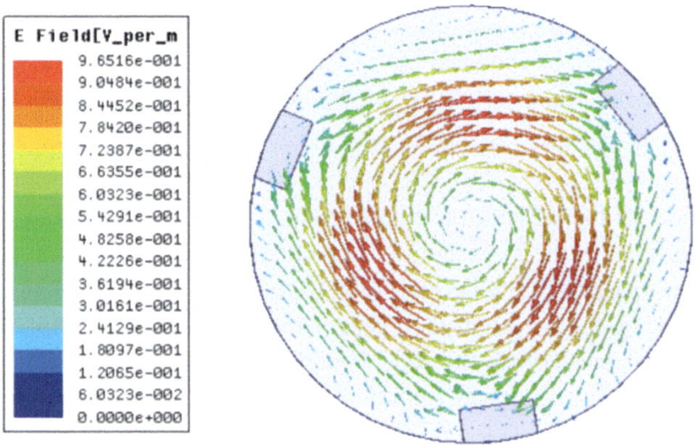

Fig. 2. Vector electric field pattern of the circular $TE_{0,1}$-mode in the presented waveguide.

Furthermore, it is necessary to examine the possibility of dispersion shaping in the given interaction structure for its feasibility to be utilized in a wideband gyro-TWT. To achieve this, the sensitivity analysis is performed for the structure parameters: radial height (h), periodicity of the structure (p), radius of cylindrical waveguide (r), axial height (l), and width (w). The dispersion characteristics are plotted taking phase delay in degree on abscissa and Eigen frequency in GHz on ordinate. The parameters: periodicity of the structure (p), radius of cylindrical waveguide (r), width (w), axial height (l) and radial height (h) are varied for different values and the graph is plotted to emphasize the opportunity of the dispersion shaping and thus enhancing the operating bandwidth of a gyro-TWT. On increasing the width of radially projected rectangular blocks inward the wall of cylindrical waveguide by keeping the other parameters constant, the curve does not emphasize the property of dispersion shaping as it is illustrative that the curve does not deviate much. (Fig. 3). The curve slightly appears to deviate with increase in axial height of the radially projected blocks by slightly lowering its lower cutoff frequency and slightly elevating its upper cutoff frequency, and thus enhances the pass band (Fig. 4).

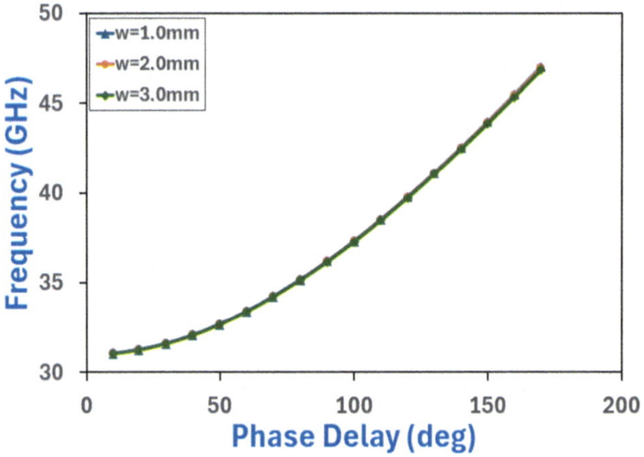

Fig. 3. Formulating the variation in width (w) of the rectangular block on the circular $TE_{0,1}$-mode dispersion characteristics of a cylindrical waveguide loaded with the rectangular blocks from the waveguide-wall keeping the parameters r = 6.0 mm, l = 2.0 mm, h = 2.0 mm, p = 4.0 mm and N = 3.

The deviation of the dispersion characteristics of the circular $TE_{0,1}$-mode in the proposed structure from the hyperbola in the smooth-wall circular waveguide (h = 0) increases with increase in radial height of the radially projected blocks in the proposed structure and by keeping the other geometric parameters constant. Thus, the dispersion shaping is observed on varying the radial height of the metallic blocks (Fig. 5). It is also observed that the lower cutoff frequency reduces with increase of radial height of the radially projected blocks (Fig. 5). Further in the next case, by varying the structure periodicity (p) and keeping the other parameters constant in the proposed waveguide, the dispersion characteristics are witnessed and a dominant dispersion shaping is noted. Increase in periodicity of the structure results in greater dispersion shaping by dominantly lowering the upper cutoff frequency and marginally increasing the lower cutoff frequency (Fig. 6).

One may also vary the number of metallic blocks in azimuth (N) while keeping the other parameters constant. By increasing the number of metallic blocks in azimuth, the dispersion characteristics of the proposed waveguide show the slight reduction in upper cutoff frequency and thus slight reduction in pass band frequency and a control

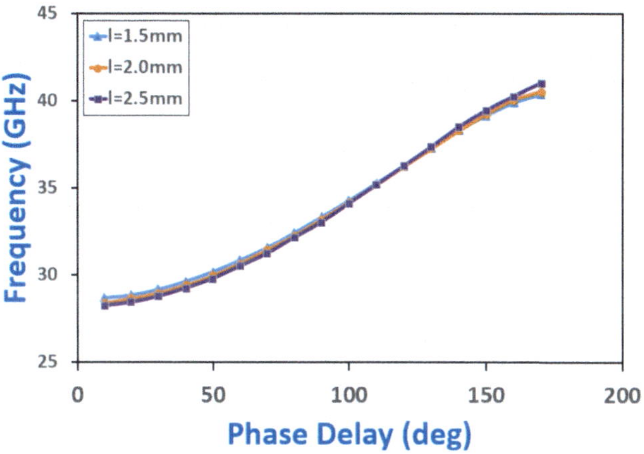

Fig. 4. Formulating the variation in axial height (l) of the rectangular block on the circular $TE_{0,1}$-mode dispersion characteristics of a cylindrical waveguide loaded with the rectangular blocks from the waveguide-wall keeping the parameters r = 6.0 mm, h = 2.0 mm, p = 4.0 mm, w = 2.0 mm and N = 3.

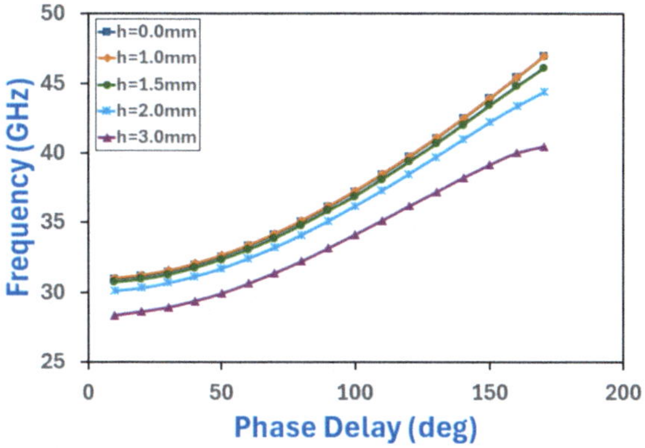

Fig. 5. Formulating the variation in radial height (h) of the rectangular block on the circular $TE_{0,1}$-mode dispersion characteristics of a cylindrical waveguide loaded with the rectangular blocks from the waveguide-wall keeping the parameters r = 6.0 mm, l = 2.0 mm, p = 4.0 mm, w = 2.0 mm and N = 3.

on dispersion shaping (Fig. 7). In the last case of varying the radius (r) of the proposed structure, both the lower and upper cut-off frequencies tend to decrease on increasing the radius, however, the upper cutoff frequency to a lower extent. Thus, the radius of the proposed structure shows a control on the pass band frequency (Fig. 8).

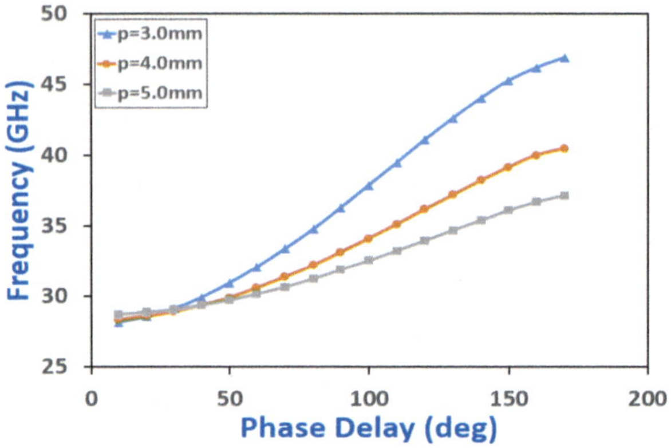

Fig. 6. Formulating the variation in structure periodicity (p) of the rectangular block on the circular $TE_{0,1}$-mode dispersion characteristics of a cylindrical waveguide loaded with the rectangular blocks from the waveguide-wall keeping the parameters r = 6.0 mm, h = 2.0 mm, l = 2.0 mm, w = 2.0 mm and N = 3.

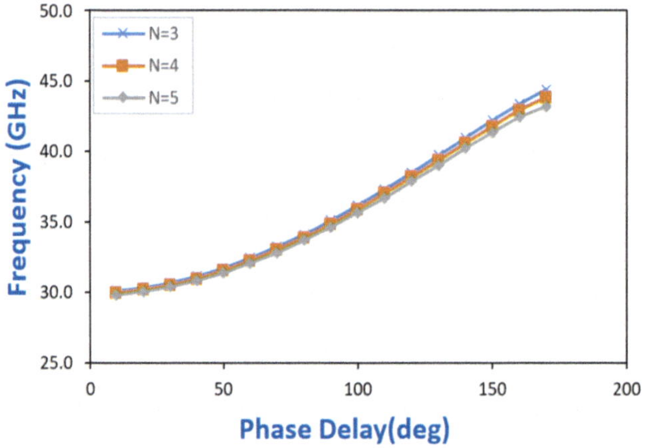

Fig. 7. Formulating the variation in number of blocks in azimuth (N) on the circular $TE_{0,1}$-mode dispersion characteristics of a cylindrical waveguide loaded with the rectangular blocks from the waveguide-wall keeping the parameters r = 6.0 mm, h = 2.5 mm, l = 2.0 mm, p = 4.0 mm and w = 2.0 mm.

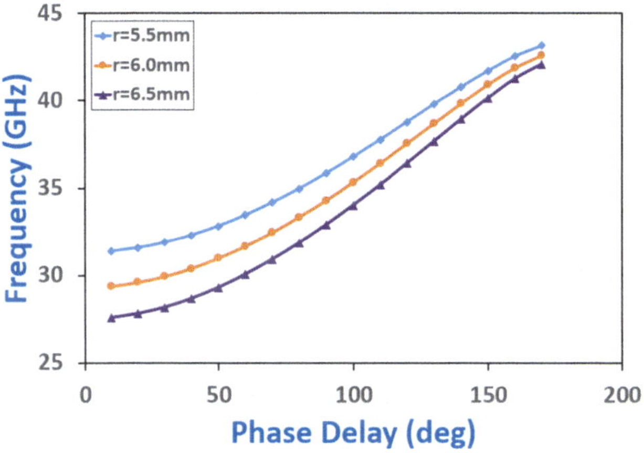

Fig. 8. Formulating the variation in waveguide radius (r) on the circular $TE_{0,1}$-mode dispersion characteristics of a cylindrical waveguide loaded with the rectangular blocks from the waveguide-wall keeping the parameters h = 2.5 mm, l = 2.0 mm, p = 4.0 mm, w = 2.0 mm and N = 3.

4 Conclusion

To emphasize its utilization in a wideband gyro-TWT, a unique radially anticipated metallic blocks derived azimuthal and axial periodic interaction structure is presented, structured, simulated and analyzed. There is an opportunity of personalizing the waveguide mode dispersion characteristics, which subsequently may yield wideband coalescence with the beam-mode dispersion line resulting in broad device bandwidth. The dispersion characteristics of the proposed waveguide is compared and validated with that of conventional smooth-wall circular waveguide in a special case. Further, the dimensional parameters: periodicity of the structure, inner radius of the waveguide, width, axial height and radial height of the metallic blocks are varied to perform the sensitivity analysis to witness the dispersion characteristics of the $TE_{0,1}$-mode in the proposed structure. It is evident that the width of radially projected metallic blocks and the structure periodicity are, respectively, the least and most impactful for controlling the dispersion shaping, pass band and hence the wideband capabilities for a gyro-TWT.

References

1. Kesari, V., Basu, B.N.: High Power Microwave Tubes: Basics and Trends, vol. 1. Morgan and Claypool Publishers, California for IOP Concise Physics, London (2018)
2. Kesari, V., Jain, P.K., Basu, B.N.: Analysis of a tapered disc-loaded waveguide for a wideband Gyro-TWT. IEEE Trans. Plasma Sci., Special issue—High-Power Microwave Generation. **34**(3), 541–546 (2006)
3. Kesari, V., Keshari, J.P.: Interwoven-disc-loaded circular waveguide for a wideband gyro-traveling-wave tube. IEEE Trans. Plasma Sci. **41**(3), 456–460 (2013)
4. Kesari, V., Jain, P.K., Basu, B.N.: Modelling of axially periodic circular waveguide with combined dielectric and metal loading. J. Phys. D: Applied Phys. **38**, 3523–3529 (2005)

5. Kesari, V.: Innovative interaction structures for wideband gyro-TWTs. In: Proc. VEDA-Kolleg-2013, pp. 1–5 (2013)
6. Singh, G., Ravi Chandra, S.M.S., Bhaskar, P.V., Jain, P.K., Basu, B.N.: Analysis of an azimuthally periodic vane-loaded cylindrical waveguide for a gyro-traveling-wave tube. Int. J. Electronics **86**(12), 1463–1479 (1999)
7. Rao, S.J., Jain, P.K., Basu, B.N.: Broadbanding of a gyro-TWT by dielectric-loading through dispersion shaping. IEEE Trans. Electron Dev. **43**(12), 2290–2299 (1996)
8. Ram, O., Shobha, H.R., Kesari, V.: Simulation based study of azimuthal and axial periodic interaction structure for a gyro-travelling-wave tube. In: IEEE Int. Conf. on Recent Innovation in Smart and Sustainable Technology, Presidency University, Bengaluru, India, 08–09 Mar. 2024 (2024)
9. User's Guide—High Frequency Structure Simulator. ANSYS, Canonsburg, PA (2016)

Design and Development of an Electromagnet for a 6.2 MW Pulsed Power S-Band Klystron

Deepender Kant(✉), Bijendra Kumar, Ayan Kumar Bandyopadhyay, and Vikram Singh Rawat

CSIR-Central Electronics Engineering Research Institute, Pilani, India
dkc@ceeri.res.in

Abstract. A pulsed power (6.2 MW) S-band klystron is being developed at CSIR-CEERI for its intended use as a high power RF source in an industrial linear accelerator. The present paper shall discuss different technological aspects involved in the design and development of an electromagnet suitable for this klystron. The simulation results for optimization of the magnetic field required for focusing of the electron beam using TRAK and CST Studio Suite will be covered. The desired electromagnet has been fabricated and being tested now. The development steps for fabrication of the desired electromagnet along with few test results for its magnetic field characterization will also be presented through this paper.

Keywords: Confined flow focusing · Electromagnet · Auxiliary Coils · TRAK · CST Studio Suite

1 Introduction

The pulsed power klystrons come into the category of microwave power amplifiers based on vacuum electronics technology. Although the semiconductor technology is continuously growing to meet the higher RF power demands but still the vacuum technology dominates where a source is needed in terms of megawatt or gigawatt power level. The high power klystrons are having a variety of applications in different strategic as well as societal areas including particle accelerators, radar and satellite communication, nuclear power reactors etc. A pulsed power (6.2 MW) S-band klystron is under development at CSIR-CEERI for its intended use in an industrial linear accelerator, as high power RF source, the desired specifications of the device are listed in Table 1.

2 Design of Electron Gun

The electrical design of electron gun is an important and primary step towards design of the required klystron, as optimization of the beam optics is essentially needed before any RF interaction mechanism. The electron gun is required to provide an electron beam of desired characteristics for interaction with the RF fields. It consists of a spherical cathode with emitting surface formed of some low work function material. The electrons

S. K. Ghosh et al. (Eds.): VEDA 2024, SPPHY 424, pp. 199–203, 2025.
https://doi.org/10.1007/978-981-96-5174-0_30

Table 1. Targeted specifications of the device

S. No.	Device Parameter	Value
1	Operating Frequency	2856 MHz
2	Power Output	6.2 MW peak, 15 kW Average
3	RF gain	45 dB (nominal)
4	Efficiency	40% (nominal)
5	Focusing	Electromagnet

emitted by the cathode through thermionic emission mechanism are accelerated towards a positively biased anode. The converging electron beam is converted into a laminar cylindrical shape with the help of a beam-focusing electrode (BFE) which is normally kept at cathode potential and the diverging lens formed by suitably shaped anode with an aperture at the center for passage of electron beam. The shape of the beam is maintained with the help of a coaxial magnetic field with flux lines parallel to axis. The schematic diagram of such an electron gun is shown in Fig. 1, which is known to be a Pierce type design [1] and the target parameters of the required electron gun are listed in Table 2.

Table 2. Parameters of the electron gun

S. No.	Target Parameter	Value
1	Beam Voltage	142 kV
2	Beam Current	110 Amp
3	Beam waist Radius	8.0 mm
4	Emission Density	5 Amp/cm^2

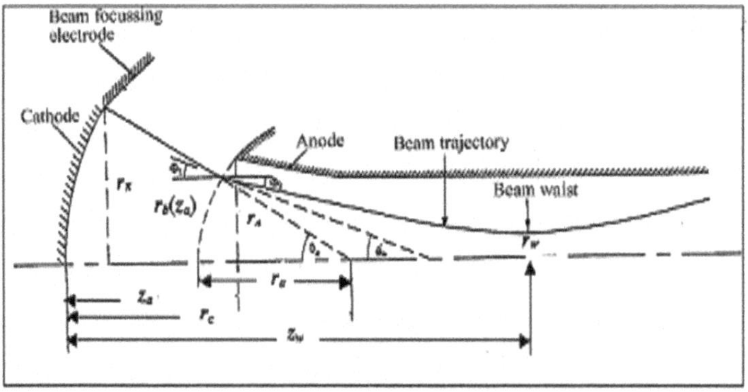

Fig. 1. Schematic of the required electron gun [1]

3 Design of Electromagnet Through TRAK Code

The design of electromagnet is required to provide an axial magnetic field suitable for focusing of the electron beam in the RF interaction region. This restricts the beam divergence while allowing the beam to drift axially without much deviations in the beam radius. The confined flow focusing scheme [2] has been used for designing axial magnetic field in the present electron gun where Brillioiun magnetic field (B_b; in Tesla) is given by following relation,

$B_b^2 = 69 \times \left({}^{I}/{V_b^{0.5}}\right) \times R_b^2$ where I = beam current, R_b = beam radius and V_b = beam voltage.

The computed value for B_b is found as 360 Gauss, thus the required value of total axial magnetic field (B_T) for the confined flow focusing will be (usually taken as 2.5–3.0 times of B_b) around 1080 Gauss [2].

A solenoid type electromagnet with suitably shaped pole pieces is designed to achieve the desired total magnetic field value (B_T) for focusing of the beam. The bore hole diameter of the pole piece decides the value of magnetic flux linked at the cathode surface (Bc) which can be computed by Busch's theorem [3].

$B_t^2 = B_b^2 + B_c^2 \times \left({}^{A_c}/{A_e}\right)^2$; Where A_c/A_e is the area compression ratio of the electron gun.

The above ratio can be calculated by knowing the cathode surface area and beam cross-section area, this ratio is 8.4 in our case, so the computed value of cathode linkage magnetic flux is about 121 Gauss. The bore hole diameter is optimized at 100 mm so as to link the magnetic field value of about 121 Gauss near the cathode surface. The profile of magnetic field after simulation of solenoid through TRAK code [4] is shown in Fig. 2, where the magnetic field in the focusing region is around 1080 Gauss (by taking NI value about 35000), thereafter in the output cavity region the magnetic field profile is designed in a peaking manner.

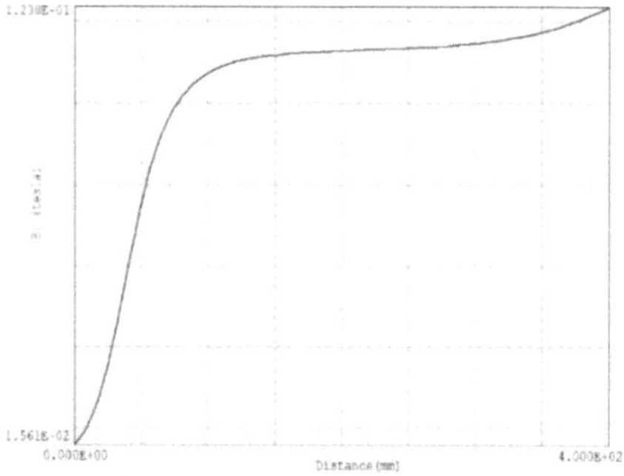

Fig. 2. Magnetic field profile in electron gun region

Electron beam focused in presence of the applied axial magnetic field is simulated in Trak code as shown in Fig. 3, where minor scalloping of the beam is observed but without any crossover of electron trajectories. A close view of the trajectories indicates that the electron beam is propagating without any interception.

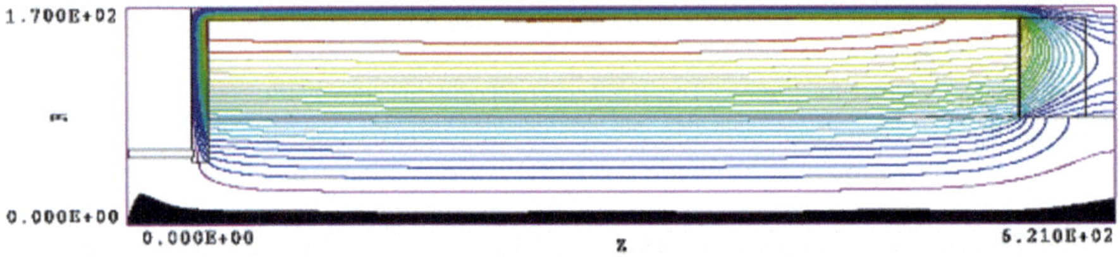

Fig. 3. Simulation of the electron beam in presence of axial magnetic field

4 Validation of Designed Electromagnet Through CST Code

The Electromagnet designed through TRAK has been cross validated by modelling through CST Particle Studio [5]. A design of electromagnet modelled in CST is shown in Fig. 4, along with simulated magnetic field profile.

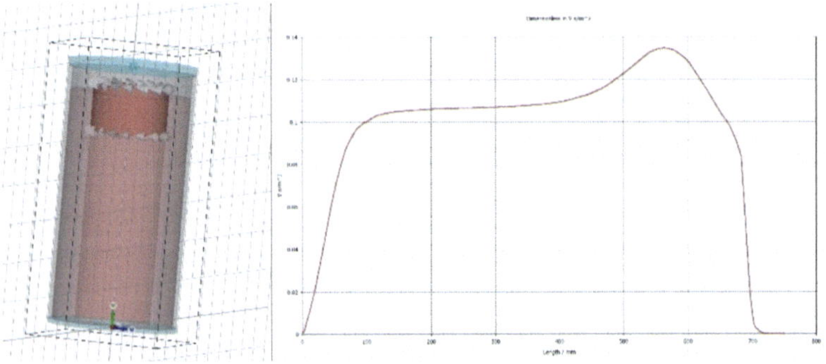

Fig. 4. Simulation of electromagnet in CST particle studio

The simulation of electron beam trajectories in presence of the optimized magnetic field profile through CST code is also shown in Fig. 5 which indicates the quite laminar electron beam with applied magnetic field, hence justifying the electromagnet design.

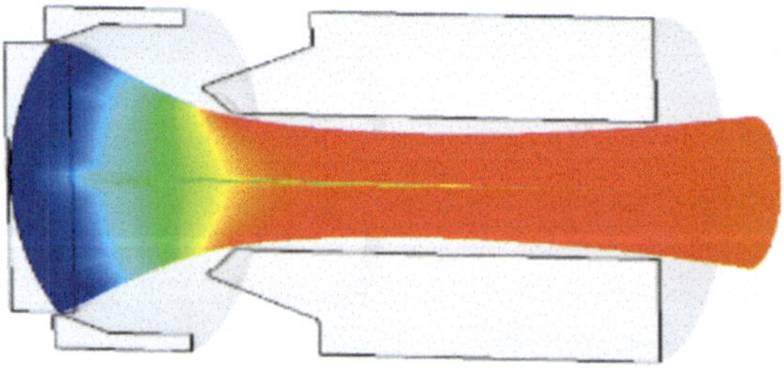

Fig. 5. Simulation of beam trajectories with applied magnetic field through CST

5 Development of Electromagnet

The optimized design of electromagnet has been developed with placement of suitable coil configurations as shown in Fig. 6. There is a main coil and an additional auxiliary coil for increasing the magnetic field near output cavity location. The coils are made of hollow copper tubes cooled with water and both coils are kept inside an envelope of mild steel while shielding with soft iron pole pieces at both ends. The fabricated electromagnet has been tested experimentally by applying 120 A current to the main coil and 200 A current to the auxiliary coil to achieve the desired magnetic field profile at the axis.

Fig. 6. The developed electromagnet

6 Conclusion

In this paper a methodology to design a suitable electromagnet required for a 6.2 MW pulsed power klystron, is presented. The optimized design of the electromagnet has cross validated through CST Studio Suite and TRAK codes. The axial magnetic field of the electromagnet has been experimentally measured confirming to the simulated results.

Acknowledgement. The authors are thankful to all colleagues of 'Vacuum electron device area' and Director, CSIR-CEERI for their support and encouragement.

References

1. Vaughan, J.R.M.: Synthesis of the Pierce gun. IEEE Trans. Electron Devices **28**(1), 37–41 (1981). https://doi.org/10.1109/TED.1981.20279
2. Gilmour, A.S., Jr.: Microwave Tubes, 1st edn. Artech House (1986)
3. Smith, M.J., Phillips, G.: Power Klystron Today, 1st edn. Wiley (1995)
4. Trak code user manual
5. CST Design Studio, 2018, user manual

Design of Applicators (2.45 GHz) for Efficient Microwave Assisted Pyrolysis (MAP) of Waste Material

Subham Chowdhury[1,2(✉)], Hasina Khatun[1,2], Nitin Kumar[1], Narendra Kumar Singh[1], and Anirban Bera[1,2]

[1] CSIR-Central Electronics Engineering Research Institute (CEERI), Pilani, Rajasthan, India
subham.ceeri22a@acsir.res.in

[2] Academy of Scientific and Innovative Research (AcSIR), CSIR-HRDG, Ghaziabad, India

Abstract. The continuous generation of waste materials, including electronics waste, plastic, and biomass waste, poses significant environmental challenges. While some of these wastes are repurposed for energy or chemical feedstock, improper disposal leads to pollution. Pyrolysis is a crucial thermochemical conversion method that transforms biomass into valuable products like heat, electricity, fuels, and chemicals. Among pyrolysis methods, microwave-assisted pyrolysis (MAP) stands out for its enhanced reaction efficiency and energy savings due to superior heat transfer profiles. This article presents microwave applicators operating at 2.45 GHz, which are capable of achieving near-uniform and efficient pyrolysis of various waste materials, transforming them into value-added products.

Keywords: Microwave applicator · Microwave Assisted Pyrolysis (MAP) · Waste to wealth

1 Introduction

The increasing buildup of waste materials, such as electronic waste, plastic, and biomass, poses serious environmental and sustainability issues [1]. Traditional methods of waste disposal, such as landfilling and incineration, contribute to pollution and resource depletion. Among the innovative solutions, microwave-assisted pyrolysis (MAP) has emerged as a highly efficient and effective technology for waste treatment. By utilizing microwave energy, MAP enhances and accelerates the breakdown of waste materials into valuable by-products like heat, electricity, fuels, and chemicals. This process offers superior heat transfer profiles, leading to faster reaction times and greater energy savings compared to conventional pyrolysis methods. Additionally, MAP's ability to achieve near-uniform heating ensures more consistent and efficient conversion of waste into high-value products. As such, MAP represents a promising advancement in waste management, addressing environmental concerns while contributing to resource recovery and energy production [2, 3].

Despite its promising advantages, MAP faces several limitations that hinder its widespread industrial adoption. One major challenge is the scalability of MAP systems,

S. K. Ghosh et al. (Eds.): VEDA 2024, SPPHY 424, pp. 204–209, 2025.
https://doi.org/10.1007/978-981-96-5174-0_31

as achieving uniform heating and efficient pyrolysis in large-scale operations remains difficult. Additionally, the design and optimization of microwave applicators to ensure consistent energy distribution across various types and quantities of waste are complex and require precise control [4–10]. Moreover, the other challenges for the system design are to control the temperature, time, and energy. In this context, efforts are being made across the globe to improve the performances of microwave-based systems. During the past few years at CSIR-CEERI, research attempts are being made in this context. As a part of that a proof-of-concept system is developing for near-uniform and efficient waste material pyrolysis.

2 Microwave Applicator Design

Many methods are proposed in different literature to address the drawback of non-uniform heating in MAP [11]. But this article deals with the cavity dimension, number of microwave launchers, number of waveguides, their positions, shapes, sizes and power of the feed to achieve a near-uniform heating.

The design of the microwave applicators (shown in Fig. 1) consist of: 1. a low reflection multimode cavity (optimized at 2.45 GHz) that can contain the microwave radiation, 2. Microwave launchers (2 numbers, which are placed strategically to the wall of the cavity) with optimized dimensions and 3. a standard waveguide (WR-340) that is compatible with microwave launchers and the feeding system with movable short.

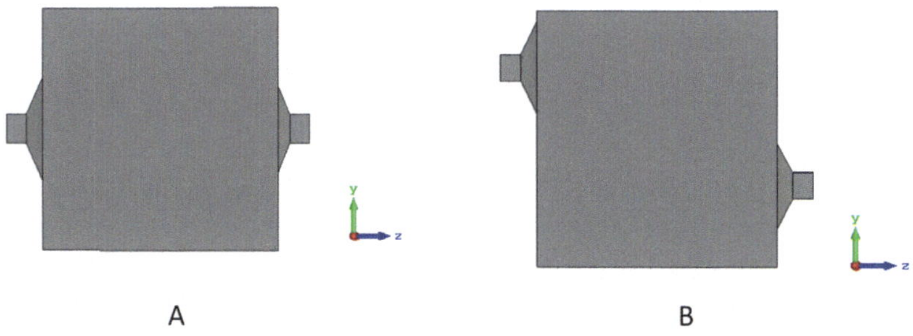

Fig. 1. 3-Dimensional simulation model of the microwave applicators A. Low volume applicator B. High volume applicator

First, the microwave launchers are designed (shown in Fig. 2). This consist of two parts which are a standard waveguide (WR-340) and a flared section. The primary advantage of this type of launcher lies in its superior impedance matching, which ensures maximum power transfer to the cavity. Consequently, the number of required microwave sources can be reduced.

The microwave launcher was optimized, resulting in a Voltage Standing Wave Ratio (VSWR) of 1.42 and S_{11} of -15 dB at 2.45 GHz, as shown in Fig. 3. This VSWR value is well below the limit of 2, and the S_{11} value is also better than the standard of -10 dB for microwave components. Additionally, it was observed that the VSWR remains below 2 across the frequency range of 2.1 GHz to 3 GHz, demonstrating the launcher's capability to operate within this bandwidth [12].

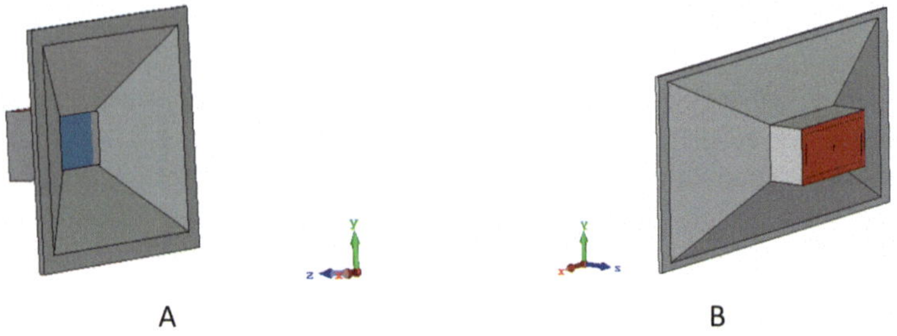

Fig. 2. 3-Dimensional simulation model of the microwave launchers **A** Front view **B** Back view

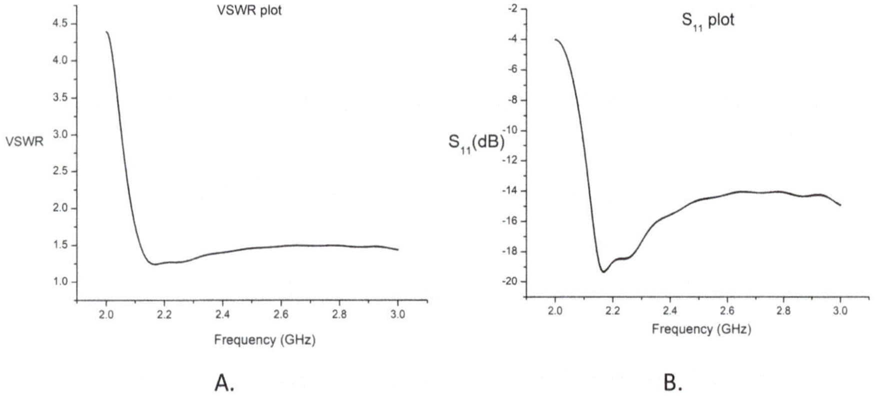

Fig. 3. **A** VSWR plot and **B** S_{11} plot of the microwave launcher at 2–3 GHz

3 Design of Low Volume (27 L) and High Volume (56 L) Applicator

Optimization of size, shape of the cavity and the position of the launcher is done to achieve uniform field distribution. The system is simultaneously fed at 2.45 GHz to observe the electric field distribution. The magnetron's input power was set to various power level to study the behavior of the system. After optimizing all parameters, an almost uniform electric field distribution is achieved, as shown in Fig. 4. This system has been fabricated and rigorous experiment has been performed. In it found that 1 kW power in each feed is sufficient for the pyrolysis process [13]. However, several issues arise with this design: 1. The capacity of the applicator is limited, making it unable to handle larger volumes of feedstock; 2. Since the feeds are positioned opposite each other, the microwave sources often get damaged after just 2–3 operations; and 3. The field uniformity is inadequate, with some areas experiencing insufficient field strength.

To address the previous issues, a high-volume applicator with a capacity of 56 L has been designed. The increased volume allows for a larger number of feedstocks to be accommodated. The microwave launchers are placed in offset positions to prevent damage to the microwave sources and to provide better field uniformity. This system is also simultaneously fed at 2.45 GHz to study the electric field distribution. The magnetron's input power was set to various levels, as in the previous setup, to observe the

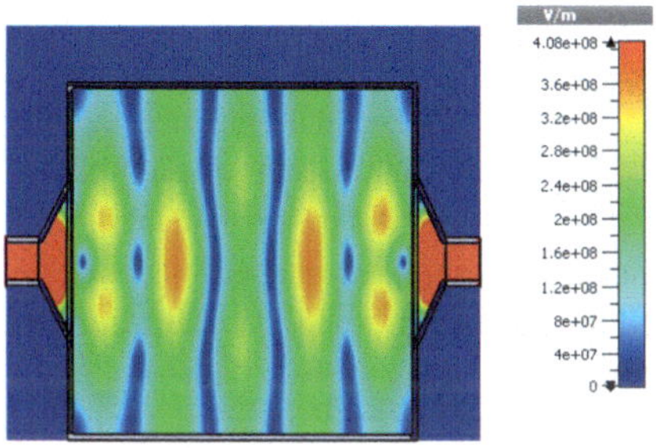

Fig. 4. Electric field distribution of the Low Volume Applicator at 2.45 GHz

system's behavior. After optimizing all parameters, a near-uniform electric field distribution, superior to the previous system, was achieved, as shown in Fig. 5. The maximum field strength of both systems has been studied and compared, as shown in Table 1.

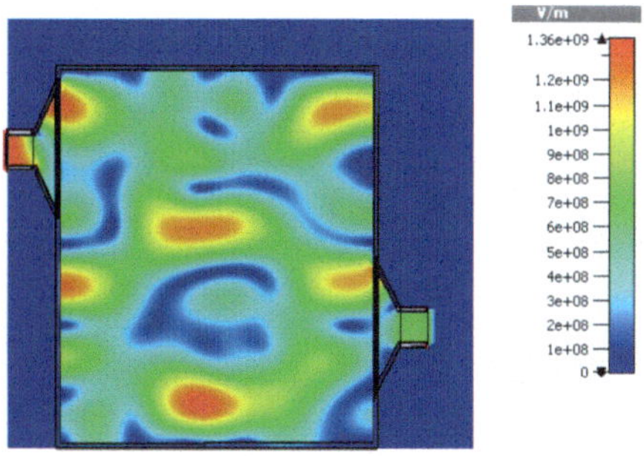

Fig. 5. Electric field distribution of the High-Volume Applicator at 2.45 GHz

Table 1. Maximum Electric field at different power level inside the applicator

Power level (Kilo-watts)	Maximum electric field (V/m)	
	Low volume applicator (27 L)	High volume applicator (56 L)
0.5	1.02×10^9	7.56×10^8
1	2.04×10^9	1.51×10^9
1.5	3.06×10^9	2.27×10^9
2	4.08×10^9	3.02×10^9

From the table above, it can be concluded that the power required for both the low-volume and high-volume applicators is nearly the same. However, the high-volume applicator offers better field distribution and a lower risk of damage to the microwave source. Therefore, the high-volume applicator is the preferred design for pyrolysis waste material.

4 Conclusion

To achieve efficient MAP, it was observed that a microwave applicator with near-uniform field distribution is essential. A couple of microwave applicator operating at 2.45 GHz was designed and fabricated to address this issue. The next step is to repeatedly test and refine the method by adjusting factors like magnetron power, operation time, process gas, output gas treatment, and collection methods. Additionally, consideration must be given to developing a higher throughput system capable of processing large quantities of waste material simultaneously, thereby facilitating the effective treatment of urban waste.

Acknowledgement. The authors would like to thank the Director, Council of Scientific and Industrial Research (CSIR)-CEERI, Pilani, Rajasthan, India, for allowing them to publish the work. This work was supported by the Council of Scientific & Industrial Research (CSIR) network program (IHP202402).

References

1. International Energy Agency (IEA).: Global energy review: CO2 emissions in 2020 (2021, December 26). https://www.iea.org/articles/global-energy-review-co2-emissions-in-2020
2. Nielsen, T.D., Hasselbalch, J., Holmberg, K., Stripple, J.: Politics and the plastic crisis: a review throughout the plastic life cycle. Wiley Interdiscip. Rev.: Energy Environ. **9**(e360) (2020)
3. Oyeyinka, S.A., Akintayo, O.A., Adebo, O.A., Kayitesi, E., Njobeh, P.B.: A review on the physicochemical properties of starches modified by microwave alone and in combination with other methods. Int. J. Biol. Macromol. **176**, 87–95 (2021)
4. Wei, W., Shao, Z.-S., Chen, W.-W., Zhang, P.-J., Yuan, Y.: A fully coupled electromagnetic irradiation, heat and mass transfer model of microwave heating on concrete. IEEE Access **9**, 1575–1589 (2020)
5. Zhu, H., et al.: A rotary radiation structure for microwave heating uniformity improvement. Appl. Therm. Eng. **141**, 648–658 (2018)
6. Chen, Y., Zhang, X., Luo, Z., Sun, J., Li, L., Yin, X., Li, J., Xu, Y.: Effects of inside-out heat-shock via microwave on the fruit softening and quality of persimmon during postharvest storage. Food Chem. **349**, 129161 (2021)
7. Zhan, L., Yang, Y., Li, W., Wang, G., Li, Y., Wang, N.: Drying kinetics and mechanical properties of low-temperature microwave dried cashmere fibers. Text. Res. J. **90**, 2745–2754 (2020)
8. Amini, A., Latifi, M., Chaouki, J.: Electrification of materials processing via microwave irradiation: a review of mechanism and applications. Appl. Therm. Eng. **193**, 117003 (2021)
9. Jie, X., Li, W., Slocombe, D., Gao, Y., Banerjee, I., Gonzalez-Cortes, S., Yao, B., Almegren, H., Alshihri, S., Dilworth, J., et al.: Microwave-initiated catalytic deconstruction of plastic waste into hydrogen and high-value carbons. Nat. Catalysis **3**, 902–912 (2020)
10. Suriapparao, D.V., Vinu, R.: Biomass waste conversion into value-added products via microwave-assisted copyrolysis platform. Renew. Energy **170**, 400–409 (2021)
11. Jeon, S., Kim, J., Yang, D.: Design of large-scale microwave cavity for uniform and efficient plastic heating. Polymers **14**(541) (2022). https://doi.org/10.3390/polym14030541
12. CST software.: Dassault Systèmes (2024). https://www.3ds.com/products/simulia/cst-studio-suite

13. Chowdhury, S., Manna, S., Kumar, K., Singh, N.K., Kumar, B., Khatun, H., Barik, R.K., Bandyopadhyay, A.K., Bera, A.: Design of pilot scale microwave cavity (2.45 GHz) for electronic-waste pyrolysis. In: National Conference on Emerging Trends in Vacuum Electronic Devices & Applications, p. 32. Rajamahendravaram, Andhra Pradesh, India. VEDA-2023-P7.2 (2023)

Preliminary Experiment on Microwave Assisted Pyrolysis (MAP) System for Plastic Waste

Anita Gahlot[1], Subham Chowdhury[1,2(✉)], Hasina Khatun[1,2], Kisan Kumar[1], Pravin Kumar[1], Narendra Kumar[1], Nitin Kumar[1], and Anirban Bera[1,2]

[1] CSIR-Central Electronics Engineering Research Institute (CEERI), Pilani, Rajasthan, India
subham.ceeri22a@acsir.res.in

[2] Academy of Scientific and Innovative Research (AcSIR), CSIR-HRDG, Ghaziabad, India

Abstract. The paper discusses a preliminary experiment of the Low-Density Polyethylene (LDPE) and High-Density Polyethylene (HDPE) plastic waste using Microwave Assisted Pyrolysis (MAP) system. The plastic waste was decomposed into hydrogen-enriched syngas and carbon black using an input microwave power of 1.2 kW at 2.45 GHz. The characterization of carbon black was carried out using Field Emission Scanning Electron Microscopy (FESEM), which confirm the traces of Carbon Nanoparticles (CNP) in HDPE. In case of LDPE, the carbon black was found to consist of Carbon Nanotubes (CNT). Energy Dispersive X-Ray (EDX) data confirmed that HDPE contains a greater amount of carbon compared to LDPE. Gas Chromatography (GC) data also confirmed the same for the syngas content.

Keywords: Microwave Assisted Pyrolysis (MAP) · Low-Density Polyethylene (LDPE) · High-Density Polyethylene (HDPE) · Syngas · Carbon Nanoparticles (CNP) · Carbon Nanotubes CNT

1 Introduction

India generated around 9.4 million tonnes (MT) of plastic waste annually during 2017–18 of this, about 5.6 MT were recycled each year, while 3.8 MT remained uncollected or were littered. Additionally, there is a constant increase in plastics waste generation and by 2050 estimated 12000MT plastic waste [1, 2]. In 2022, government launched the Waste to Wealth Mission aims to discover, evaluate, confirm, and implement technologies that transform waste into energy, recycle materials, and recover valuable resources. In India, 60% of plastic waste are recycled out of which 70% are registered recycling facilities. And all these registered plastic recycling facilities use conventional methods i.e. chemical, thermal and catalytic processes to recycle the plastic and produce large amounts of toxic chemical and gases to the environmental [3, 4].

Microwave assisted pyrolysis (MAP) technology is the effective alternate method to overcome the issues of conventional methods. The MAP can efficiently heat up materials, for better heating efficiency and easily control the process temperature and its byproducts fraction [5–8]. At CSIR CEERI Pilani, recently started research activity on microwave

S. K. Ghosh et al. (Eds.): VEDA 2024, SPPHY 424, pp. 210–214, 2025.
https://doi.org/10.1007/978-981-96-5174-0_32

assisted pyrolysis (MAP) system to extract value added products from waste plastic [9]. The paper discussed decompositions of LDPE and HDPE plastic waste into char, tar and syngas and the characterizations of the value-added output products.

2 Experimental Set Up

Based on our electromagnetic simulation, developed a MAP system shown in Fig. 1. In the MAP system, the SS chamber made of waveguide with microwave launchers, ports for inlet and outlet gas and a view port. The LDPE/HDPE plastic waste sample shredded into small pieces of size 10–20 mm mixed with iron powder in a ratio of 2:1 and keep inside the chamber. The SS chamber is microwave leak tight. The microwave power is radiated to the sample through microwave launcher and magnetron of 2 KW/2.45 GHz. The coupling of maximum input power to the sample is carried out by tuner.

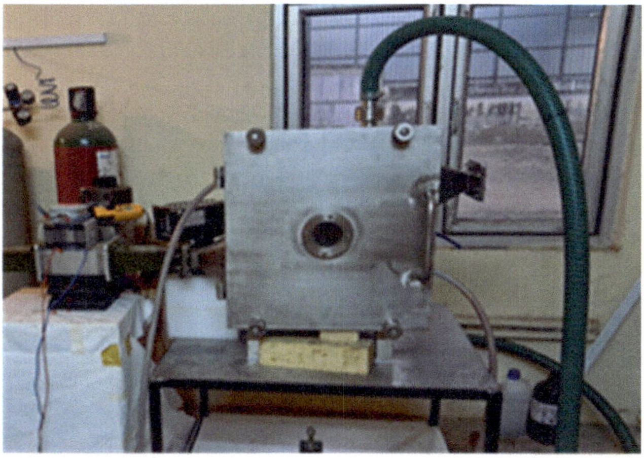

Fig. 1. Microwave assisted pyrolysis system consist of 1. SS chamber, 2. Microwave launcher with 2 kW microwave source, 3. Port for gas inlet, 4. Port for outlet gas and view port.

3 Characterization of Syngas and Char

Under control environment i.e. in the presence of nitrogen gas (0.5–1 L/m) at atmospheric condition and catalyst like iron powder, the shredded plastic waste inside the chamber is heated by magnetron and arcing occurred. At 700–1000 °C, the plastic sample decompose into carbon black and syngas with tar. The syngas collected in a 500 ml tedlar bag through the PTFE pipe and carbon black settled on the bottom of the chamber. The syngas and carbon black are characterized using gas chromatography (GC) and Field emission scanning electron microscopy (FESEM), respectively.

Table 1 shows the syngas composition and its percentage concentration. It may conclude that the HDPE plastic convert into more H_2 enriched syngas than LDPE.

Table 1. Syngas composition and its percentage concentration

Sl.no	Type of plastic waste	Syngas composition	% concentration
1	LDPE	n-propane	0.48
		Hydrogen	**2.747**
		Nitrogen	96.773
2	HDPE	n-propane	0.5
		Hydrogen	**5.54**
		Nitrogen	93.96

The FESEM shows the HDFE carbon black consists of carbon nanoparticles (CNP) of size 70–100 nm as shown in Fig. 2. The LDPE carbon black may consist of carbon nanotubes (CNT) of size 120–150 nm as shown in Fig. 3. The Energy Dispersive X-Ray (EDX) results show the presence of iron, 35% and carbon, 58% in LDFE. The EDX also ensure the presence of iron, 5% and carbon, 88% in HDPE.

Table 2 shows EDX data of the both sample (LDPE and HDPE) which indicate element composition of the output material. It may conclude that the HDPE plastic contents more carbon than LDPE.

Fig. 2. FESEM image of carbon achieved by MAP of HDPE

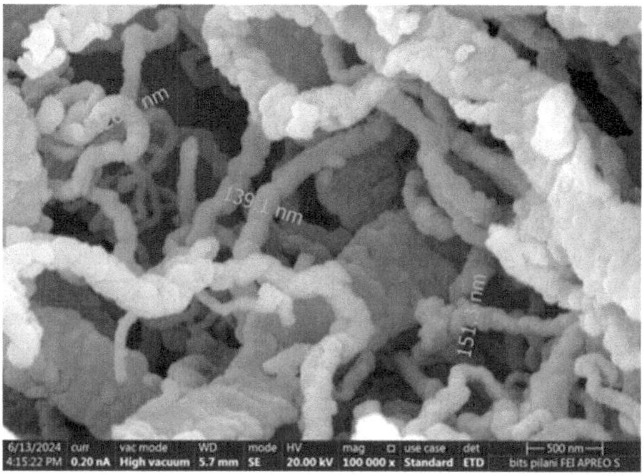

Fig. 3. FESEM image of carbon achieved by MAP of LDPE

Table 2. EDX analysis of the output products (LDPE and HDPE)

Sl. no	Type of plastic waste	Composition of output solid	% concentration
1	LDPE	**Carbon**	**58.81**
		Oxygen	4.12
		Silicon	1.64
		Iron	35.43
2	HDPE	**Carbon**	**88.30**
		Oxygen	6.43
		Silicon	0.20
		Iron	5.07

4 Conclusion

The preliminary experiment on microwave assisted pyrolysis system for LDPE and HDPE shows the promising outputs. Thus, the MAP technology helps to reduce environmental pollution and energy consumption. And the use of final products also helps to achieve carbon neutrality. In future, the optimization of the input and output parameters needs to be carried out.

Acknowledgement. The authors would like to express their gratitude to the Director of CSIR-CEERI Pilani for the ongoing support provided to conduct this research work. Also acknowledges CSIR HQ, New-Delhi, for funding (IHP202402) the research work.

References

1. GRID-Arendal.: Global plastic production, accumulation and future trends (n.d.)
2. OECD.: Plastic pollution is growing relentlessly as waste management and recycling fall short (n.d.). https://www.oecd.org/environment/plastic-pollution-is-growing-relentlessly-as-waste-management-and-recycling-fall-short.htm
3. Bose, P.: Converting plastic waste into carbon nanotubes. AZO Network (2021, January)
4. Centre for Science and Environment.: Managing plastic waste in India (n.d.). http://cdn.cseindia.org/attachments/0.34437100_1599806161_managing-plastic-waste-in-india.pdf
5. Yang, C., Shang, H., Li, J., Fan, X., Sun, J., Duan, A.: A review on the microwave-assisted pyrolysis of waste plastics. Processes **11**(1), 123 (2023)
6. Vijayakumar, A., Sebastian, J.: Pyrolysis process to produce fuel from different types of plastic—A review. IOP Conf. Series: Mater. Sci. Eng. **376**(1), 012007 (2018)
7. Ren, S., Xu, X., Hu, K., Tian, W., Duan, X., Yi, J., Wang, S.: Structure-oriented conversions of plastics to carbon nanomaterials. Carbon Res. **3**(2), 203–216 (2022)
8. Jie, X., Li, W., Slocombe, D., Gao, Y., Banerjee, I., Gonzalez-Cortes, S., ... Edwards, P.: Microwave-initiated catalytic deconstruction of plastic waste into hydrogen and high-value carbons. Nat. Catal. **3**(1), 37–44 (2020)
9. Chowdhury, S., Manna, S., Kumar, K., Singh, N.K., Kumar, B., Khatun, H., Barik, R. K., Bandyopadhyay, A.K., Bera, A.: Design of pilot scale microwave cavity (2.45 GHz) for electronic-waste pyrolysis. In: National Conference on Emerging Trends in Vacuum Electronic Devices & Applications, p. 32. Rajamahendravaram, Andhra Pradesh, India. VEDA-2023-P7.2 (2023)

Advances in Development of Microfabricated Sub-Terahertz Vacuum Electron Devices

Nikita M. Ryskin[1,2(✉)]

[1] Saratov Branch, Kotelnikov Institute of Radio Engineering and Electronics, Saratov, Russia
ryskinnm@gmail.com

[2] Saratov State University, 83 Astrakhanskaya St., Saratov 410012, Russia

Abstract. This paper presents a review of recent advances in development of sub-THz miniaturized vacuum electron devices (μVEDs). We focus on O-type μVEDs, which operation principle is based on interaction of a linear electron beam with a slow electromagnetic wave. Traveling-wave-tubes (TWT) are of the most interest as high-gain and wide-bandwidth amplifiers. The most attention is paid to sub-THz TWTs with folded-waveguide slow-wave structures (SWS), TWTs with grating-type SWS and sheet electron beam, and TWTs with planar microstrip SWSs. Backward-wave oscillators (BWO) are used as low-power radiation sources with broadband frequency tuning. BWOs provide power of several mW at frequencies up to 1 THz. Extended-interaction klystron amplifiers and extended-interaction oscillators offer high power at sub-THz frequency range but at the cost of narrow bandwidth. Best performances achieved by various research teams are summarized.

Keywords: Terahertz band · Traveling-wave tube · Backward-wave oscillator · Klystron · Microfabrication

1 Introduction

Many applications, such as novel high-data-rate wireless communication systems, security systems, high-resolution spectroscopy, radar and remote sensing, radio astronomy, etc. require high-power, compact table-top radiation sources operating at sub-THz (0.1–0.3 THz) frequency range [1]. Miniaturized vacuum electron devices (μVEDs) still deliver higher power than other kinds of sub-THz sources. The advancement of vacuum microwave electronics devices in frequency range above 0.1 THz began in the late 1950s. However this research field has been developing most actively in the last two decades. Although several more or less recent reviews of the topic have been published (see e.g. [2–7]), the field is rapidly developing, which makes it necessary to provide an overview of the most recent results. This paper presents a review of recent advances in development of sub-THz μVEDs with focus on miniaturized O-type linear-beam devices such as TWT, BWO, and klystron.

S. K. Ghosh et al. (Eds.): VEDA 2024, SPPHY 424, pp. 215–226, 2025.
https://doi.org/10.1007/978-981-96-5174-0_33

2 Traveling-Wave Tubes

2.1 Folded-Waveguide TWTs

The most developed as sources of the THz range are folded-waveguide (FW) TWTs, which began to attract a serious interest in the early 2000s [8, 9]. FW SWS is a periodically deformed rectangular waveguide in which a tunnel for an electron beam is cut out (Fig. 1). This SWS has a number of advantages. First of all, the slow electromagnetic wave propagating in a FW is not a surface wave but has a structure close to the TE_{10} mode of a rectangular waveguide. Therefore, an electron beam of a relatively large radius can be used. Input and output of the RF signal of the FW SWS can be easily matched with a rectangular waveguide of standard cross-section. Being an all-metal structure, FW SWS has high mechanical and thermal stability and high power-handling capacity. Finally, this SWS is compatible with modern MEMS-fabrication technologies [10]. Initially, LIGA technology based on X-ray or ultraviolet lithography was mainly used, as well as deep reactive ion etching (DRIE) [5]. Currently, the most widespread technology is computer-numerical-control (CNC) micro- and nanomilling [10, 11], at least for the manufacture of structures in the 0.1–0.3 THz range.

In Table 1, main output parameters of FW TWT amplifiers are presented. The power of several hundred watts in the W-band and more than 50 watts in the G-band has been obtained.

Table 1. Performance of sub-THz FW TWTs.

Bandwidth (GHz)	Gain (dB)	Max. Power (W)	Efficiency	Duty cycle	References
88–98	28	285	9.8%	0.1%	[12]
90–96	37	647	13.4%	10%	[13]
94–110	35	143	3–4%	1%	[14]
206–244	15	63	5%	0.1%	[15]
215.4–219	39	56.7	3.5%	5%	[16]
217–219.4	35	22	5%	CW	[17]
210–216	29.5	27	2.5%	CW	[18]
235–239	35.7	8.9	<1%	10%	[19]
258–264	25.5	11.9	<1%	10%	[20]
336–340	31.5	3	1%	10%	[21]

Based on the FW TWTs, sub-THz microwave power modules (MPM) have also been developed. An MPM is a complex device combining a gain equalizer, a miniaturized TWT, and a solid-state preamplifier, which have a common power supply source. MPMs with power over 65 W in the W-band (91–97 GHz) [22] and 32 W in the G-band (231.5–235 GHz) [23] were reported.

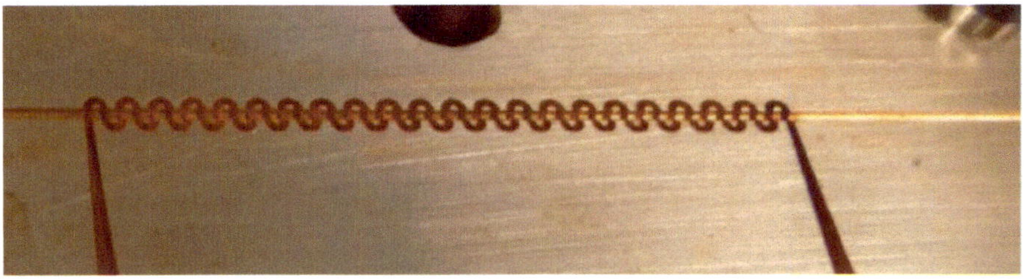

Fig. 1. FW slow-wave structure.

At Northrop Grumman, an exclusive series of THz-band FW TWT amplifiers has been developed. Their main parameters are summarized in Table 2 [24]. Miniaturized SWS circuits for these devices were fabricated by Si-DRIE process and subsequent metallization by copper electroplating.

Table 2. Parameters of Northrop-Grumman THz TWTs.

Frequency (THz)	0.64	0.67	0.85	1.03
Beam voltage (kV)	9.7	9.6	11.4	12.1
Beam current (mA)	4.8	3.1	2.8	2.3
Gain at saturation (dB)	22	17	22	20
Peak power (mW)	259	71	39	29
3-dB bandwidth	15	15	15	5
Duty cycle (%)	10	0.5	11	0.3

2.2 Sheet-Electron-Beam TWTs with Staggered Dual-Grating SWS

Despite important advantages, the possibility for increasing the power in FW TWT is limited by the use of a small-diameter pencil electron beam with ultrahigh current density. For most devices listed in Table 1, the current density is ~500 A/cm^2. In this regard, devices with sheet electron beams have attracted attention. Using of high-aspect-ratio sheet beams with large cross-section makes it possible to increase the current while maintaining the current density available for modern thermionic cathodes.

Staggered dual-grating (SDG) SWS is most commonly used in sheet-beam TWTs (Fig. 2a). In the half-period staggered SDG SWS, two modes coalesce at the π-mode frequency providing wide bandwidth [25, 26] as is illustrated by Fig. 2b. This SWS is capable to adopt a high-aspect-ratio sheet beam and compatible with MEMS-fabrication technologies.

On the other hand, the SDG SWS also has a number of disadvantages. Due to relatively low slow-down factor, an electron beam is synchronous with the +1st spatial harmonic at phase advance per SWS period $2\pi < \varphi < 4\pi$ (Fig. 2b). At this harmonic,

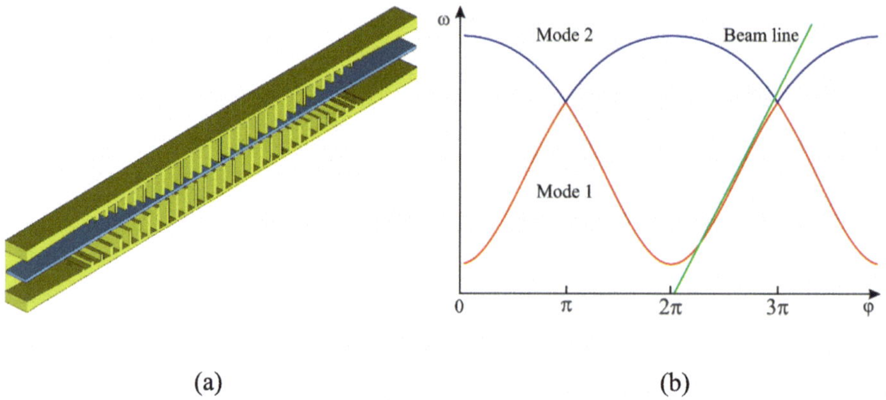

Fig. 2. Schematic cut view of the sheet-beam TWT FW with SDG SWS (**a**) and qualitative picture of its dispersion diagram (**b**).

Pierce interaction impedance is relatively low. Furthermore, when a sheet beam propagates in the uniform magnetic field, the diocotron instability arises that causes tilt and filamentation of the beam [27]. Therefore, for transportation of an intensive sheet beam at distances of several centimeters, high magnetic field of ~1 T is required. Thus, weight and size of the magnetic system may be large.

In [28, 29], development and hot test of the 0.2-THz sheet-beam TWT with SDG SWS was reported. Over 100 W power and 30-dB gain with 6-GHz –3 dB bandwidth was obtained. The TWT was driven with a 0.25-A sheet beam with 12.5:1 aspect ratio and nearly 100-μm thickness. However, the current density was as high as 438 A/cm^2. For beam focusing, a permanent-magnet focusing system with weight over 100 lbs was developed. To compare, the weight of FW TWTs with cylindrical pencil beam and focused by periodic permanent magnets (Sect. 2.1) was less than 2 kg.

Development of periodic-cusp-magnet (PCM) systems for focusing of sheet beams is a challenging task. Only recently a compact 0.2-THz TWT with a PCM-focused sheet electron beam was developed [30]. Power of 115 W at 212.4 GHz was demonstrated and weight of the tube was 1.8 kg. However, the problems with beam transmission were not fully resolved and collector current was 81.5% of the cathode current. As a result, the interaction length was limited to 47 circuit pitches and the small-signal gain was less than 15 dB. Therefore, to drive the tube in the saturated mode, a FW TWT pre-amplifier was used to amplify the input signal to the level of 30 W, i.e., power over 100 W was actually achieved with help of two cascaded TWT amplifiers. Nevertheless, in the nearest future one can expect a substantial progress in development of PCM-focused sheet-beam devices.

2.3 TWTs with Planar Microstrip SWS

Microstrip planar SWSs on dielectric substrate have attracted attention because they provide high slow-down factor and low operating voltage that is of primary importance for miniaturization. Sub-THz microstrip SWSs developed by photolithography were successfully demonstrated in [31–34]. In [35–37], we proposed a new technology based on laser micromachining and magnetron sputtering, which has a number of advantages in comparison with photolithography. Among them, there are high speed, low cost, and high flexibility. In addition, thickness of microstrip structures made by photolithography is ~1 μm, while magnetron sputtering allows formation of coatings with up to 10-μm thickness, which has a positive effect on thermal stability. Using the proposed technology, we fabricated microstrip meander-line (MML) SWSs on quartz substrates in V, W, and D band (Fig. 3) for tubes with 3–6 kV operating voltage that is much lower than for FW and SDG SWSs. The manufacturing error does not exceed 3–4 μm.

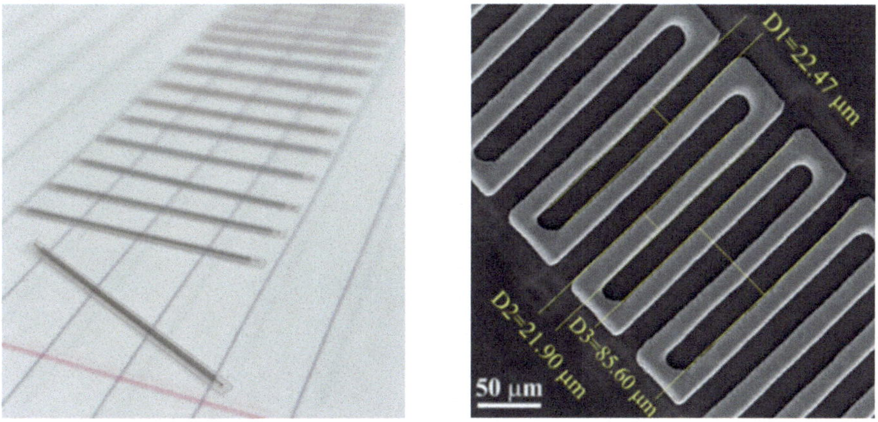

a b

Fig. 3. Photo of the batch of fabricated MML SWSs (**a**) and SEM image of a D-band SWS (**b**).

Clod test measurements of electromagnetic parameters demonstrate good transmission characteristics. For example, in the V-band reflection loss below –10 dB in the wide frequency range and transmission loss about 0.2–0.4 dB per pitch were measured.

3 Backward-Wave Oscillators

BWO is an oscillator that can be tuned electronically over a wide frequency range. Miniaturized CW-mode BWOs operating at sub-THz and THz frequencies with 1–10 mW power level find applications in measurement equipment, heterodyne receivers for radioastronomy, remote sensing, and imaging. Low-power, compact, and lightweight BWOs has been developed at Istok (Russia). Main parameters of the series of tubes that cover a range of frequencies from 0.258 to 1.4 THz are summarized in Table 3 [38]. Photo of one of such tubes is presented in Fig. 4. These devices utilize original multi-row interdigital or grating SWSs and a large electron beam scraped into a multiple sheet beam.

THz-band BWOs with similar parameters were developed at Thomson CSF, France (now Thales Electron Devices) [39]. These devices used single-grating grooved SWS

Table 3. Parameters of THz BWOs manufactured by Istok.

Model	Frequency (GHz)	Power (mW)	Voltage (kV)	Current (mA)	Magnetic field (T)
OB-65	258–375	1–10	1–4	25–40	0.7
OB-67	370–535	1–5	1–4	25–40	0.9
OB-80	530–714	1–5	2–6	30–45	1.0
OB-81	690–850	1–5	2–6	30–45	1.0
OB-82	790–970	0.5–3	2–6	30–45	1.1
OB-83	900–1100	0.5–3	2–6	30–45	1.1
OB-84	1070–1200	0.5–2	2–6	30–45	1.1
OB-85	1170–1400	0.5–2	2–6	30–45	1.1

Fig. 4. Photo of a sub-THz Istok BWO

and cylindrical pencil beam. For example, in [40] a 0.85–1.0 THz BWO with 1–2 mW output power was reported.

Using a high-aspect-ratio sheet electron beam allows further increase of power. The BWO with a 200-mA, 5-kV, 2.5 mm × 0.14 mm sheet electron beam reported in [41] delivers up to 180 mW at 0.34 THz operating in a CW mode.

In a conventional BWO, an electron beam propagating parallel to the SWS interacts with a surface wave, which magnitude rapidly decreases with distance from the SWS. Therefore, it is necessary to use electron beams with small thickness. However, current density of available electron guns is limited, which, in turn, limits the output power of THz BWOs. To overcome these limitations, a modification of a BWO with an inclined electron beam (clinotron) was proposed. In a clinotron, a relatively thick electron beam propagates at a small angle α to the grating SWS as shown in Fig. 5. In the beginning of interaction space, lower layers of the beam, which move in a strong field, give their energy to the wave and settle on the SWS surface. They are replaced by the following layers, which gradually approach the SWS, and so on [42].

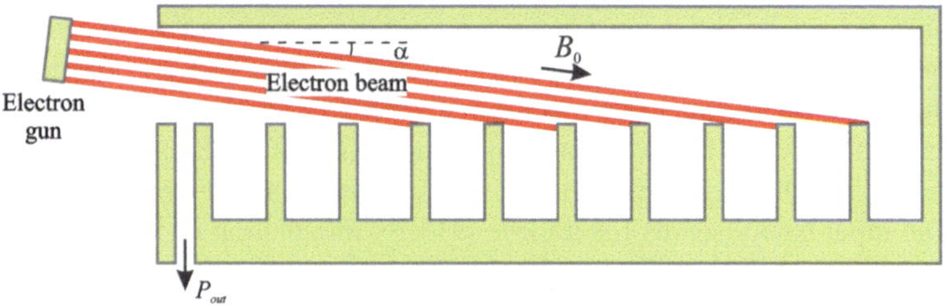

Fig. 5. Schematic of a clinotron oscillator.

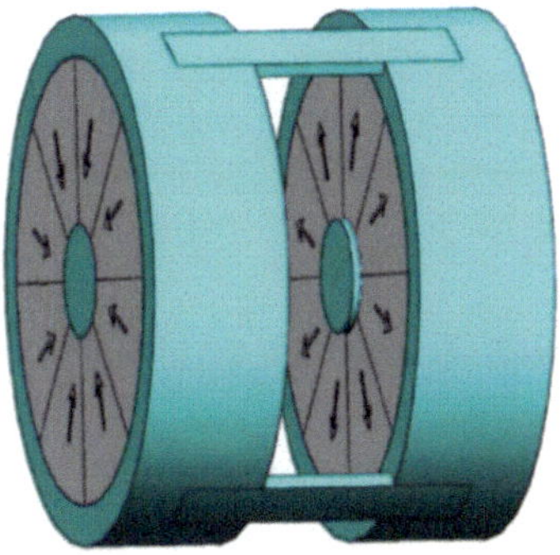

Fig. 6. Schematic of the multiple-segment radially magnetized magnetic focusing system.

CW clinotrons developed at Usikov IRE NASU, Kharkiv, Ukraine can provide power of 0.7–3.0 W at 0.1–0.2 THz, 0.05–0.7 W at 0.2–0.3 THz, and 0.02–0.1 W at 0.3–0.4 THz [42–44]. Based on clinotrons with stabilized frequency, compact radiation modules for spectroscopy are developed [44]. For beam focusing, original compact magnetic focusing systems with up to 1 T magnetic field were developed. These systems comprise of Nd-Fe-B multiple-segment permanent magnets magnetized in radial direction and fixed in an iron cage (Fig. 6). Such a design allows for a substantial reduce of weight and size.

It worth noting that clinotron is a resonant-type oscillator as it operates at to strong reflections of the wave from the SWS ends. Therefore, frequency tuning is stepwise since with variation of the beam voltage successive excitation of various axial modes of the resonator occurs.

4 Extended-Interaction Klystrons

Unlike the slow-wave devices such as TWT and BWO, in klystrons interaction of electrons with the RF field occurs discretely in the cavity gaps. However, in the THz band, it becomes difficult to use conventional reentrant cavities, since the coupling of the beam with the field weakens with the decrease of size. Therefore, extended-interaction klystron (EIK) are used instead standard klystrons. EIK cavities have multiple gaps, which raises the cavity shunt-impedance, i.e., increases the beam-wave coupling. The multi-gap cavities are actually shorted sections of the ladder SWS and RF field in such cavities has the form of a standing wave. Usually either π-mode or 2π-mode is selected as an operation mode, for highest shunt-impedance. As a rule, EIKs have a large number of cavities (5–7), which can be divided into input, output and intermediate (idler), see Fig. 7. The input cavity serves for impose velocity modulation on the beam, the idler cavities serve for improve beam bunching, and the output cavity serves to extract RF power from the bunched beam.

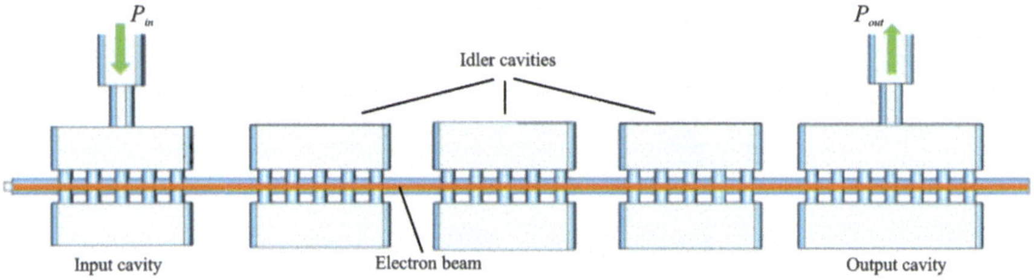

Fig. 7. Schematic of the EIK amplifier.

Sub-THz EIKs are used for radar, spectroscopy, and remote sensing. The main manufacturer of EIK is CPI Canada [45]. Summary of their parameters is presented in Table 4. W-band EIKs can provide 1–2 kW peak power, 40–45 dB gain and up to 1 GHz–3dB bandwidth. The 94-GHz EIK developed for CloudSat space-borne radar for measurements of cloud structures has 2-kW power, 250-MHz bandwidth, 56-dB gain, and 32% efficiency [46]. CPI also reported development of a 263-GHz amplifier for NMR spectroscopy with dynamic nuclear polarization, which demonstrated 72-W peak power, 31-dB gain, and 1-GHz bandwidth at 1% duty cycle [47]. In addition, CPI has developed W-band EIKs operating in a CW mode provide 20 W power and gain of about 30 dB.

At Aerospace Information Research Institute, Chinese Academy of Sciences, a 0.22-THz EIK with power over 100 W was recently reported [48].

Same as for TWT or BWO, using a sheet electron beam allows a substantial increase of the oscillation power. A sheet-beam 94-GHz EIK with 7.7 kW power, 17.2% efficiency, and 40 dB, gain has been developed [49]. In this device, a 20 kV/4 A beam with 4 mm × 0.32 mm cross section was focused by 0.76–0.86 uniform magnetic field, which slightly increased along the system. Since interaction length of EIK is much shorter than that of a TWT, difficulties in focusing a sheet beam are less pronounced.

Finally, an extended-interaction oscillator (EIO) worth mentioning. EIO is a self-excited oscillator, which interaction structure is a multiple-gap cavity with large number

Table 4. Parameters of pulse sub-THz EIKs manufactured by CPI.

Model	Frequency (GHz)	Peak power (W)	Gain (dB)	Duty cycle	–3 dB Bandwidth (GHz)
VKB2459	80–93	1000	45	10%	0.3
VKB2461	93–97	1500	45	10%	0.1
VKB2475	93–97	1000	40	10%	1.0
VKT2467	110–145	300	30	4%	0.3
VKY2462	220	100	26	5%	0.35
VKY2492	263–264	0.5–3	24	1%	0.8

of gaps. The main difference from a BWO is that beam-wave synchronism is attained at π- or 2π-mode, not at the backward harmonic. Therefore, there is no possibility for wideband electric frequency tuning. However, EIO can provide high power and efficiency. CPI produces oscillators with pulsed power of 1 kW in W-band and up to 40 W in G-band. In CW mode, the achieved power levels are tens and units of watts, respectively. Mechanical frequency tuning within 1–2 GHz is possible.

In [50], a 95-GHz sheet-beam EIO with 10.2 kW/2.2 kW peak/average power and 18% efficiency was presented. Weight of the device was less than 10 kg.

5 Conclusion

In the past two decades, a significant advancement in the development of sub-THz μVEDs has been made. Rapid progress of microfabrication technologies have made it possible to manufacture miniaturized interaction structures with dimensions of order of hundreds of microns and with accuracy of 1–2 um. Broadband TWT amplifiers deliver power approaching 1 kW in W band and 100 W in G band with 5–10 GHz bandwidth. Miniaturized CW BWOs deliver mW-level power at frequencies of 1 THz and above with wideband electric frequency tuning. Clinotrons provide 1–2 orders of magnitude higher power than BWOs but with stepwise frequency tuning. EIK amplifiers deliver kilowatts of power in W band and over 100 W in G band with bandwidth up to 1 GHz. These devices have a great potential for application in wireless communication, radar, spectroscopy, imaging, remote sensing and non-destructive evaluation.

Acknowledgment. This work was partly supported by the Russian Science Foundation (grant # 22-12-00181).

References

1. Leitenstorfer, A., et al.: The 2023 terahertz science and technology roadmap. J. Phys. D Appl. Phys. **56**, 223001 (2023)
2. Rozhnev, A.G., Ryskin, N.M., Sokolov, D.V., Trubetskov, D.I., Han, S.T., Kim, J.I., Park, G.S.: Novel concepts of vacuum microelectronic microwave devices with field emitter cathode arrays. Phys. Plasmas **9**(9), 4020–4027 (2002).
3. Booske, J.: Plasma physics and related challenges of millimeter-wave-to-terahertz and high power microwave generation. Phys. Plasmas **15**, 055502 (2008)
4. Srivastava, V.: THz Vacuum Microelectronic Devices. J. Phys.: Conf. Ser. **114**, 012015 (2008)
5. Booske, J.H., Dobbs, R.J., Joye, C.D., Kory, C.L., Neil, G.R., Park, G.S., Park, J.H., Temkin, R.J.: Vacuum electronic high power terahertz sources. IEEE Trans. Terahertz Sci. Technol. **1**(1), 54–75 (2011)
6. Srivastava, A.: Microfabricated terahertz vacuum electron devices: technology, capabilities and performance overview. Europ. J. Adv. Eng. Technol. **2**(8), 54–64 (2015)
7. Paoloni, C., Gamzina, D., Letizia, R., Zheng, Y., Luhmann, N.C.: Millimeter wave traveling wave tubes for the 21st century. J. Electromag. Waves Appl. **35**(5), 567–603 (2021)
8. Han, S.T., Jang, K.H., So, J.K., Kim, J.I., Shin, Y.M., Ryskin, N.M., Chang, S.S., Park, G.S.: Low–voltage operation of Ka–band folded waveguide traveling–wave tube. IEEE Trans. Plasma Sci. **32**(1), 60–66 (2004)
9. Bhattacharjee, S., Booske, J.H., Kory, C.L., van der Weide, D.W., et al.: Folded waveguide traveling–wave tube sources for terahertz radiation. IEEE Trans. Plasma Sci. **32**(3), 1002–1014 (2004)
10. Ives, R.L.: Advanced fabrication of vacuum electron devices. IEEE Trans. Electron Devices **70**(6), 2693–2701 (2023)
11. Gamzina, D., Himes, L.G., Barchfeld, R., Zheng, Y., Popovic, B.K., Paoloni, C., Choi, E.M., Luhmann N.C.: Nano-CNC machining of sub-THz vacuum electron devices. IEEE Trans. Electron Devices **63**(10), 4067–4073 (2016)
12. Cook, A.M., et al.: Demonstration of a W-band traveling-wave tube power amplifier with 10-GHz bandwidth. IEEE Trans. Electron Devices **68**(5), 2492–2498 (2021)
13. Zhang, X., Feng, J., Cai, J., Du, Y., Dong, R., Wu, X.: Power enhancement in W-band pulsed folded waveguide TWT. IEEE Trans. Electron Devices **68**(5), 2504–2508 (2021)
14. Du, Y., Cai, J., Pan, P., Dong, R., Zhang, X., Liu, S., Wu, X., Feng J.: Experimental investigation of an ultrawide bandwidth W-band pulsed traveling-wave tube with microfabricated folded-waveguide circuits. IEEE Trans. Plasma Sci. **47**(1), 219–225. (2019)
15. Bian, X., Pan, P., Tang, Y., Lu, Q., Li, Y., Zhang, L., Wu, X., Cai, J., Feng, J.: Demonstration of a pulsed G-band 50-W traveling wave tube. IEEE Electron Device Lett. **42**(2), 248-251 (2021)
16. Joye, C., Cook, A., Calame, J., Abe, D., Vlasov, A.N., Chernyavskiy, I.A.: Demonstration of a high power, wideband 220-GHz traveling wave amplifier fabricated by UV-LIGA. IEEE Trans. Electron Devices **61**(6), 1672–1678 (2014)
17. Pan, P., Tang, Y., Bian, X., Zhang, L., Lu, Q., Li, Y., Feng, Y., Feng, J.: A G-band traveling wave tube with 20 W continuous wave output power. IEEE Electron Device Lett. **41**(12), 1833–1836 (2020)
18. Jin, Z., Liu, W., Wang, J., Zhang, Z., Zhang, F., Zhao, K., Ou, Y., Zhang, Z., Zhao, C., Zhang, Z.: Development of G-band continuous-wave folded waveguide traveling-wave tube. IEEE Trans. Plasma Sci. **51**(2), 598–604 (2023)
19. Li, Y., Pan, P., Song, B., Zhang, L., Feng, J.: A 237 GHz traveling wave tube for cloud radar. Electronics **12**(10), 2153 (2023)

20. Pan, P., Zheng, Y., Li, Y., Song, X., Feng, Z., Feng, J., Britt, R.D., Luhmann, N.C.: Demonstration of a 263-GHz traveling wave tube for electron paramagnetic resonance spectroscopy. IEEE Trans. Electron Devices **70**(11), 5897–5902 (2023)
21. Hu, P. Lei, W., Jiang, Y., Huang, Y., Song, R., Chen, H., Dong, Y.: Demonstration of a watt-level traveling wave tube amplifier operating above 0.3 THz. IEEE Electron Device Lett. **40**(6), 973–976 (2019)
22. Chen, J., Wang, Z., Wang, F., Gao, C., Li, J., Xie, Q., Zi, Z., Cai, J., Feng, J.: Demonstration of a broadband W-band microwave power module with improved gain flatness. IEEE Electron Device Lett. **43**(8), 1335-1338 (2022)
23. Armstrong, C.M., et al.: A compact extremely high frequency MPM power amplifier. IEEE Trans. Electron Devices **65**(6), 2183–2188 (2018)
24. Tucek, J.C., Basten, M.A., Gallagher, D.A., Kreischer, K.E.: Operation of a compact 1.03 THz power amplifier. In: 2016 IEEE International Vacuum Electronics Conference (IVEC), pp. 1–2. IEEE, Piscataway NJ (2016)
25. Shin, Y.-M., Barnett, L.R., Luhmann, N.C.: Phase-shifted traveling-wave-tube circuit for ultrawideband high-power submillimeter-wave generation. IEEE Trans. Electron Devices **56**(5), 706–712 (2009)
26. Rozhnev, A.G., Ryskin, N.M., Karetnikova, T.A., Torgashov, G.V., Sinitsyn, N.I., Shalayev, P.D., Burtsev, A.A.: Studying characteristics of the slow-wave system of the traveling-wave tube with a sheet electron beam. Radiophys. Quantum Electron. **56**(8-9), 542–553 (2014)
27. Antonsen, T.M., Ott, E.: Velocity shear driven instabilities of an unneutralized electron beam. Phys. Fluids **18**(9), 1197–1208 (1975)
28. Baig, A., Gamzina, D., Kimura, T., Atkinson, J., Domier, C., Popovic, B., Himes, L., Barchfeld, R., Field, M., Luhmann N.C.: Performance of a nano-CNC machined 220-GHz traveling wave tube amplifier. IEEE Trans. Electron Devices **64**(5), 2390-2397 (2017)
29. Field, M., Kimura, T., Atkinson, J., Gamzina, D., Luhmann, N.C., et al.: Development of a 100-W 200-GHz high bandwidth mm-wave amplifier. IEEE Trans. Electron Devices **65**(6), 2122–2128 (2018)
30. Zhang, C., Pan, P., Cai, J., Chen, X., Tian, H., Su, S., Zhou, K., Meng, W., Li, Y., Song, B., Gong, Y., Feng, J.: Demonstration of a PCM-focused sheet beam TWT amplifier at G-Band. IEEE Trans. Electron Devices **70**(6), 2798-2803 (2023)
31. Sumathy, M., Augustin, D., Datta, S.K., Christie, L., Kumar, L.: Design and RF characterization of W-band meander-line and folded-waveguide slow-wave structures for TWTs. IEEE Trans. Electron Devices **60**(5), 1769–1775 (2013)
32. Socuéllamos, J.M., Dionisio, R., Letizia, R., Paoloni, C.: Experimental validation of phase velocity and interaction impedance of meander-line slow-wave structures for space traveling-wave tubes. IEEE Trans. Microwave Theory Tech. **69**(4), 2148–2154 (2021)
33. Guo, G., Zhang, T., Zeng, J., Yang, Z., Yue, L., Wei, Y.: Investigation and fabrication of the printed microstrip meander-line slow-wave structures for D-band traveling wave tubes. IEEE Trans. Electron Devices **69**(9), 5229–5234 (2022)
34. Guo, G., Zeng, J., Zhao, Q., Zhang, T., Zhou, T., Lu, P., Tan, H., Wei, Y.: Investigation and experiment of a novel chamfered V-shaped microstrip slow-wave structure for W-band traveling-wave tube. J. Infrared Millim. Terahertz Waves **45**, 629–644 (2024)
35. Ryskin, N.M., Rozhnev, A.G., Starodubov, A.V., Serdobintsev, A.A., et al.: Planar microstrip slow-wave structure for a low-voltage V-band traveling-wave tube with a sheet electron beam. IEEE Electron Device Lett. **39**(5), 757–760 (2018)
36. Torgashov, R.A., et al.: Theoretical and experimental study of a compact planar slow-wave structure on a dielectric substrate for the W-band traveling-wave tube. Tech. Phys. **65**(4), 660–665 (2020)

37. Ryskin, N.M., et al.: Development of microfabricated planar slow-wave structures on dielectric substrates for miniaturized millimeter-band traveling-wave tubes. J. Vacuum Sci. Technol. B **39**, 013204 (2021)
38. Borisov, A.A., Budzinsky, U.A., Bykovsky, S.V., Galdetskiy, A.V., et al.: The development of vacuum microwave devices in Istok. In: 2011 IEEE International Vacuum Electronics Conference (IVEC), pp. 437–438. IEEE, Piscataway, NJ (2011)
39. Boissiere, J., Epsztein, B., Teyssier L.: Advances in submillimeter-wave carcinotrons. In: 1981 International Conference on Submillimeter Waves and Their Applications, pp. 1–2. IEEE, Piscataway, NJ (1981)
40. Garcin, P., Grauleau, D., Gerber, R., Teyssier, L.: New technologies used for the 1 THz backward-wave oscillator. In: Technical Digest, International Electron Devices Meeting, pp. 850–853. IEEE, Piscataway, NJ (1988)
41. Xi, H., Wang, J., He, Z., Zhu, G., Wang, Y., Wang, H., Chen, Z., Li, R., Liu L.: Continuous-wave Y-band planar BWO with wide tunable bandwidth. Sci. Rep. **8**, 348 (2018)
42. Lukin, K., Khutoryan, E., Kuleshov, A., Ponomarenko, S., Sattorov, M., Park, G.-S.: THz oscillators based on Cherenkov, Smith-Purcell and hybrid radiation effects. In: Park, G.-S., Tani, M., Rieh, J.-S., Park, S.Y. (eds.) Advances in Terahertz Source Technologies, pp. 487–524. Jenny Stanford Publ, Singapore (2024)
43. Ponomarenko, S.S., Kishko, S.A., Zavertanniy, V.V., Khutoryan, E.M., Lopatin, I.V., Yefimov, B.P., Kuleshov A.N.: 400-GHz continuous-wave clinotron oscillator. IEEE Trans. Plasma Sci. **41**(1), 82–86 (2013)
44. Likhachev, A., Danik, A., Kovshov, Yu., Kishko, S., Ponomarenko, S., Martseniak, O., Khutoryan, E., Ogawa, I., Idehara, T., Kuleshov A.: Compact radiation module for THz spectroscopy using 300 GHz continuous-wave clinotron. Rev. Sci. Instrum. **90**, 034703 (2019)
45. Berry, D., Deng, H., Dobbs, R., Horoyski, P., Hyttinen, M., Kingsmill, A., MacHattie, R., Roitman, A., Sokol, E., Steer B.: Practical aspects of EIK technology. IEEE Trans. Electron Devices **61**(6), 1830-1835 (2014)
46. Roitman, A., Berry, D., Steer, B.: State-of-the-art W-band extended interaction klystron for the CloudSat program. IEEE Trans. Electron Devices **52**(5), 895–898 (2005)
47. Hyttinen, M., Roitman, A., Horoyski, P., Deng H.: High power pulsed 263 GHz extended interaction amplifier. In: 2020 IEEE 21st International Conference on Vacuum Electronics (IVEC), pp. 1–2. IEEE, Piscataway, NJ (2020)
48. Qu, Z., Zhang, Z., Guo, N., Jiang, C., Wang, S., Li, Q., Zhao, D., Xue, Q., Yin, S., Ding, Y.: Demonstration of a G-band high-power extended interaction klystron. IEEE Electron Device Lett. **44**(10), 1736–1739 (2023)
49. Pasour, J., Wright, E., Nguyen, K.T., Balkcum, A., Wood, F.N., Myers, R.E., Levush B.: Demonstration of a multikilowatt, solenoidally focused sheet beam amplifier at 94 GHz. IEEE Trans. Electron Devices **61**(6), 1630–1635 (2014)
50. Wang, J., Wan, Y., Xu, D., Li, X., Dai, Z., Li, H., Jiang, W., Wu, Z., Liu, G., Yao, Y., Luo Y.: Performance and experimental progress of a compact W-band high average power sheet beam extended interaction oscillator. IEEE Electron Device Lett. **44**(1), 144-147 (2023)

Design of an Equiangular Spiral Antenna for Efficiency Enhancement in Tokamak for Fusion Reactors

Dimple Yadav[1], Poonam Tiwari[1], Vaishali[1], Meenu Kaushik[2], Raj Singh[3], and Vishant Gahlaut[1(✉)]

[1] Department of Physical Sciences, Banasthali Vidyapith, Rajasthan 304022, India
vgceeri@gmail.com
[2] School of Automation, Banasthali Vidyapith, Rajasthan 304022, India
[3] RF and Fundamental Plasma Division, Institute for Plasma Research, Gandhinagar 382428, India
raj@ipr.res.in

Abstract. The proposed spiral antenna serves as an alternative structure to the traditional setup for launching fast waves in the SST-1 tokamak, operating at a frequency of 350 MHz. The study becomes essential to determine the impedance matching in the middle of antenna and the coaxial transmission line, considering the significant several-hundred-kilowatt continuous wave power supplied to the spiral antenna as RF input. The design of the spiral antenna adheres to the port space restrictions of SST-1. The design of the antenna is simple and compact in size. Bidirectional radiation is observed within the specified frequency band in antenna prototype. The modeling findings show that the antenna exhibits an inherent resonance frequency approximately 350 MHz, with respective reflection coefficient of ~−15 dB respectively.

Keywords: Impedance matching · Fusion research · Plasma parameters · SST-1 · Tokamak · Two-arm resistively loaded spiral antenna

1 Introduction

In the forefront of advanced fusion research and plasma physics, SST-1 tokamaks are propelling efforts towards sustained nuclear fusion as an environmentally friendly and renewable source of energy. Antenna coupling impedance is investigated in various operating conditions taking into account variations in frequency spectra, antenna designs, and plasma characteristics. Reactor-grade plasmas have higher temperatures and electron densities, which make slower or lower hybrid waves less accessible and lead to power deposition at the plasma's edge [2]. Strategies for driving current using fast waves as well as helicon waves are thought to be better suitable for these kinds of situations. The plasma persists for many hundred milliseconds, and several hundred kilowatts of RF power are

S. K. Ghosh et al. (Eds.): VEDA 2024, SPPHY 424, pp. 227–235, 2025.
https://doi.org/10.1007/978-981-96-5174-0_34

fed into it. By applying the fast wave current driving technology, it is possible to inject fast waves into plasma at such short times and high power levels, which prolongs the current's life. Consequently, an innovative technique is currently being used to measure the impedance within a tokamak. Despite the spiral antenna's long history, this study examines its design within the limitations of the SST-1 port to see whether it has the ability to produce fast waves.

The researcher recently conducted a large-scale modeling research to examine the viability of fast waves for non-inductive plasma current drive with sst-1 plasmas. The sst-1 machine comprises 1.1 m major radii and 0.2 m minor radii. Their toroidal magnetic field strength is 3 T at the minor axis [3]. Fwcd typically uses off-axis current driving, while sst-1 has distinct behaviors, with a significant amount of power being deposited within and around its centre.

This study's antenna incorporates the device parameters and plasma properties particular to the SST-1 tokamak in order to meet the system requirements. Using a novel method for a tokamak, this study aims to investigate the different spiral antenna performance characteristics.

2 Spiral Antenna Design and Proposed Technique

A circularly polarized two-arm equiangular spiral antenna has been designed, achieving reflection coefficient of ~−15 dB and a VSWR close to (VSWR < 2) at 350 MHz. The antenna maintains unchanged polarization, radiation pattern, and also impedance matching over a broader range of frequencies. The two-arm equiangular helical antenna uses a novel design approach to achieve circular polarization. It operates over the frequency range about 0.1–1 GHz. Using CST microwave studio; the performance characteristics of the antenna were analyzed to assess its capabilities.

Spiral antennas are radio frequency and frequency independent antennas, typically characterized as Archimedean and equiangular spirals. Archimedean spiral antennas exhibits a uniform increase in size, maintaining a constant gap between them, as depicted below.

Fig. 1. **a** Two arm Archimedean spiral antenna, **b** Two arm equiangular spiral antenna, **c** Two arm equiangular spiral antenna

Fig. 1. (*continued*)

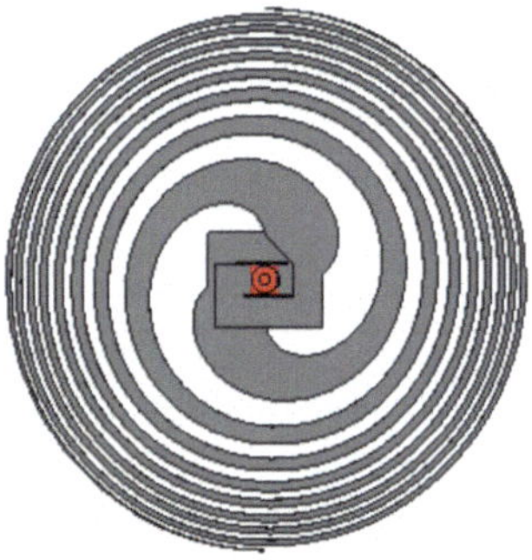

Fig. 1. (*continued*)

Figure 1a demonstrate the uniform increase in the antennas radius within the change of angle, described by formula (1)

$$r = r_0 + a(\varphi - \varphi_0) \tag{1}$$

In this context the variables are defined as follows: r_0 represents the spiral starting radius; φ denotes the azimuth; φ_0 indicated the starting angle; a represents the spiral growth rate; (r, φ) represents the polar coordinates of any point on the curve. Zoned radiation is produced by the arms of the Archimedean antenna, the antenna's spiral band, whose diameter is approximately comparable to one wavelength, produces the effective radiation.

Figure 1b indicates that the angle, which is given by Eq. (2), directly determines the geometry for an equiangular spiral antenna.

$$r = r_0.e^{a(\varphi - \varphi_0)} \tag{2}$$

Here, r_0 represents spiral starting radius; φ denotes the azimuth; $\varphi_0$indicated the starting angle; a represents the spiral growth rate; (r, φ) represents any position on the curve polar coordinated. An ultra-wideband antenna is referred to as being isometric. When the spiral arm's length exceeds one wavelength, the antenna starts showing non-frequency characteristics.

This paper introduces an innovative logarithmic spiral antenna design crafted from PEC material. The antenna's intricate performance analysis was executed through CST simulations. The transmission line boasts a characteristic impedance of 50 Ω, and the antenna is elegantly fed using a coaxial feed mechanism. Notably, no phase shifter, stub tuner, or balun was used for matching, which adds to the antenna's novelty. The antenna was naturally tuned to resonate, and the chosen dimensions taking into account the spatial constraints of the SST-1 tokamak.

3 Results and Discussions

Within the confines of this paper, the antenna was designed using CST-MWS and its dimensions were carefully tailored to accommodate the spatial limitations of the SST-1 port. The simulated outcomes demonstrate a natural resonance occurring at 350 MHz frequency, yielding the reflection coefficient of approximately −15 dB. The subsequent section outlines the comprehensive set of results performed on the antenna, highlighting its observed attributes and characteristics. Figure 2 displays the cut view of coaxial feed of spiral antenna

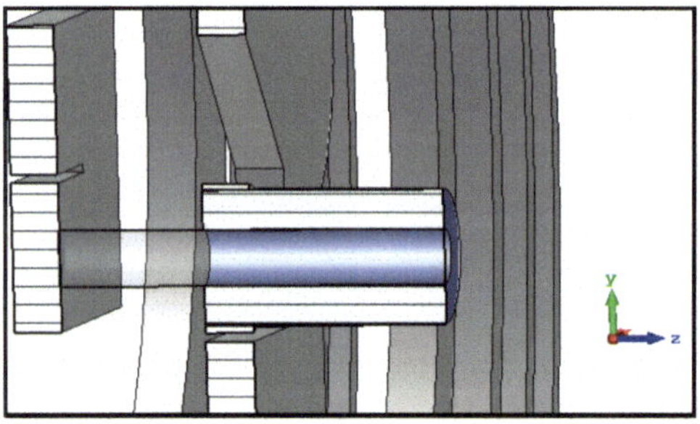

Fig. 2. Cut view of the coaxial feed of spiral antenna

A. ***Reflection coefficient***

A reflection coefficient of –15.14 dB (in Fig. 3) wavelength indicates that the antenna achieves best impedance matching at the given frequency of 350 MHz.

B. ***Voltage standing wave ratio (VSWR)***

At 350 MHz, the antenna exhibits a VSWR value that is approaching (VSWR > 2) as shown in Fig. 4, indicating almost perfect impedance alignment via the transmission line. This alignment leads to efficient power transmission to the antenna.

C. ***Gain***

The gain pattern of the antenna is graphically depicted in Fig. 5. It is clearly discernible that the antenna's gain fluctuates within the interval of 1.5–4.2 dBi, signifying an enduring and progressive augmentation.

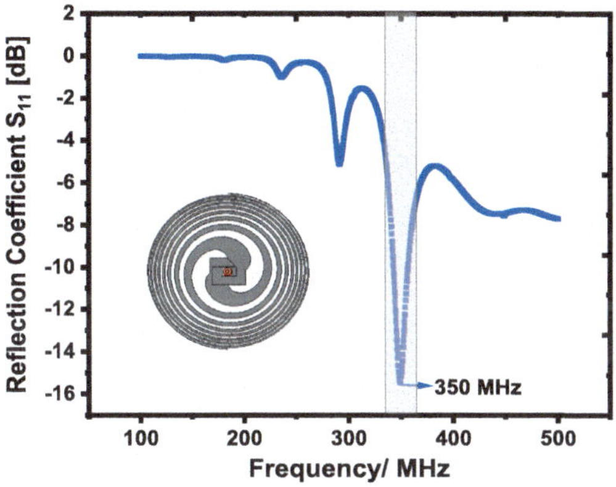

Fig. 3. Simulated S_{11} of spiral antenna

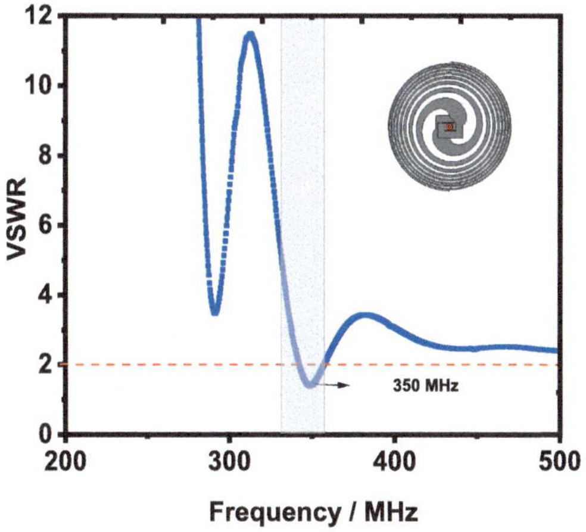

Fig. 4. VSWR of spiral antenna

D. ***Z-matrix***

The Z-matrix portrays the impedance characteristics of the envisaged antenna, as depicted in Fig. 6. Imparting 50 Ω input impedance to the antenna reveals its self-alignment closely within the vicinity of 50 Ω, thus yielding minimal power reflection.

E. ***Surface current***

The current distribution of the surface current is graphically represented in Fig. 7, the region of maximum current density predominantly converges around the vicinity of the coaxial feeding structure.

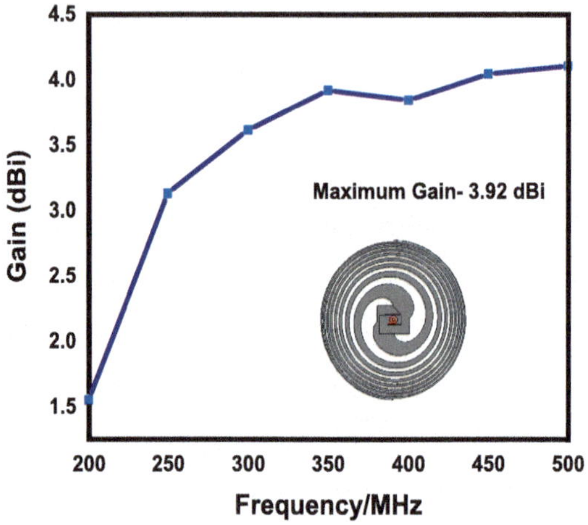

Fig. 5. Gain of spiral antenna

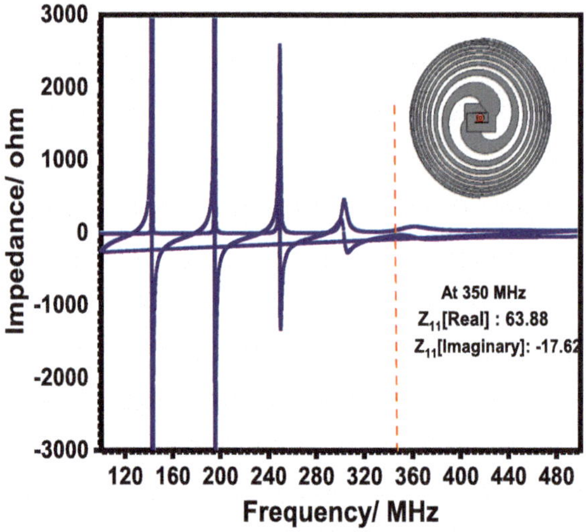

Fig. 6. Z-matrix of spiral antenna

F. ***Radiation pattern***

Radiation pattern graphs show the many directions in which energy is transmitted by an antenna or other electromagnetic device. Figure 8a, b show the antennas polar as well as 3D radiation patterns at 350 MHz, respectively. The proposed spiral antenna displays a bidirection radiation pattern (Table 1).

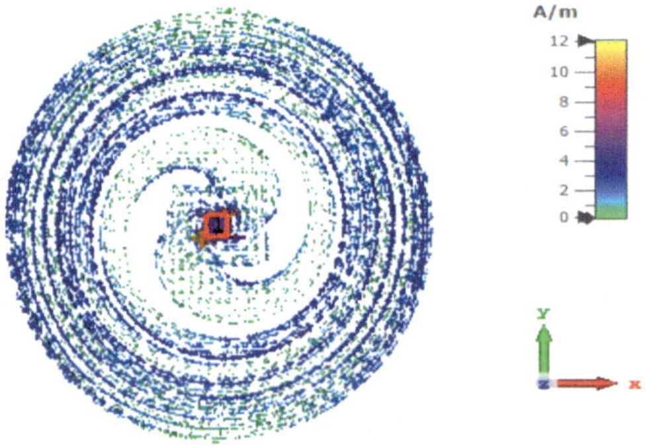

Fig. 7. Current distribution curve for spiral antenna

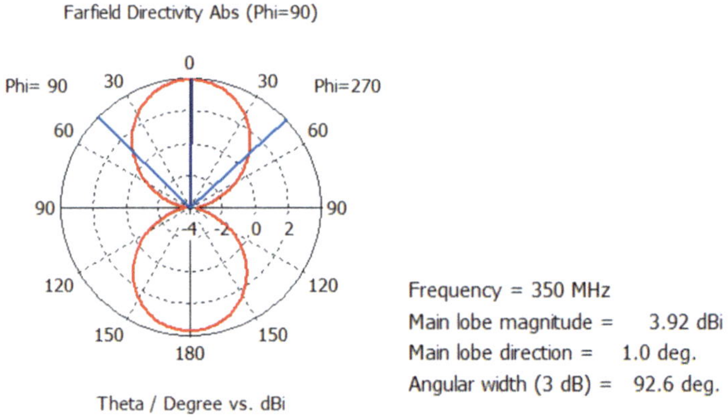

Fig. 8. **a** Radiation pattern for spiral antenna, **b** 3D radiation pattern for spiral antenna

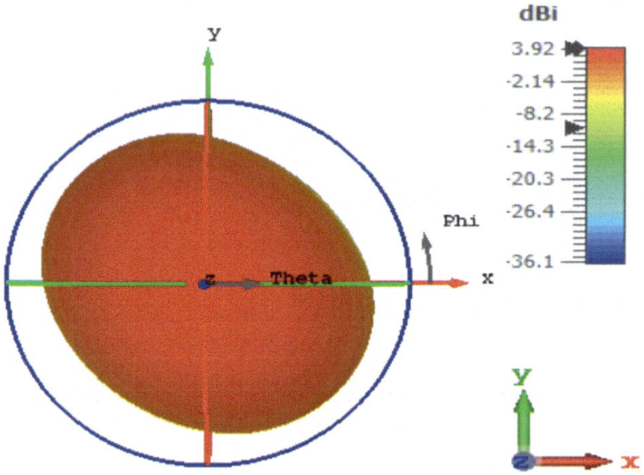

Fig. 8. *(continued)*

Table 1. Result of the proposed equiangular spiral antenna

Antenna profile	Compact size and simple
Reflection coefficient	−15.14dB
VSWR	1.08
Efficiency	High (>90%)
Impedance matching	Z1,1 (Re) = 63.88 Z1,1 (Im) = −17.62
Radiation pattern	Bidirectional
Gain	3.92 dBi
Directivity	18.08 dB

4 Conclusion

This research introduces an innovative antenna design aimed at achieving impedance matching as well as effective feeding in the middle of the antenna and coaxial transmission line. The simulation outcomes demonstrate excellent VSWR, closely approaching ~1, and a reflection coefficient of less than −15.14 dB. The CST simulations indicate a close agreement between the results of the two-arm logarithmic spiral antenna and the performance requirements specified for the SST-1 tokamak. Upon subjecting the antenna to a power input of several watts during simulation, it is evident that there was no occurrence of arcing between the spiral arms. This implies that in the times ahead, experimental investigations can be performed of this spiral antenna.

References

1. Oikawa, T., Kamada, Y., Isayama, A., Fujita, T., Suzuki, T., Umeda, N., Itoh, T.: Reactor relevant current drive and heating by N-NBI on JT-60U. Nuclear Fusion **41**(11), 1575 (2001)
2. Kumar, S.R., Jeevitha, E., Keerthi, V., Kiruthika, S., Manimakalai, V.: Design of Archimedean spiral antenna for radar applications. Int. J. Pure Appl. Math. **119**(15), 315–319 (2018)
3. Sharma, P.K., Rao, S.L., Bora, D., Trivedi, R.G., Mishra, K., Ambulkar, K., Rao, C.V.S.: Commissioning of 3.7 GHz LHCD system on ADITYA tokamak and some initial results. Fusion Eng. Des. **82**(1), 41–53 (2007)
4. Gormezano, C.: Review of lower hybrid wave heating and current drive. Plasma Phys. Control. Fusion **28**(9A), 1365 (1986)
5. Bora, D., Sharma, P.K., Rao, S.L., Trivedi, R.G., Gupta, M.K., Ambulkar, K., LHCD Group.: Design criteria and current status of LHCD system on SST1 machine. Fusion Eng. Des. **82**(2), 141–152 (2007)
6. Chen, D., Zhang, H., Xu, J.: A broadband planar spiral antenna design for electromagnetic signal monitoring. IEEE Access **8**, 73451–73456 (2020)
7. Ferron, J.R., Casper, T.A., Doyle, E.J., Garofalo, A.M., Gohil, P., Greenfield, C.M., Wade, M.R.: Optimization of DIII-D advanced tokamak discharges with respect to the β limit. Phys. Plasmas **12**(5), 056126 (2005)
8. Alhayali, A.N.R.J.: Design of a planar spiral antenna in the 5.5–6.8 GHz frequency range using moment method. Tikrit J. Pure Sci. **17**(4) (2012)
9. Cvetnic, E.E., Adams, R.S.: Impedance matching of an Archimedean spiral antenna. In: 2020 IEEE International Conference on Electro Information Technology, pp. 344–352. IEEE (2020)

Design of a Planar Cathode Electron Gun for High-Power Inductive Output Tubes Operating in the UHF Band

Vaishali[1], Dimple Yadav[1], Poonam Tiwari[1], Meenu Kaushik[2], A. Mercy Latha[3], and Vishant Gahlaut[1(✉)]

[1] Department of Physical Sciences, Banasthali Vidyapith, Rajasthan 304022, India
vgceeri@gmail.com
[2] School of Automation, Banasthali Vidyapith, Rajasthan 304022, India
[3] CSIR-CEERI, Chennai Centre, Chennai, India

Abstract. Inductive Output Tube (IOT) represents the latest generation of amplification technology employed in high-power applications. Owing to their electrical efficiency and established reliability, IOTs are gaining acknowledgment as RF sources within industrial, plasma heating, research and scientific domains. In an IOT, electron gun is not an independent component but is integrated with input cavity. This study focuses on the design of a planar cathode electron gun used in an IOT. The integration of electron gun with the input cavity within the IOT is highlighted, with its significance in efficient RF signal amplification. The design optimizes electrode geometry and dimensions using the Pierce Synthesis approach, with specific attention to grid dimensions and the cathode-grid gap to enhance density modulation and electron bunching. An electron gun for an IOT operating at a frequency of 350 MHz with a 4 A beam current is presented along with the fine-tuning of grid dimensions to facilitate effective modulation and the gap between the cathode and grid deliberately minimized to improve efficiency. The research provides valuable insights and ways for further advancements in increasing the applications of high-power IOTs.

Keywords: RF signal · UHF band · Planer cathode electron gun and grid

1 Introduction

The Inductive Output Tube (IOT) represents the latest generation of amplification technology employed in high-power transmitter applications [1]. IOTs are also known as klystrodes [2] and have utility in high-power contexts; including serving as drivers for Radio Frequency (RF) accelerators [3–5]. It was invented by Andrew V. Haeff in 1938. However, because of specific technological constraints; IOT did not gain significant prominence until the 1980s. Owing to their electrical efficiency and established reliability, IOTs are gaining acknowledgment as RF sources within industrial, plasma heating, research and scientific domains [6].

S. K. Ghosh et al. (Eds.): VEDA 2024, SPPHY 424, pp. 236–242, 2025.
https://doi.org/10.1007/978-981-96-5174-0_35

The key components of an IOT include the electron gun and the input resonant cavity, input RF coupler, collector, focusing structure, output cavity, output waveguide and output coupler [7]. The electron gun stands as the pivotal component within vacuum electron devices, serving as the active mechanism that generates the primary electron stream which underpins their fundamental performance. It combines a powerful electron gun with resonant cavities and magnetic fields to efficiently amplify RF signals [8]. In an IOT, signal is applied to the input cavity using input coupler. Cathodes are essential components of electron guns because they act as the origin of electrons within the device. In the electron gun setup, electrons are discharged from the cathode and then propelled towards an anode. Electron emission encompasses various mechanisms, including photoelectric emission, thermionic emission, secondary electron emission, and field emission [9]. The stream of electron is emitted from the cathode, when heated to a specific temperature. The cathode and control grid are placed very close to each other. When this stream of electrons interact with RF input signal then the formation of bunch takes place based on density modulation. Inside an input cavity, there is a space known as interaction space where interaction between the electron gun and RF signal takes place which amplifies the input signal. Electron guns play a crucial role in the efficiency of high-power devices like IOTs. By precisely controlling the emission and trajectory of electrons, electron guns ensure that a large portion of the input power is converted into RF output power, minimizing energy loss and maximizing efficiency. Electron guns are integral components of high-power devices such as IOTs, enabling efficient RF signal generation, amplification, and transmission across a wide range of applications in the UHF band.

The schematic of IOT has been shown in Fig. 1. This density modulated beam is the accelerated towards the output cavity. In the output cavity, the kinetic energy of electrons is converted into the RF signal [10]. IOTs gained prominence in broadcasting and communication systems due to their high power capabilities, with ongoing refinements in electron gun design, materials, and electromagnetic simulation techniques further enhancing their performance and applicability in modern high-frequency applications [11]. The IOT represents a remarkable synthesis of advanced vacuum tube technology and electromagnetic principles, designed to amplify radio frequency signals with exceptional efficiency and high-power [12]. In this introduction, we delve into the significance of the electron gun within the context of an inductive output tube, exploring its design, operation, and pivotal role in enabling the IOT to achieve high-power microwave amplification for diverse applications in broadcasting, communication, and beyond. The UHF band is vital in space applications for communication, telemetry, tracking, remote sensing, navigation, and more. It facilitates communication between satellites and ground stations, allowing for data transmission and command functions.

Using advanced electromagnetic simulations and precise engineering techniques, the study seeks to optimize the design parameters of the planar cathode electron gun to enhance its performance and efficiency. This study is undertaken in order to efficiently design a planar cathode electron gun for high-power IOTs operating in the UHF band. Previous generations of IOTs have utilized cathodes which are capable of producing RF power, but have poor efficiency and short lifespan. With these facts in mind, a new generation of IOTs is being developed to achieve the same RF power output as a

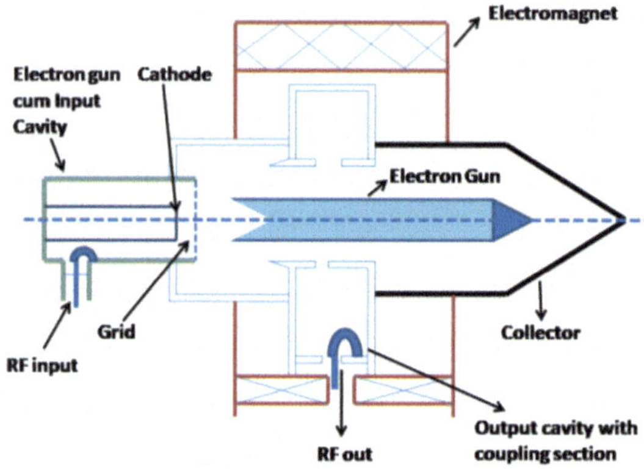

Fig. 1. Schematic diagram of IOT

conventional IOT, but at a higher efficiency and with a longer life. Since the new tube is required to have the same RF power as the older generation IOT, it is necessary to design an electron gun which can deliver the same space charge and produce equivalent RF power in the collector circuit. This creates an extremely challenging task, as reducing the voltage of the triode tube will increase the space charge density and make it difficult to generate the same RF power in the collector. An electron gun design with a successful outcome will be of great significance to the researchers, as they will have an IOT which can produce the same RF power at a lower cost due to energy savings, and will have to replace it less frequently because of the longer life.

Section 1 of the paper introduces the background, necessity, and prospective approach in designing the planar cathode electron gun for the IOT. In Sect. 2, the methodology for designing this electron gun is detailed. Section 3 explores the results and discussions pertaining to particle trajectory and beam current. Finally, Sect. 4 concludes the paper.

2 Design Methodology

The quality of the electron beam generated by the electron gun directly impacts the performance of the high-power device. A well-designed electron gun produces a focused, stable beam with uniform characteristics, leading to improved signal quality and reliability in the output. Electron guns can be optimized for specific frequency bands, making them versatile for various applications. In the case of IOTs, electron guns are designed to operate efficiently in the UHF band, allowing for the generation of high-power RF signals in this frequency range. The design of electron gun based on Computer Simulation Technology (CST) software utilizes the Finite Integrated Method for solving equations. The electron gun is mainly consisting of a cathode for emitting the stream of electrons, Beam Focusing Electrode (BFE) for shaping the beam, and anode for accelerating the electrons. Mainly this design includes the cathode that is able to produce the beam current required to perform desired tube operation. Also, designing the electrodes is important part of electron gun which is used for shaping electrostatic field between the cathode and the beam tunnel extracts electrons emitted by the cathode and maintains them in a laminar flow with the desired radius [13]. For the initial estimation of gun parameters, a Pierce-type gun is used, following an initial design developed through Vaughan's synthesis method, which employs a series of analytical equations [14]. The electron gun has been designed with operating voltage of 40 kV and beams current of 4 A. The beam waist radius is the radius where the beam diameter converges to the minimum, the beam radius at this point is the beam waist radius. The beam waist radius's importance stems from its impact on beam characteristics. This parameter is of significance because, when the magnetic field captures the beam at this specific point, it results in a seamless transition into the interaction region with a laminar flow. In an IOT, input cavity is not an independent component but is integrated with electron gun. Consequently, the design of the electron gun holds critical significance as certain limitations need to be factored in. Characteristics of the IOT System are shown in Table 1.

Table 1. Characteristics of the IOT.

Parameter	Value
Beam current	4 A
Beam voltage	40 kV
Resonant frequency	350 MHz
Output power	112 kW
Efficiency	70%

The approach of Pierce Synthesis has been embraced for the purpose of electron gun design. The integration of the planar electron gun with UHF IOTs opens doors to improved signal amplification, transmission, and communication across a wide array of applications, including telecommunications, radar systems, and satellite communication. Parameters for planar cathode, gridded gun structure has been shown in Table 2. Analysis of gridded gun has been performed and the electron gun has been designed using CST software.

3 Result and Discussion

An analytical method has been successfully employed to calculate the dimensions of the electron gun, and the outcomes have been compared to numerical estimations derived from CST Particle Studio. The geometry has been simulated, and an analysis has been performed to optimize the dimensions of all electrodes. A sectional cut view of the electron gun with control grid is illustrated in Fig. 2. Furthermore, the dimensions of the grid are fine-tuned to enable effective modulation. To ensure seamless alignment with the curvature of the cathode, the spherical radius of the grid is set to match that of the cathode. The space between the cathode and the grid is deliberately kept minimal, approximately 0.37 mm, ensuring that only a small fraction, approximately 1–2% of the RF cycle interacts directly with the electron beam. This deliberate gap facilitates efficient density modulation, enabling the appropriate bunching of electrons.

Table 2. Design parameters for planar cathode, gridded gun structure.

Parameter	Value
Planar cathode radius	37 mm
Anode tunnel radius	50 mm
Cathode anode spacing	70.39 mm
Anode aperture radius	50 mm
Beam radius	37 mm
Grid radius	37 mm
Cathode grid spacing	0.37 mm
No. of grid segments	6

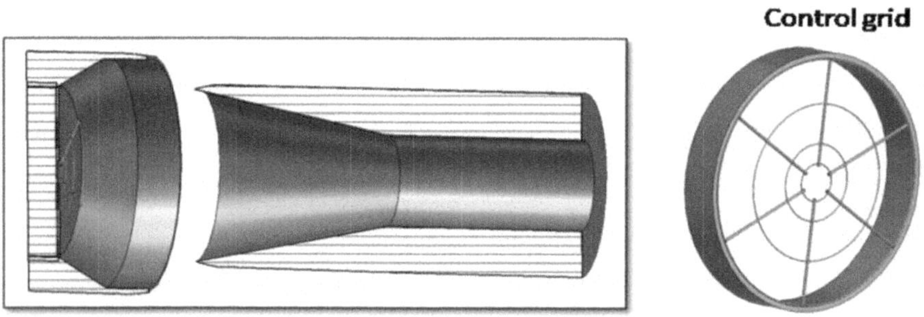

Fig. 2. Cross-sectional view of gun geometry with control grid

Figure 3 illustrates the beam trajectories observed through CST, accompanied by the beam current, obtained at the full beam voltage. An electromagnetic analysis of the electron gun has been carried out, which has yielded a beam current of 4 A. This aligns perfectly with the specifications required for the device. This means that the electron gun meets the desired performance criteria, as outlined in the device specifications. The plot between the beam current and the iteration has been shown in Fig. 4.

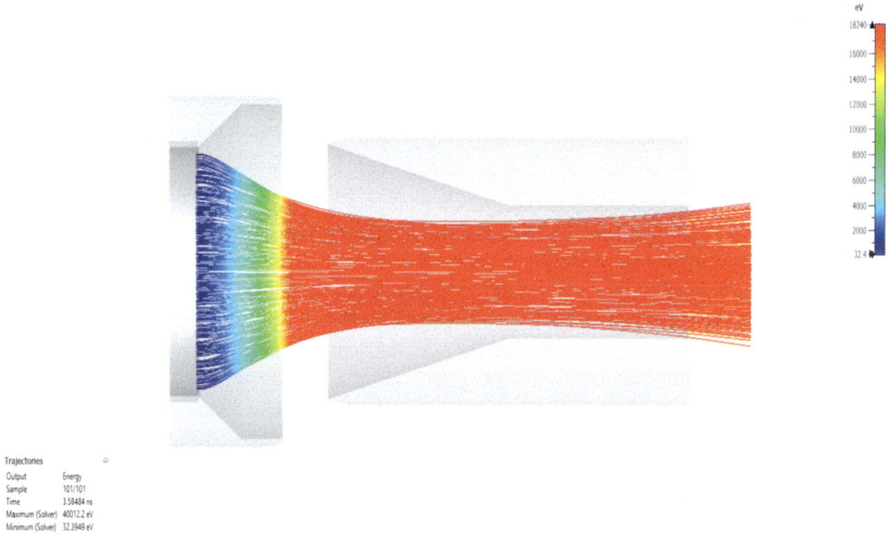

Fig. 3. Beam trajectory of electron gun

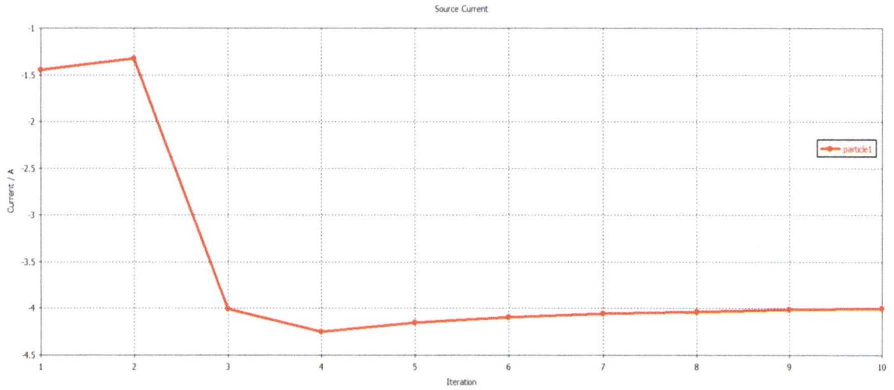

Fig. 4. Beam current is approximatetly 4 A

4 Conclusion

In conclusion, the paper has illuminated the crucial role of electron guns in the efficient generation and control of electron streams, pivotal for achieving high-power microwave amplification. Through the seamless integration of advanced electromagnetic simulations and precise engineering, CST software has emerged as an indispensable tool in shaping electron gun configurations. An electron gun has been meticulously designed for an IOT operating at a frequency of 350 MHz with 112 kW continuous wave (CW) power output. This configuration has successfully achieved the desired beam characteristics.

This research contributes valuable insights to the field of electron gun technology, particularly in the context of high-power IOTs, and paves the way for further advancements in increasing the area application in the field of particle accelerator, plasma heating etc.

References

1. Heppinstall, R., Clayworth, G.T.: The Inductive Output Tube. EBU Technical Review, pp. 43–52 (1997)
2. Freund, H.P., Ives, R.L., Sessions, W.D.: High-efficiency inductive output tubes using a third harmonic drive on the grid. IEEE Trans. Plasma Sci. **48**(11), 3827–3831 (2020)
3. Preist, D.H., Shrader, M.B.: The Klystrode—An unusual transmitting tube with potential for UHF-TV. Proc. IEEE **70**(11), 1318–1325 (1982)
4. Whitaker, J.C.: Power Vacuum Tubes Handbook (Electronics Handbook Series) (1999)
5. Gilmour, A.S.: Klystrons, Traveling Wave Tubes, Magnetrons, Crossed-Field Amplifiers, and Gyrotrons. Artech House (2011)
6. Li, Y., Bohlen, H., Tornoe, B.: Development and deployment of 50 kW high efficiency IOTs for industrial and scientific services. In: 4th IEEE International Conference on Vacuum Electronics, pp. 18–19 (2003)
7. Kaushik, M., Bandyopadhyay, A.K., Joshi, L.M.: RF design of output resonator and conjoined input cavity with gridded electron gun for UHF inductive output tube (2017)
8. Ju, J., Zhang, J., Shu, T., Zhong, H.: An Improved X-Band triaxial Klystron amplifier for Gigawatt long-pulse high-power microwave generation. IEEE Electron Device Lett. **38**(2), 270–272 (2016)
9. Ahmadian, S., Davani, F.A., Ghasemi, F., Shafiee, M.: A design of thermionic electron gun for traveling wave electron-Linac in order to inject beam into booster synchrotron accelerator, pp. 286–288 (2016)
10. Kaushik, M., Joshi, L.M.: Design of planar electron gun for UHF range, CW power inductive output tube. In: Proceedings of the Seventh DAE-BRNS Indian Particle Accelerator Conference: Book of Abstracts, vol. 47(1) (2015)
11. Bohlen, H.: Inductive output tubes—Status and future direction. In: AIP Conference Proceedings American Institute of Physics, vol. 625(1), pp. 35–44 (2002)
12. Grudiev, A.V., Myakishev, D.G., Yakovlev, V.P., Luetgert, S., Krueger, S.: Numerical simulation of a gridded inductive output amplifier. In: Proceedings of the 1996 European Particle Accelerator Conference (EPAC'96), Barcelona, pp. 1244–1246 (1996)
13. Kaushik, M., Joshi, L.M.: Design of spherical electron gun for ultra high frequency, CW power inductive output tube. In: AIP Conference Proceedings AIP Publishing, vol. 1715(1), pp. 020047 (2016)
14. Vaughan, J.R.M.: Synthesis of the pierce gun. IEEE Trans. Electron Devices **28**(1), 37–41 (1981)

Coupled Multi-physics Analysis of a S Band Klystron for Improved Efficiency and Compactness

Janvin Itteera(✉), Mahima, Vikas Tiwari, Elina Mishra, Himanshu Bisht, Kumud Singh, and Sanjay Malhotra

Electromagnetic Applications and Instrumentation Division, B.A.R.C, Mumbai, India
janvin@barc.gov.in

Abstract. This paper proposes the coupled multi-physics analysis of a S-band klystron for electron linac applications. The proposed klystron is in pulsed mode of operation and is designed to deliver average power of 24 kW with the peak power of 6 MW at the desired frequency of 2856 MHz. This paper presents the 1D/3D analysis results of the klystron with various configurations having higher order harmonic intermediate buncher cavities for improved gain and bandwidth performance. The entire model is solved for the Maxwell's equation in time domain to arrive at the coupling factor, bunching, gain and bandwidth. The position of the cavity, interaction gap length, resonant frequency of each cavity and the number of cavities has been varied to obtain an optimal design having desired gain and bandwidth. Sub-harmonic cavities have been placed to shorten the RF interaction section and make the design compact. A study of the klystron beam wave interaction has been carried out using AJDISK. AJDISK is a one-dimensional simulation tool for round and sheet beam klystrons. As first order tool, it provides insight into the klystron design by running fast large signal design computations prior to running more accurate and time consuming 2D and/or 3D particle in cell analysis. Moreover, 1-D solutions offers a fairly good estimate of various electrical and dimensional parameters of RF section to get desired tube performance and compare reasonably with experimentally results of the klystron, particularly, 5-cell cavity klystron (SLAC-5045). The RF gain, phase and velocity profile of particles, energy distribution and RF current are plotted along the travel of the particle. Subsequently, the validation and optimization of design parameters of complete RF section using 3D CST—Particle in Cell code.

Keywords: First keyword · Second keyword · Third keyword

1 Introduction

Design of a S band klystron is being pursued for integration with e-linacs in industrial applications. Klystron are the major technological bottlenecks for the indigenization of industrial e-linacs. A high power pulsed 2856 MHz klystrons are being proposed to be developed to deliver peak power of 6 MW with a PRF of 400 Hz for 10 μs flat top period.

S. K. Ghosh et al. (Eds.): VEDA 2024, SPPHY 424, pp. 243–247, 2025.
https://doi.org/10.1007/978-981-96-5174-0_36

Klystron amplifier is composed of three functional section namely, an electron gun, RF circuit and collector system. The electron gun manifests the wall plug power onto kinetic energy of the electron beam. The RF circuit primarily is derived from several resonating cavities that serially modulates the velocity of the electron beam from low level input RF signal on microscopic scale. For enhanced conversion, the partially modulated electron beam which is further loaded to achieve spatial bunching and the high RF power is extracted from the resultant electromagnetic fields induced on the output part of circuit. Final phase of the spent beam having sufficient kinetic energy is then captured in the output collector, wherein substantial portion of the energy is manifested as thermal.

Key parameters to be factored in during the design conceptualization of the klystron begins with the electron gun perveance [1]. It is known empirically, efficiency of klystron falls linearly with increasing perveance, as higher perveance gun produces higher charge density in the interaction region of the RF circuit. The higher volumetric charge density limits the penetration of RF fields through the beam cross-section leading to poor conversion efficiencies.

2 Design Consideration

Based on the past experience with a similar klystron designed by CSIR-CEERI, a reasonable micro-perveance of 1.86 was selected to initiate the design [3]. A practical achievable beam current of 90 A under applied voltage of 130 kV has been considered to be the power source for the RF circuit.

Key parameters of RF circuit are the R/Q, resonant frequencies, accelerating gap of the respective accelerating cavities, coupling coefficient, number of cavities and their respective gap centre [6]. Klystron having more intermediate cavities are known to have higher efficiencies [4].

For designers of RF power devices such as klystron, effective accelerating voltage governed by R/Q (on axis) is shown in Eq. (1).

$$\frac{R}{Q} = \frac{V^2}{2\omega_0 W} \tag{1}$$

where V represents for on-axis interaction gap, ω_0 is the resonant frequency and W is the stored energy of RF fields. The cavity dimensions were optimized with due regards to the resonant frequency ω_0.

The gain bandwidth of the amplifier to a large extent, is proportional to the R/Q of the output cavity in the RF circuit.

Another parameter of interest to design, is the coupling of the modulated RF beam current across the electrodes in the cavity interaction region. The RF voltage, V_k are induced across the gap of a kth resonating cavities by a driving beam current I_j modulated by the preceding section of the circuit. This voltage and current measured in the circuit are related for evaluating the efficacy of the bunching by respective transconductance

$$G_{kj} = \frac{I_j}{V_k} \tag{2}$$

The transconductance parameters modifies the feedforward currents from preceding cells across the interacting gap of the output cavity, wherein amplified power is drawing out. Practically, intermediate cavities are tuned to resonate at higher frequencies, serving as inductive load to the electron beam which is modulated to the operating frequency.

A five-cavity design is envisaged for this klystron, wherein, intermediate cavities are staggered tuned for achieving the desired gain-bandwidth. Considering the project requirements, the key design parameters of the proposed klystron are summarized in Table 1.

Table 1. Major parameters of the proposed S band klystron

Design parameters	Values
Operating frequency	2856 MHz
Bandwidth	8 MHz
Peak output power	6 MW (min)
Pulse width/Pulse repetition Frequency	10 μs (Flat top)/400 Hz
Average RF output power	24 kW
Beam voltage/current	130 keV/90 A

2.1 1D Electromagnetic Design

For a multi cavity klystron, the Eq. (2) is far more complex summation of several feed-forward currents from the preceding cavities on to the output cavities through their respective transconductances. This is more often challenging to compute analytically and optimize the parameters. However, there are few RF dependent geometric features that may be analytically estimated, based on which, design can evolve further.

Among them, following parameters can be set analytically are the radius of the drift tube (R_d) and envelope of the DC beam (R_b) [6].

The drift tube radius is inversely proportional to the electron wave number (K_e)

$$R_d = \frac{1}{K_e} \tag{3}$$

While, radius of the beam is so chosen to ensure a reasonably high transmission through the drift space. In RF cavities, since, electric field is maximum near the gap electrodes, the beam is allowed to expand and fill nearly 70–80% of the tube radius for optimal RF coupling.

For the proposed klystron, all the cavities are operated in proximity of the fundamental harmonic, due to challenges in fabrication and processing of the RF surface.

With the above parameters evaluated, basic design of the klystron is carried out in 1D AJDISK, wherein the electron beam is modelled as series of discs in one dimension (longitudinal). The RF circuit parameters can be reasonably designed conveniently with the help of 1D model, wherein, following input parameters pertinent to RF cavities were derived from 2D eigen mode solution of the cavities using Superfish software, shown in Fig. 1.

The R/Q, intrinsic quality factor (Q_0), transit time factor (Gap coupling coefficient, M) and resonant frequency were estimated while incorporating requisite correction as applicable for tube design. This 1D code can accurately and quickly predict saturated gain, reduced plasma wavelength, cavity voltages, and beam loading using FDTD technique. It offers an optimal design parameters, which may be tested further with more time consuming 2D or 3D particle in cell (PIC) simulations.

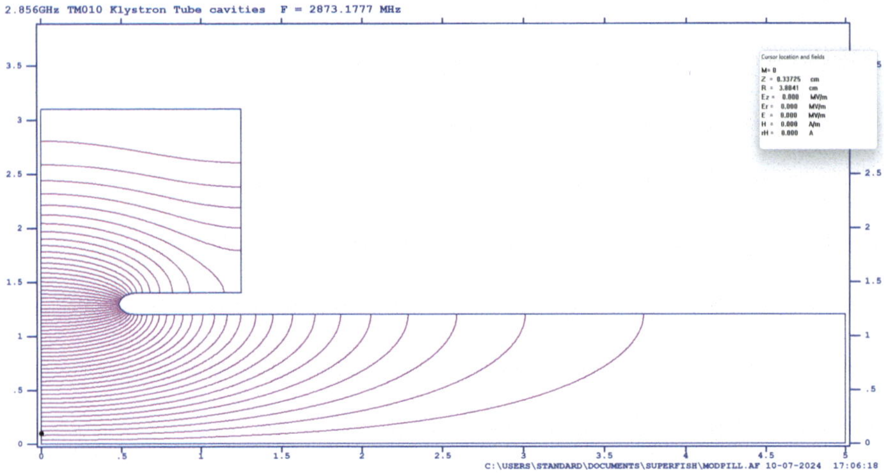

Fig. 1. 2D design of the Re-entrant cavities

3 Results

The lengths of the drift sections between cavities and the cavity characteristic parameters such as frequency, R/Q and coupling coefficients are optimized iteratively to obtain the desired saturation gain and reasonable efficiency. The results of AJDISK predicts the design to have an efficiency of more than 55% and the gain of 48.2 dB(6.48 MW) can be achieved. The proposed design envisages a RF circuit length of ~56 cm with cylindrical re-entrant cavities (Fig. 2).

Nearly 45 discs with finite thickness accounting for space charge forces were used for achieving sufficient convergence and the phase of the particle are nearly in bunch phase. The analysis has also considered effect of reduced space charge potential depression imposed by the intense DC beam in the tunnel.

Since the 1D analysis does not consider the effect of the focusing magnetic field, the 3D computations (particle in cell) are in progress to validate this optimal design.

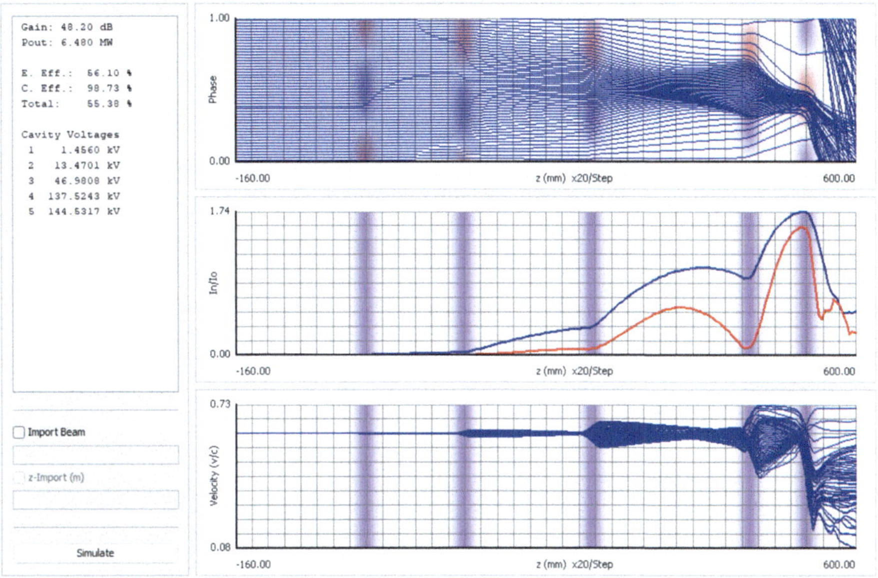

Fig. 2. Results of 1D modelling of the RF circuit

Acknowledgement. The authors express their heart-felt gratitude to the scientific support extended by the klystron design Team from CSIR -CEERI, Pilani for their valuable scientific contributions during the design and testing phase of earlier S-band klystron.

References

1. Chin, Y.H.: Modeling and Design of Klystron. KEK
2. Clerk Maxwell, J.: One-dimensional disk model simulation for klystron design, SLAC
3. Joshi, L.M., et al.: Design and development of a 6 MW peak, 24 kW average power S-band Klystron. In: IEEE Internal Conference on Vacuum Electronics (2011)
4. Catalan Lasheras, N., et al.: High-efficiency klystrons from a dream to a reality. In: 15th International Particle Accelerator Conference, Nashville, TN. ISBN 978-3-95450-247-9
5. Bandyopadhyay, A.K., et al.: Design and development aspects of Klystrons. In: IEEE Internal Conference on Vacuum Electronics (2021)
6. Santosk Kumar, M. et al.: Design studies of Re-entrant square cavities for V-band klystrons. Electronics **12** (2023). https://doi.org/10.3390/electronics12010006

LTCC-Based Rectangular DRA Operating in Millimeter-Wave Range

Poonam Tiwari[1(✉)], Dimple Yadav[1], Vaishali[1], Meenu Kaushik[2], Anshuman Shastri[3], and Vishant Gahlaut[1]

[1] Department of Physical Sciences, Banasthali Vidyapith, Rajasthan 304022, India
poonamrakeshtiwari2596@gmail.com
[2] School of Automation, Banasthali Vidyapith, Rajasthan 304022, India
[3] Centre of Artificial Intelligence, Banasthali Vidyapith, Rajasthan, India
anshumanshastri@banasthali.in

Abstract. This paper deals with the feasibility of using low-temperature co-fired ceramics (LTCC) as a dielectric substrate in the design of a rectangular dielectric resonator antenna (DRA). The primary objective of this paper is to design an antenna that efficiently uses higher-order modes within a wide frequency range for an increase in efficiency and directivity. LTCC material is chosen for its good dielectric properties and excellent thermal stability, and it is also suitable for multilayer circuits; thus, it is perfectly appropriate for compact designs of high-performance antennas. A substrate-integrated waveguide with a circular slot is used to realize efficient feeding to the resonator and suppress back radiation. The reason for choosing this feeding technique instead of the conventional microstrip feed line is that this offers better performance in reducing back radiation and enhancing the antenna efficiency. This designed antenna realizes a quite high radiation efficiency, greater than 90%, very critical for applications that require significant power transfer and fewer losses. Added to this, the antenna has been carefully designed for broadside radiation patterns with linear polarization, hence making it useful in various wireless communication and sensing applications. It demonstrates that LTCC-based DRAs make possible whole classes of advanced antenna designs and provide a route to accomplish high-efficiency, wideband antennas having excellent directivity with low back radiation.

1 Introduction

DRAs may provide the solution for mm-wave bands because of the low conduction losses. In these antennas, operation is usually in the low-order modes, and again, approximately 5 dBi may be expected; then, the emission patterns are similar to those from microstrip antennas or small magnetic dipoles. Arrays are usually used to generate more focused radiation patterns. Various techniques to increase the gain of certain features of DRA have been seen coming up in the recent past. These include those that incorporate the investigation of excitation of higher-order modes in simply shaped DRA, the use of superstrates and electromagnetic band-gap structures, and modifying the ground plane [2]. Among these strategies, higher-order modes are found to be the most interesting

S. K. Ghosh et al. (Eds.): VEDA 2024, SPPHY 424, pp. 248–256, 2025.
https://doi.org/10.1007/978-981-96-5174-0_37

ones since they do not add a lot of new features and do not significantly complicate the fabrication process [3]. One of the special features of DRAs is that there is a special part to operate at higher-order modes with an unbelievable maximum gain of about 9.6 dBi [4]. This can open a completely new window regarding DRA performance since its virtues may make it possible to attain more functional and effective mm-wave communication systems [5]. This paper attempts to design a mm-wave rectangular DRA using LTCC technology. This paper, therefore, investigates the suitability of the technology for the realization of highly radiation-efficient mm-wave band antennas [6]. This LTCC technology will also exhibit a clear advantage when designing complex resonator structures for very high frequency bands, which overcomes problems that the conventional production methods often have when meeting the required standards. It has been shown that increased geometric complexity, resonator shapes similar to the lower mode DRA performance operating in the higher modes, can increase impedance bandwidth. However, small errors in the final dimensions of the resonator can result because of the complicated manufacturing process inherent in LTCC technology during the fabrication of a multilayer structure on the LTCC tape [8]. LTCC ceramic is greatly stable and also provides good mechanical and dielectric strength. As a result, the use of LTCC ceramic makes it high-performance for mm-wave applications. The chosen LTCC material for this design is Green Tape 9K7. It comes in many different thicknesses, so it's a material that is easy to use. This LTCC material is replete with versatile properties and performance characteristics, which makes it find wide applications in multichip multilayer modules, single chip packages, ceramic printed wiring boards, and mm-wave modules. As such, a slot within a substrate-integrated waveguide (SIW) feeds the antenna because of the spurious radiation-related problems that result from this feeding mechanism. As a result, the paper has been broken down into sections to enable a reader to grasp an overview of the design and functionality of the antenna concept.

Section 1 explains the properties of the material used for design. This forms basic background material and is important for the following sections. Section 2 explains or further discusses the methodology adopted for the design that obtained those dimensions of the antenna. It also describes how the performance characteristics of the antenna were obtained. Like frequency of operation, radiation pattern, and impedance matching, are achieved. Final results are presented in Sect. 3, hence giving insight into the real performance of the antenna. Section 4 concludes the paper with the main findings and insights derived from the study. It also gives a synthesis of the presented research, its main implications, and which possible lines can be followed for further research on the design of mm-wave antennas. Table 1 compares the proposed work to the existing contributions in the literature, showing that the current study is ahead in terms of various performance metrics and a few design aspects.

2 Antenna Design Methodology

The antenna is designed to be a simple rectangular DRA made from the LTCC material—9K7 Green Tape, characterized by its electrical properties with $\varepsilon r = 7.1$ and $\tan \delta = 0.0009$. The proposed antenna measures $(16 \times 16 \times 1.6)\ \text{mm}^3$, as shown in Fig. 1. Since substrate dimensions are given for length and breadth, the conducting ground plane

Table 1. Performance Evaluation of the Suggested Antenna Compared to Previous Work

References	Substrate (ϵ_r) (Thickness)	Size (mm^2)	Shape of DR (ϵ_r)	Bandwidth (GHz)	Gain (dBi)	Complexity of the design
[11]	Rogers5880 (2.2) (0.254)	17.77 × 16	Rectangular (9.8)	25–27 (7.69)	6	Complex
[12]	Rogers3003 (3) (0.13)	20 × 20	Cylindrical (10)	29 GHz (6.7)	6.5	Complex
[13]	Rogers6002 (2.94) (0.254)	35 × 35	Rectangular (9.8)	24.50–27.50 (24) 33–37 (12)	9.3	Complex
[14]	Rogers5880 (2.2) (0.254)	20 × 20	Rectangular (9.8)	27.25–28.59 (4.8)	9	Complex
[15]	Rogers4003C (3.55) (0.1)	10 × 30	Rectangular (45)	26.41–30.42 (14.11) 36.05–40.88 (12.56)	6.8	Complex
[16]	Rogers4003C (3.55) (0.305)	46.5 × 22.5	Rectangular (3.55)	13.3–19 (35.3) 36.3–40 (9.7)	14	Complex
[17]	Rogers5880 (2.2) (0.254)	27.5 × 30	Rectangular (6.15)	25.41–31.18 (11.6)	5	Complex
This work	**Arion 25N (2.2) (1.6)**	**16 x 16**	**Rectangular (7.1)**	**30.3 to 54.5 GHz**	**9.57**	**Simple**

of the antenna will be in the shape of a square. In this work, a CST tool utilizing the third-order electric dispersion model to extrapolate LTCC properties in the simulation of fields is used. The principle of operation for the antenna results from the excitation of higher-order modes in a rectangular DRA. From this setup, the resulting radiation pattern was of a relatively high radiation efficiency. Nevertheless, it can be noted that edge diffraction in the E-plane is responsible for the major effect of the directivity in the wider side direction, which results in the maximum gain of the antenna as a periodic function of the size of the ground plane. A resonator with fed through an aperture within an SIW to prevent back radiation. A piece of Arlon 25N microwave substrate with $\tan \delta = 0.0025$ and $\varepsilon r = 3.38$ is used for the feeding structure. The linearly polarized radiation resulted in an antenna having fields inside the aperture that run parallel to it. The feeding is performed by a segment of the grounded coplanar waveguide ended by a linear taper connected to the SIW. This kind of setup has already demonstrated the conversion effectiveness of the transverse electric TE10 mode of the SIW from the QTEM of the GCPW line. The antenna is initially designed to cover a wide frequency of 30.3–54.5 GHz, with an expected relatively wide impedance bandwidth. Even using the low relative permittivity materials for the resonator, expanding the bandwidth is not easy because higher order modes, like TE211 and TE221, have higher Q than their lower order counterparts. This makes it challenging to broaden the bandwidth due to the nature of those modes in the resonator. The major objectives of the design are high radiation efficiency and enlarged impedance bandwidth. First, by scaling, RL × RW × RH dimensions are found through scaling down the DRA antenna. Now, in the design process, it is used which simulates dielectric-air interactions as perfect magnetic barriers-a magnetic wall model; for our purpose, this approximation is good enough. As this model does not include the imposition of the Dirichlet boundary condition at the dielectric-air interfaces, Ht = 0 A/m, as lossy walls do, the projected resonant frequency is slightly higher. The exact measurements of the antenna are outlined in Table 2. The width on an Arlon 25N substrate of the SIW feed defines a cutoff frequency of 42.4 GHz at the initial design level.

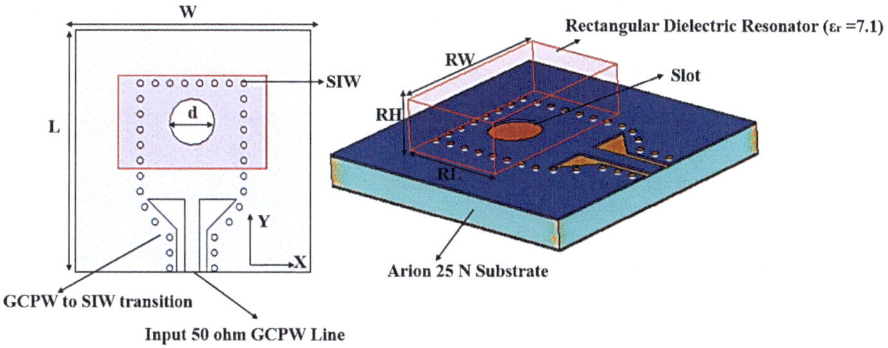

s

Fig. 1. The LTCC-based dielectric resonator antenna configuration

Table 2. Antenna Dimensions

Symbol	L	W	d	RW	RH	RL	h
Values (mm)	16	16	3	10	4	6	1.6

At initial design, the SIW is approximated and regarded as a rectangular waveguide with vertical walls to reduce complications in calculations to save time. In the later stage of design, where accuracy is required, real simulations consist of the model of SIW with vias. Equations (1–3) are used to determine the length (l), width (w) and distance from the shorted end of the SIW (s) for the resonator as given below:

$$l = \frac{0.4\lambda_0}{\varepsilon_e} \tag{1}$$

$$w = 0.2l \tag{2}$$

$$s = \frac{\lambda_g}{4} \tag{3}$$

In the equation, ε_{effr} represents the effective relative permittivity, which is determined by averaging ε_s (the relative permittivity of the substrate) with εr (the relative permittivity of the Dielectric Resonator Antenna, DRA). λg denotes the guided wavelength within the slot of the SIW. To prevent half-wave resonance in the slot, the length parameter is chosen according to a specific equation. A linear taper of predetermined length and width is inserted between the SIW and the input 50 Ω GCPW line, with the taper dimensions optimized using CST. The size of the ground plane is the final design parameter. Research has shown that edge diffraction effects significantly impact gain in the broadside direction of SIWs. Therefore, optimizing the ground plane dimensions is essential to achieve maximum antenna performance, particularly in terms of gain and radiation pattern. The goal is to balance the ground plane's minimum area with the highest possible directivity. In this study, a square ground plane footprint is used, though a circular profile could further enhance directivity. The next section presents the simulation results.

3 Simulated Outcomes

Figure 2 gives the reflection coefficient (S11) of the antenna and its performance over a wide frequency span of 30.3–54.5 GHz. It can be seen that the resonant peaks occur at 37 GHz and 45 GHz, which correspond to the excitation of the TE211 and TE221 modes at these resonating frequencies.

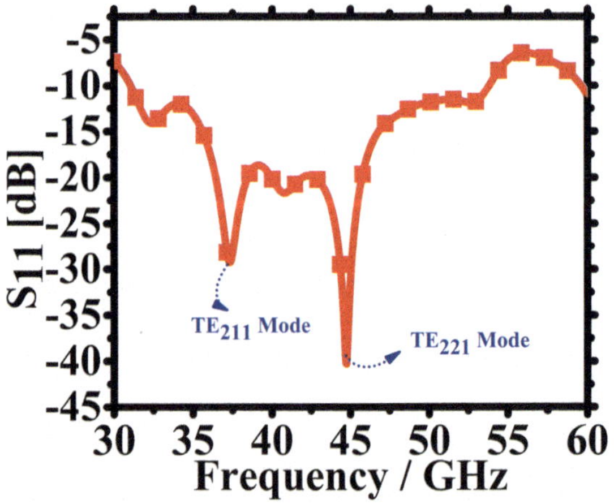

Fig. 2. Reflection Coefficient S_{11} of LTCC DRA

These resonant modes are therefore of importance because they directly affect the features of operation and the effectiveness of the antenna within the whole frequency band. According to Fig. 3, the antenna realizes around 88.2% radiation efficiency. A substrate having less loss tangent compared to that of the Arlon 25N will further reduce losses and hence have higher radiation efficiency. Figure 4 (a) E-plane and, (b) field distributions: Electric-color and Magnetic-arrows inside the resonator in cross-section. Mainly these fields show the TE221 mode. This paper shows the radiation patterns of

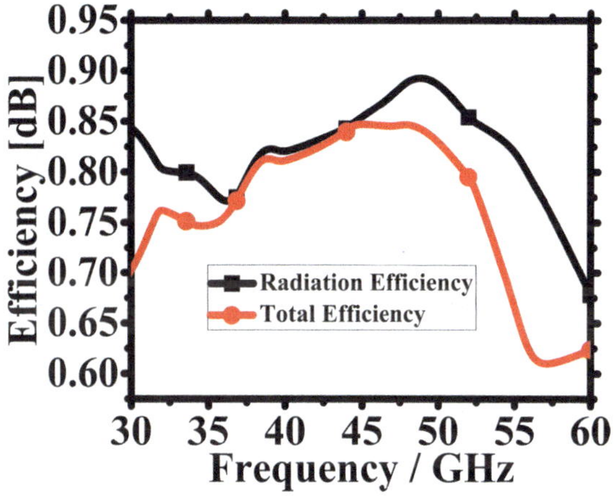

Fig. 3. Efficiency of LTCC DRA

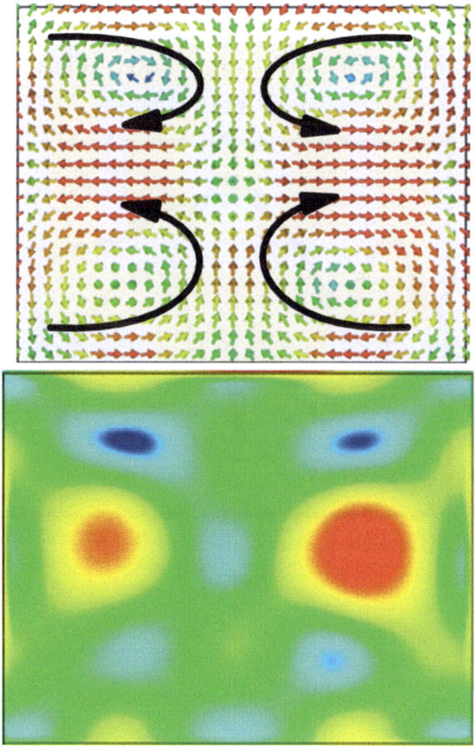

(a) (b)

Fig. 4. **a** Electric field distribution in the E-plane (xz plane) and **b** Magnetic field distribution in the E-plane (xz plane) corresponding to the TE_{221} mode the antenna in the E-plane that is in the xz plane as shown in Fig. 5a. It is shown that broadside gain has 9.57 dBi. We now turn to Fig. 5b to view the radiation patterns in the H-plane, which is in the yz plane. In that plane, however, the sidelobe level turns out to be very sensitive to differences in the antenna design.

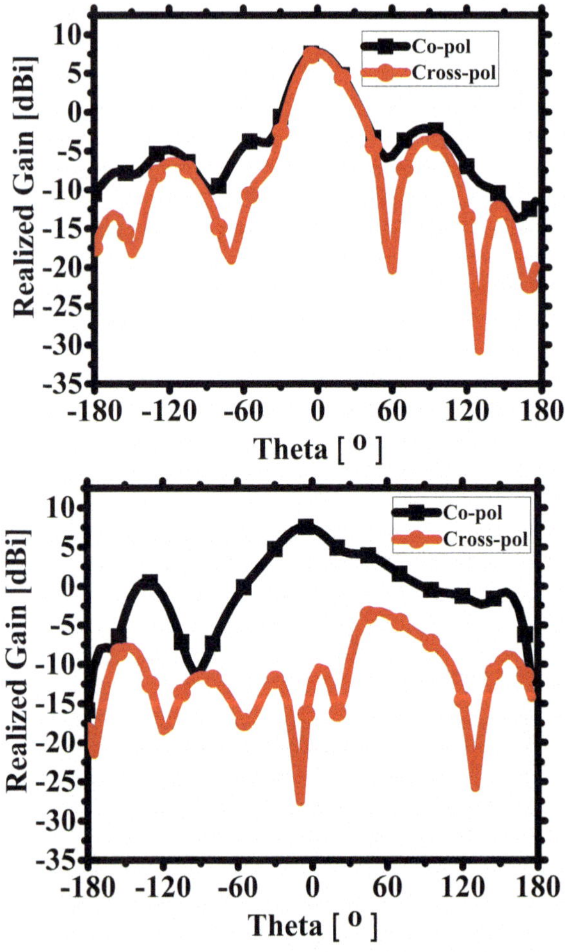

Fig. 5. Radiation patterns of the antenna; **a** E-plane (*xz* plane), **b** H-plane (*xz* plane)

4 Conclusion

It presents the design of an ultra-efficient mm-wave antenna for excellent performance. Initially targeted to be in a wide frequency band of 30.3–54.5 GHz and primarily working with the TE211 mode and TE221 Mode, this antenna has an overall bandwidth of

24.2 GHz with 57.07% and a gain of 9.57 dBi. Shockingly, it realizes 88.2% in radiation efficiency, to a great extent because LTCC has almost negligible dielectric losses. Comparing the performance of our antenna with [11–17], oriented to the improvement of the antenna gain respectively, we can formulate the following: there are considerable advantages although our results are obtained at a much higher frequency. However, this increased gain comes at the expense of bandwidth, which is reduced when compared to both referenced designs.

References

1. Tiwari, P., Kaushik, M., Shastri, A., Gahlaut, V.: Two element microstrip-fed slot loaded millimeter wave MIMO antenna. In: IEEE International Conference for Advancement in Technology (ICONAT), pp. 1–5 (2023)
2. Tiwari, P., Kaushik, M., Shastri, A., Rai, J.K., Gahlaut, V.: Simulated design and analysis of highly isolated 5G millimeter-waves MIMO antenna with wideband characteristic. In: IEEE First International Conference on Microwave, Antenna and Communication (MAC), Prayagraj, India, pp. 1–6 (2023)
3. Tiwari, P., Gahlaut, V., Kaushik, M., Rani, P., Shastri, A., Singh, B.: Advancing 5G connectivity: a comprehensive review of MIMO antennas for 5G applications. Int. J. Antennas Propagat. 19 (2023). Article ID 5906721
4. Tiwari, P., Gahlaut, V., Narayan Mishra, U., Shastri, A., Kaushik, M.: Polarization diversity configuration of millimeter wave MIMO antenna for Ka-band application in 5G wireless networks. J. Electromagn. Waves Appl. 1–22 (2024)
5. Tiwari, P., Kaushik, M., Shastri, A., Gahlaut, V.: Design of a MIMO rectangular dielectric resonator antenna for 5G millimeter-wave communications. In: 2023 IEEE Wireless Antenna and Microwave Symposium (WAMS), Ahmedabad, India, pp. 1–6 (2023)
6. Tiwari, P., Gahlaut, V., Kaushik, M., Shastri, A., Siddiqui, G., Singh, B.: A high-frequency planar-configured millimeter-wave MIMO antenna for fifth-generation NR operations. Int. J. RF Microwave Comput.-Aided Eng. (2023)
7. Mrnka, M., Raida, Z.: Enhanced-gain dielectric resonator antenna based on the combination of higher-order modes. IEEE Antennas Wireless Propag. Lett. **15**, 710–713 (2016)
8. Jain, R., Tiwari, P., Praveena, A., Bhardwaj, A., Rai, J.K., Ranjan, P.: A novel modified P-shaped microstrip antenna for 5G mm-wave applications. In: 2023 First International Conference on Microwave, Antenna and Communication (MAC), March 24, pp. 1–4. IEEE (2023)
9. Tiwari, P., Rai, J. K., Gahlauta, V., Ranjan, P.: Compact quad element dual band high gain MIMO rectangular dielectric resonators antenna for 5G millimeter wave application. AEU-Int. J. Electron. Commun. 155280 (2024)
10. Tiwari, P., Gahlaut, V., Kaushik, M., Shastri, A., Arya, V., Elfergani, I., Zebiri, C., Rodriguez, J.: Enhancing performance of millimeter wave MIMO antenna with a decoupling and common defected ground approach. Technologies **11**(5), 142 (2023)
11. Pan, Y.M., Qin, X., Sun, Y.X., Zheng, S.Y.: A simple decoupling method for 5G millimeter-wave MIMO dielectric resonator antennas. IEEE Trans. Antennas Propag. **67**(4), 2224–2234 (2019)
12. Hussain, H.M.T., Sharawi, M.S., Podilchack, S., Antar, Y.M.: Closely packed millimeter-wave MIMO antenna arrays with dielectric resonator elements. In: 10th European Conference on Antennas and Propagation (EuCAP), April 10, pp. 1–4 (2016)

13. Girjashankar, G.P.R., Upadhyaya, T.: Substrate integrated waveguide fed dual-band quad-elements rectangular dielectric resonator MIMO antenna for millimeter wave 5G wireless communication systems. AEU-Int. J. Electron. Commun. **137**, 153821 (2021)
14. Zhang, Y., Deng, J.Y., Li, M.J., Sun, D., Guo, L.X.: A MIMO dielectric resonator antenna with improved isolation for 5G mm-wave applications. IEEE Antennas Wirel. Propag. Lett. **18**(4), 747–51 (2019)
15. Cui, L.X., Ding, X.H., Yang, W.W., Guo, L., Zhou, L.H., Chen, J.X.: Communication compact dual-band hybrid dielectric resonator antenna for 5G millimeter-wave applications. IEEE Trans. Antennas Propag. **71**(1), 1005–1010 (2022)
16. Wang, Y.-X., Chen, Z., Yuan, T.: Design of a dual-band dielectric resonator antenna array for millimeter-wave communication. IEEE Antennas Wirel. Propagat. Lett. (2023)
17. Xu, M., Zhang, J.: Circularly polarized multiple-input multiple-output dielectric resonator antenna for 5G millimeter-wave application. Electronics **12**(20), 4258 (2023)

Materials and Their Machining Characteristics for TWT Applications: A Review

H. S. Karthik, Santosh Kumar Jha, P. Srikrishna, K. Santhosh Kumar(✉), and S. K. Datta

Microwave Tube Research and Development Centre, DRDO, Ministry of Defence, Bengaluru, India
santosh.mtrdc@gov.in

Abstract. Material selection is crucial for the functionality and performance of Traveling Wave Tubes (TWTs) in high frequency applications. Internal components such as an electron gun and slow-wave structure require materials with low vapor pressure and high thermal stability to maintain ultrahigh vacuum. Vacuum tube envelopes made from ceramics or glass provide mechanical strength and vacuum integrity, with ceramic-to-metal seals preferred for their thermal resistance. External components must withstand environmental stresses such as high-voltage gradients and corrosion. The machinability of materials such as tungsten, molybdenum, ferrous alloys and copper alloys is influenced by their mechanical properties, which in turn affect factors such as tool wear, cutting forces, precision and surface finish. Material Properties such as hardness and ductility directly affect the machinability, which determines the tool life, cutting speed and quality of the final product. In addition, cutting tool, cutting fluid and appropriate machining parameters are critical for achieving the precision required for high-performance TWT devices. Advanced fabrication and microfabrication techniques further ensure the accuracy of these applications.

Keywords: Traveling wave tubes (TWT) · Electron gun · Slow wave structure · Microfabrication and Machinability

1 Introduction

The traveling wave tube works under ultra-high vacuum conditions; therefore, the selection of materials for the construction of such vacuum tube electronics device is critical for reliable performance. The commercial and defence applications of vacuum devices have expanded to millimeter-wave and terahertz frequencies. As the frequencies increase into the millimeter-wave and terahertz ranges, the components must be manufactured with smaller and more precise features to ensure desired performance. Microwave tubes require high or ultrahigh vacuum to be maintained throughout their lifespan, which demands specialized techniques, processes and materials. Material choice varies by TWTs type and service environment with distinct requirements for internal and external components such as the electron gun, slow-wave structures, collectors and RF windows,

S. K. Ghosh et al. (Eds.): VEDA 2024, SPPHY 424, pp. 257–270, 2025.
https://doi.org/10.1007/978-981-96-5174-0_38

housed in sealed envelopes where ultrahigh vacuum must be upheld [1]. Materials such as ferrous and non-ferrous metals and alloys, ceramics, magnets and refractory metals are essential for the functional requirements of TWTs, such as thermal management, air-vacuum interfaces, controlling electromagnetic instabilities and focusing electron beams. These materials must have low vapor pressure to ensure vacuum-tight seals for a considerable life span of the TWTs. Unique materials with desirable thermal, mechanical, electrical and magnetic properties enhance vacuum tube efficiency, compactness and power [1]. Machinability is the ease with which materials can be machined using various cutting tools and techniques, it is a key factor in the manufacturing process. Tungsten and molybdenum require specialized fabrication techniques because of their hardness, whereas glass and ceramics require delicate handling due to their brittleness. Balancing and optimizing the machining processes are crucial for ensuring the reliability and performance of vacuum tube devices.

1.1 Materials Functionality in Traveling Wave Tubes

See Fig. 1.

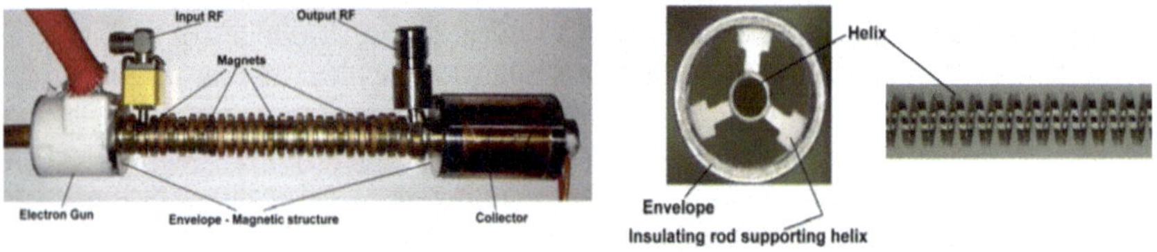

Fig. 1. Helix Traveling Wave Tube. Source: [2]

In a TWT, the selection of materials is very important for the vacuum seal envelope which preserves vacuum environment required for its reliable operation. This vacuum is essential for enabling the free electrons emitted from the cathode to interact effectively with electric and magnetic fields as they travel along the beam axis towards the collector. The integrity of the envelope and the appropriate choice of materials play significant role for TWTs developed for high-frequency applications.

Internal Components

To ensure the efficient operation of a TWT, a high vacuum (~10^{-7} mm Hg) within the envelope of the tube is essential. The envelope and its internal components must be made of materials that exhibit low vapor pressures at elevated temperatures to prevent vacuum deterioration. The materials used for joining components, such as brazing fillers must also have low vapor pressure to prevent the emission of molecules.

During manufacturing processes, use of high vapor pressure material such as Sulphur, Silicon, Carbon, etc. should be avoided. Sulfur-free lubricants are recommended to prevent contamination, as even minute amounts of residue can outgas and damage the vacuum. Sustainable mechanical strength and material compatibility is essential for preserving component integrity. Prolonged high-temperature contact between dissimilar materials can cause chemical reactions or form brittle phase-weakening interface in the components, and may affect the structural integrity even at room temperature [3].

The Tube Envelope

The tube envelope serves as a vacuum-tight container and must withstand mechanical and thermal stress due to the differential pressure and high processing temperatures. Tube envelopes are usually cylindrical, combining metallic and non-metallic sections with either glass-to-metal or ceramic-to-metal seals, depending on the insulation needed. While glass-to-metal seals are more common, ceramic-to-metal seals are increasingly used in high-frequency applications such as UHF tubes, including TWTs, Klystrons, Magnetrons and X-ray tubes. Ceramic envelopes are stronger than glass envelopes, allowing for higher processing temperatures. Alumina ceramics are preferred material that offer better resistance to thermal shock and lower dielectric losses at high frequencies compare to glass [3].

External Components

The key considerations for vacuum tube external components, include high-voltage breakdown, corrosion resistance and contact resistance. Materials must be carefully selected based on these factors. For instance, ceramic-insulated bases are preferred over plastics for high-voltage and low-temperature solders in high-altitude or sub-zero environments to prevent standard solders from losing strength. To ensure the reliability of a tube, attention must be paid to each and every component and even a minor issue such as a corroded contact pin can compromise the system. In critical applications, gold-plated pins and socket contacts are used to enhance durability, thereby emphasizing the importance of precise material selection and quality [3].

1.2 Structural Materials for Electron Devices

See Figs. 2 and 3.

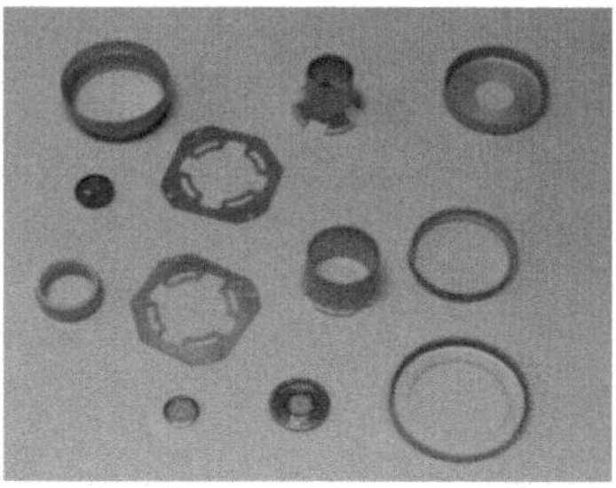

Fig. 2. Electron gun components. Source: [1]

Kovar, an Iron-Nickel-Cobalt controlled expansion alloy (CEA), is a suitable material of choice for ceramic-metal or glass-metal seals due to its matching thermal expansion properties and is utilized in electron gun and structural components [1]. Typical Gun and fixture components are shown in Fig. 1. Kovar is machined at low speeds using tungsten

Fig. 3. SS components & fixtures (300 series). Source: [1]

carbide or high-speed tools with sharp and smooth edges to obtain the desired component quality. Inadequate lubrication can cause overheating and loss of cutting efficiency [4]. Nickel is widely used in electron guns, cathode bases and structural components in vacuum tubes owing to its excellent mechanical strength, relatively low vapor pressure, good electrical and thermal conductivities and ease of outgassing and fabrication [1].

Nickel is extensively used in vacuum technology because of its high melting point, low vapor pressure and resistance to corrosion by substances such as water, saltwater and alkalis. It serves as a base material for components such as coated cathodes, grids, anodes, getters and heat shields and can be used in its pure form, electroplated or as part of alloys. Nickel 200 with 99.5% nickel is known for its high ductility and high tensile strength. Nickel UNS N02201 with very low carbon and sulfur contents, is preferred owing to its softness and ductility. During machining, nickel resembles AISI 1010 steel, requiring well-supported tools to prevent issues such as chatter and backlash [5]. Its Young's modulus compared to that of other metals is illustrated in Fig. 4 showcasing its performance at various temperatures.

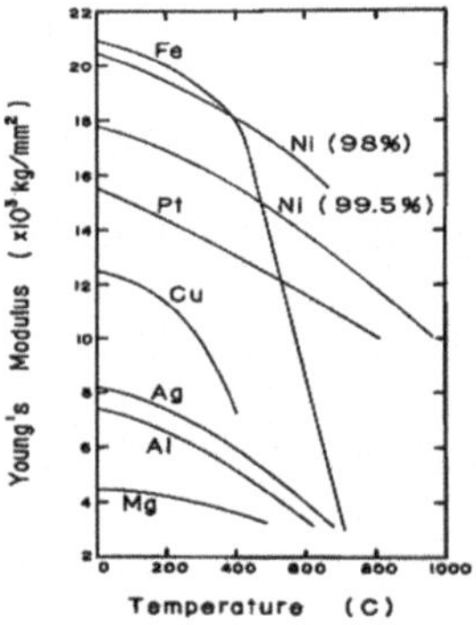

Fig. 4. Young's Modulus of metals v/s temperature. Source: [6]

Figure 5 shows the effect of temperature (ranging from -196°C to 834°C) on the mechanical properties of annealed 99-85% nickel, specifically focusing on yield strength, tensile strength, elongation and reduction in area. Figure 6 presents the approximate relationship between various mechanical properties, such as yield strength, tensile strength,

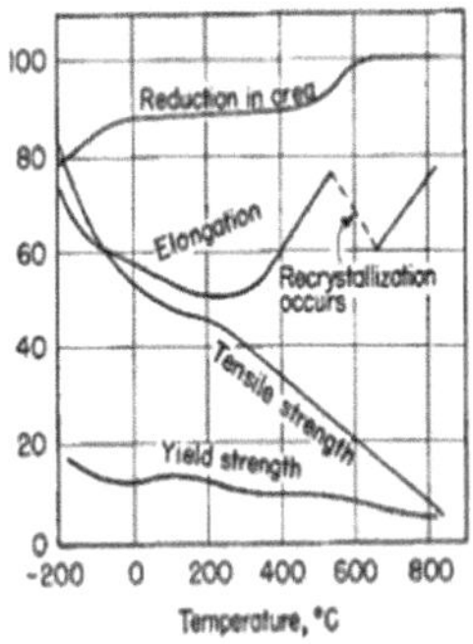

Fig. 5. Mechanical properties variation with temperature. Source: [2]

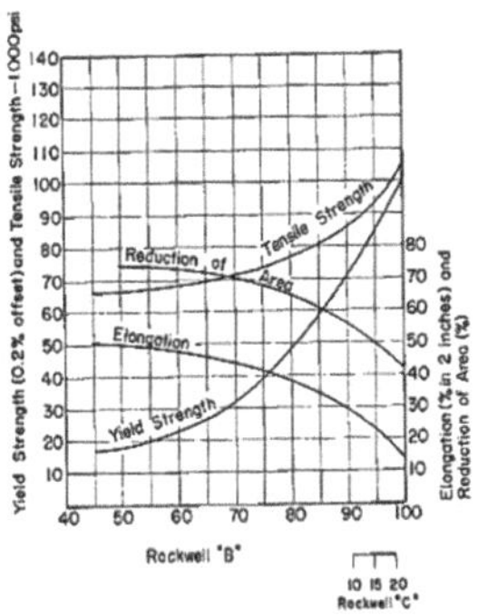

Fig. 6. Relationship of mechanical properties. Source: [2]

elongation and reduction in area and hardness (Rockwell "B" scale) for hot-rolled, forged and cold-drawn nickel.

***Oxygen-Free High Conductivity (OFHC)* copper** is a type of high-purity copper used extensively in high-vacuum technology. With a copper content of at least 99.98%, this material is oxygen-free and is known for its excellent electrical and thermal conductivity, ductility and non-porosity. These qualities make them ideal for various vacuum applications such as gasket seals, pumping stems, tubing and coolers [4]. Although annealed copper is relatively strong and machinable using conventional methods, its softness presents challenges when machining to close tolerances and its tensile strength weakens significantly above 200 °C, restricting its use in high-temperature structural applications. Although copper is generally avoided in applications exceeding 500 °C because of its relatively high vapor pressure compared to nickel, OFHC copper remains widely used for efficient heat and electrical conduction in vacuum systems owing to its cost-effectiveness and superior properties [2].

Stainless Steels (SS), such as austenitic CrNiFe alloys, particularly SS 316 are widely used in ultra-high vacuum (UHV) systems and pumps because of their excellent corrosion resistance and mechanical strength. Among these alloys, SS 304 stands out for its high strength and superior resistance to corrosion, although this may be more challenging to machine [4]. Unlike some alloys that can be hardened through heat treatment, SS 304 can be hardened through cold working. Its remarkable corrosion resistance is primarily due to the formation of a thin, self-healing chromium oxide layer on freshly machined surfaces. When machining SS 304, it is essential to use sharp tools with narrow cutting edges and to apply cutting lubricants to ensure smooth processing. Figure 7 shows the temperature-dependent thermal conductivity of SS 304, highlighting its comparison with other metals. Although SS 304 has good machinability and weldability, it is somewhat less favorable in these areas compared to SS 303. Despite this, SS 304 remains the material of choice for UHV applications owing to its low vapor pressure, nonmagnetic properties and exceptional corrosion resistance. These qualities make it ideal for fabricating critical components, such as UHV chambers, flanges, ion pumps and various fasteners such as nuts and bolts [2].

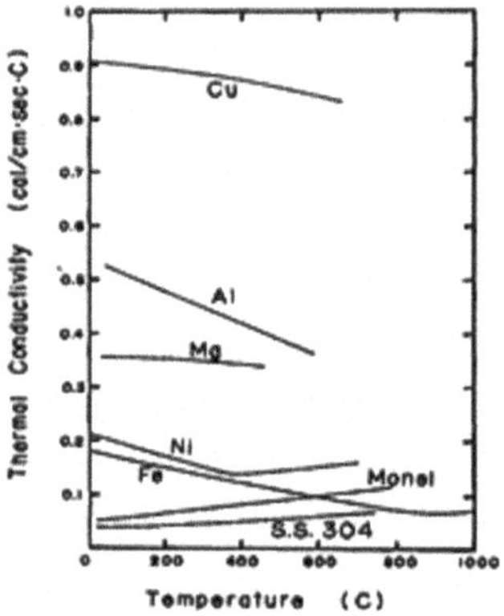

Fig. 7 Temperature dependent thermal conductivity of metals. Source: [6]

Monel alloy 404 is another suitable material with high strength, ductility, weldability, exceptional corrosion resistance and has good machinability. Its low nickel content ensures that it remains non-magnetic at and above room temperature. The standard specification for Monel 404 allows a maximum of 0.12% manganese (Mn). However, it is preferable to use material with Mn contents less than 0.05% to avoid poor braze flow when brazed in wet hydrogen [2].

Figure 8 depicts the variation in tensile and yield strength of annealed Monel 403 that is examined across different temperatures and the results are compared with SS 302 and Copper (Fig. 9).

Titanium (Ti) *is* a strong, lightweight metal that is highly valuable in vacuum applications. It can be machined and formed like steel and its nonmagnetic nature makes it well suited for structural purposes. However, because of its rarity and high cost, similar to zirconium (Zr), titanium is primarily used for gettering, ion-pump and cathode applications. It can be machined with high-speed steel tools and welded using various methods. However, welding must be performed in a protective atmosphere to prevent gas adsorption, which can cause embrittlement as well as warping and deformation in thin films [2].

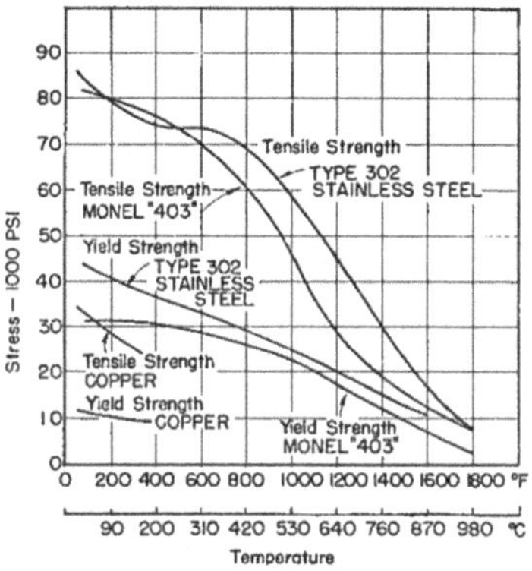

Fig. 8. Comparison of mechanical properties of metals at elevated temperature. Source: [2]

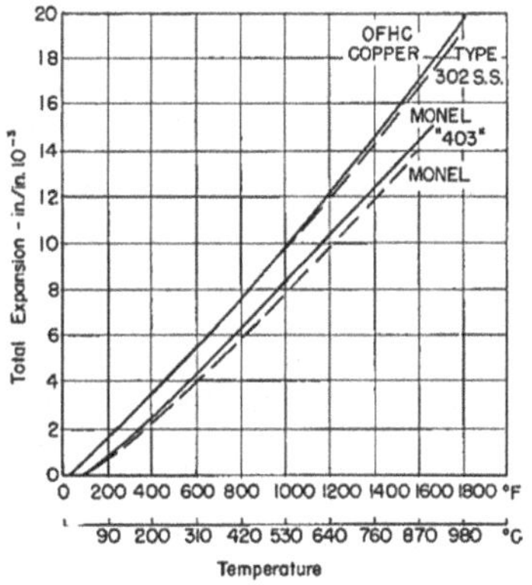

Fig. 9. Thermal expansion of Monel 403, OFHC Copper and SS 302 at elevated temperature. Source: [2]

Ceramics/Dielectric plays a crucial role in microwave tubes, but require consistent dielectric properties and precise geometric tolerances for optimal performance. They are categorized as metallized and lossy ceramics, each serving specific purposes. In applications such as collectors, ceramics require high dielectric strength and thermal conductivity, while electron gun components require high-quality hermetically sealed metallized ceramics, with aluminum oxide being the most commonly used material. Ceramics also function as spacers in tube structures to manage voltage differences and serve as transmission windows for RF energy, sometimes forming a part of the vacuum envelope walls [2].

Aluminum oxide, also known as alumina or sapphire ceramics is a highly durable material widely used in high-temperature and corrosive environments. Composed mainly of Al_2O_3, alumina exhibits excellent electrical resistivity, thermal conductivity and chemical inertness even under harsh conditions such as exposure to hydrofluoric acid [4]. In vacuum technology it is commonly used for insulation, support structures, spacers and ceramic-to-metal seals. Its extreme hardness makes machining difficult and requires grinding or diamond tools for processing. Outside of vacuum technology, alumina is used in producing firing and assembly jigs, as well as electrical components such as insulators, coil forms, condenser shafts and cores for capacitors and resistors due to its hardness, chemical resistance and wear resistance highlighting its versatility in various technical applications [2].

Zirconium (Zr) is a rare, highly reactive metal commonly used as a getter for absorbing gases such as hydrogen, nitrogen and oxygen making it valuable in vacuum applications. One of its main advantages is its excellent corrosion resistance, owing to the thin oxide layer on its surface. Zr efficiently absorbs gases at moderate temperatures, trapping them even at very high temperatures. Its low neutron cross section makes it suitable for neutron windows and it has a low secondary electron emission yield, which is useful for coatings on other materials. Zirconium is machinable like brass and can be spot-welded to molybdenum or tungsten; however, it cannot be brazed with silver. Alloying Zr with small amounts of Mo enhances its mechanical strength [2].

Macor, a glass-mica ceramic which is notable for its machinability can be fabricated with tight tolerances using conventional metalworking tools. It retains the typical hardness and strength of ordinary ceramics without the need for firing after machining. Macor maintains its mechanical and electrical properties at temperatures of up to 1000 °C, making it suitable for a variety of high-temperature applications. The ease with which Macor can be machined to precision tolerance makes it an ideal choice for fabricating accurate fixtures with complex geometries. Combined with its exceptional properties, Macor is expected to be widely adopted as a ceramic material in a variety of vacuum technology applications [2].

Refractory Metals are favored because of their exceptional mechanical strength and high melting points (> 1900 °C), making them ideal for use near cathodes in the tube industry. Tungsten, Molybdenum and Tantalum are commonly employed as structural components of vacuum tubes [2].

Tungsten and Its Alloys: Pure tungsten is widely used in vacuum tube applications, especially as a cathode or direct emitter/helices, owing to its high melting point (2000–3000 °C) and its ability to retain mechanical strength at elevated temperatures. Unlike most metals, tungsten is difficult to work with conventional metal working techniques because it cannot be melted in a furnace and poured into molds. Instead, they are fabricated using powder metallurgy, where tungsten powder is formed into solid ingots by pressing and sintering. It requires no activation, remains largely unaffected by gases and contaminants and has an extremely low vapor pressure, making it ideal for use in ultrahigh vacuum (UHV) systems [2]. The performance of tungsten particularly in wire form depends on its purity, additives used and the mechanical and thermal treatment it undergoes during manufacturing. Tungsten-Rhenium (W-Re) and Tungsten-Molybdenum (W-Mo) alloys are employed in microwave tubes as electron tube heaters to enhance both the tensile strength and ductility. Owing to its hardness and high melting point, tungsten is challenging to machine and form [2].

Molybdenum and Its Alloys: Mo is extensively used in microwave tubes for grid and gun components because of its ability to withstand high stress at elevated temperatures in vacuum environments. However, it is prone to oxidation, starting at 250°C and accelerating rapidly at 600°C, which necessitates storage in an inert atmosphere. To address these limitations, alloys such as Mo-Re, Mo-W, and Mo-Cu have been employed as heat shields and other components offering improved ductility, strength and electrical resistivity. Molybdenum has superior formability and lower cost than tungsten, along with its excellent high-temperature properties compared to more common metals, making it a preferred material for specific high-stress applications [2].

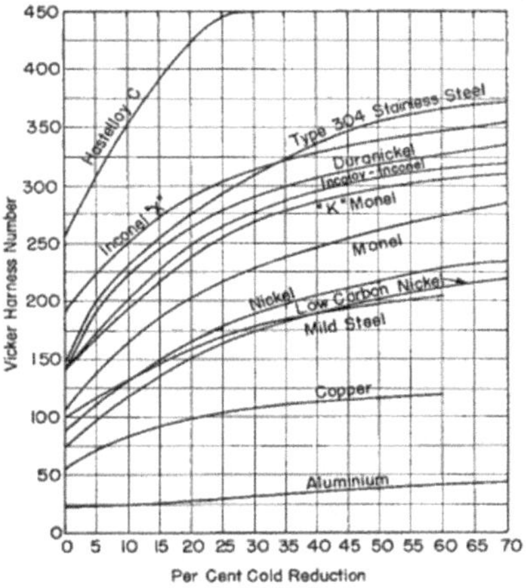

Fig. 10 Effect of cold working on the hardness of materials. Source: [2]

Figure 10 depicts the effect of cold working of various metals and alloys, it can be observed that as the percentage of cold reduction during a cold metal working increases the hardness also increases significantly. During cold working, metals harden and exhibit an increase in tensile strength due to strain hardening, while concurrently experiencing a reduction in ductility.

2 Influence of Work-Piece Material Properties on Chip Formation and Machinability

Copper-based materials are generally easier to machine than other structural materials such as steels or aluminum alloys because their mechanical and physical properties lead to lower cutting forces during the machining processes. However, despite its high conductivity pure copper can be challenging to machine because of the formation of long, tubular or tangled chips. Alloying with elements such as like tellurium, sulfur or lead produces shorter, more manageable chips, although they slightly reduce the conductivity [7].

The key machinability aspects include the following.

Tool Wear: The softness of copper leads to lower tool wear but can cause built-up edge (BUE) formation if not controlled. Alloys such as brass and bronze exhibit varying tool wear based on hardness.

Chip Formation: Pure copper tends to form long, continuous chips, whereas alloys such as brass produce shorter ones. Therefore, effective chip management is crucial.

Cutting Forces: Machining copper involves moderate cutting forces owing to its softness, whereas copper alloys require varying forces based on hardness. Proper control of machining parameters minimizes the strain on tools and machinery.

Surface Quality: Copper can achieve high-quality surface finishes, but care is needed to prevent scratches. Copper alloys can also achieve good finishing, depending on their hardness.

Optimizing machining parameters, tool selection and using appropriate lubricants and coolants are essential for balancing the tool wear, chip formation, cutting forces and surface quality. In rough machining, tool wear and cutting forces are key concerns, whereas surface quality is the focus in finishing operation. In automatic lathe operations managing chip formation is crucial for maintaining productivity and safety [7].

2.1 Impact of Work-Piece Material's Mechanical Properties on Chip Formation

When a cutting tool engages with a work material, the material initially deforms elastically, before deforming plastically. Once the shear stress exceeds the material's shear strength the material flows and forms a chip. The shape and flow of the chip are influenced by the design of the tool. Effective chip formation and removal are critical in machining operations such as drilling, reaming, milling and turning especially on automatic lathes, where precise control is needed owing to the limited cutting area [7].

2.2 Cutting Fluids in Machining Copper-Based Materials

Cutting fluids are essential in machining processes and serve as either coolants or lubricants based on the specific tools and operations involved. For high-speed steel (HSS) tools, they help to maintain optimal temperatures and prevent overheating, as HSS loses hardness at temperatures between 550 and 600 °C. For tools that are less affected by temperature like carbide tools cutting fluids mainly serve as lubricants in operations where parts of the tool do not cut material (such as with reamers and taps). The selection of the cutting fluid is influenced by its ability to remove chips effectively, dissipate heat, lubricate machine components and facilitate chip breakage [7].

2.3 Mechanical Micro-Machining

The use of end mills for component fabrication is a conventional milling method, where a rotating cutting tool advances against a stationary workpiece. Micro-machining has gained traction due to advancements in machinery, including high-precision linear tables, accurate axis positioning, high-speed spindles and innovations in micro-end mills. Ultra-precision machining requires a tolerance of $\pm 1\ \mu m$, achievable with linear drive systems that reduce friction and backlash errors. While conventional CNC milling spindle speeds range from 2000 to 6000 rpm, higher-speed options are available. As end mill sizes decrease to micro-end mills (see Fig. 11), both the material removal rate and linear cutting speed are significantly reduced. To ensure good surface integrity, spindle speeds often reach 40,000 to 100,000 rpm, depending on the application.

Advancements in cutting tool materials that maintain hot hardness at elevated temperatures are crucial for precision machining. Carbide and ceramic tools exhibit good hot

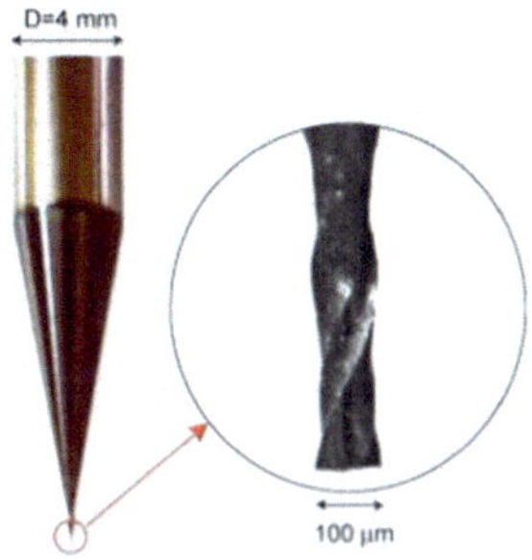

Fig. 11. Micro-end mill with two flutes

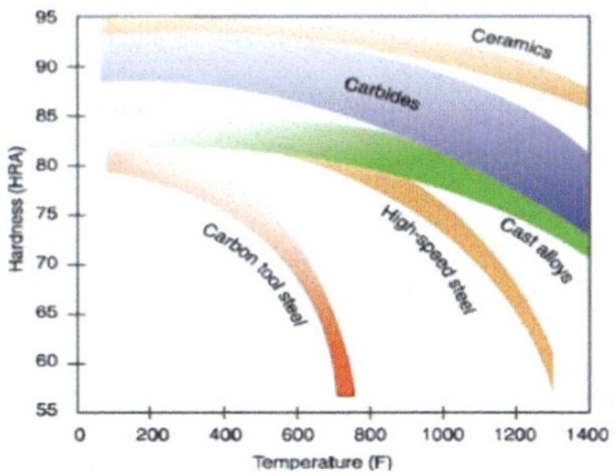

Fig. 12. Hardness of tool material as function of temperature

hardness (see Fig. 12). Diamond tools, the hardest known material, is particularly suitable for machining non-ferrous materials. These developments in machinery, tooling and processes ensure that the accuracy of machining small structures is not compromised, establishing the process as an ultra-precision material removal technique comparable to the accuracy found in precision engineering. While achieving complex 3D geometries and intricate micro-features with micro-machining can be challenging, it remains essential for the development of micro-components used in millimeter-wave structures.

3 Ultra-Precision Machining Techniques for Copper

Ultra-precision machining, particularly diamond turning, is a specialized technique used to achieve extremely high-quality surfaces, particularly for materials such as copper. This process allows for the production of optical components with mirror like surfaces and exceptional dimensional accuracy. The key factor in this technique is the use of monocrystalline diamond cutting tools, which have very sharp edges with nose radii of approximately 50 nm, thereby ensuring minimal wear and precise cutting (Fig. 13).

The cutting speed is an important factor in determining surface quality. For example, when turning, the spindle speed is usually constant which means the cutting speed varies along the work-piece at the center, the speed is theoretically zero. Despite the low cutting forces (typically less than 1 N) involved, the machining process requires extremely rigid machines to avoid vibrations and maintain the desired precision often within 0.1 μm.

Fig. 13. Ultra-precision machining of brass component. Source: [7]

This allows the surface roughness (Ra) achieved to be only a few nanometers making the process ideal for applications requiring high optical quality [8].

3.1 Diamond as a Cutting Material

Diamond, which is composed of pure carbon, is widely recognized as the hardest known material in the world. However, its extreme hardness also results in brittleness, rendering diamonds sensitive to both impact and thermal stress. These intrinsic properties critically influence their applications in cutting operations. In machining processes, both natural and synthetic diamonds are employed, with monocrystalline diamonds (DIN ISO 513 code: DM) and polycrystalline diamonds (DIN ISO 513 code: DP) being the most prevalent. The term PCD is frequently used to denote tools manufactured from polycrystalline diamond. Monocrystalline diamonds are particularly effective for precision machining, especially in ultra-precision applications. Conversely, PCD tools are versatile, accommodating both precision and roughing operations. Importantly, in certain applications, the rough and finish machining stages can be consolidated into a single operation, thereby enhancing efficiency and streamlining manufacturing processes. This adaptability underscores the significant role of diamonds in the advancement of machining technologies [7] (Table 1).

Table 1. Comparison of different machining processes for the manufacture of micro-devices.

Machining method	Minimum feature (μm)	Tolerance (μm)
Mechanical micromachining	10	1
Focused ion beam	100	10
Femto-second laser	1	0.5
Micro-EDM	10	1
Diamond milling and micromachining	10	1

3.2 Material Properties and Their Influence on Ultra-Precision Machining

The outcome of an ultra-precision machining process is influenced not only by the cutting tool and machine characteristics but also by the properties of the work material. Copper and its alloys often used in engineering typically have a polycrystalline microstructure. Studies on polycrystalline copper have shown that the back effect (compression of individual grains during machining) can cause level differences of up to 40 nm between neighboring crystals. Despite these challenges, diamond cutting of OFHC copper can still produce surfaces with roughness parameters (Ra) as low as 3 nm [9].

4 Advanced Micro-Fabrication Technologies

Micro-fabrication techniques are essential for the production of micro-scale devices and structures with precise dimensions and high surface quality. Table 2 summarizes the accuracy and surface roughness of various advanced micro-fabrication methods.

Table 2. Accuracy and Surface Roughness—Micro-fabrication technologies. Source: [10]

Technique	Accuracy (μm)	Surface roughness (nm)
UV LIGA	3–5	> 30
Deep X-ray LIGA	3–5	> 30
DRIE	1 μm	> 30
Nano CNC	0.5 μm	> 10

Acknowledgement. Authors are very much grateful to Centre Head, MTRDC for providing an opportunity to present our work in the conference also acknowledge to DRDO, New Delhi for funding sources.

References

1. Devi, P.D.: Overview of Materials and Processes in Microwave Tubes. Microwave Tube Research & Development Centre, DRDO, Bengaluru.
2. Kohl, W.H.: Handbook of Materials and Techniques for Vacuum Devices, 345–423. American institute of physics, New York (1995)
3. Kohl, W.H.: Materials and Techniques for Electron Tubes, 83–332. Reinhold, New York (1960)
4. Bharat Electronics Limited (BEL): Skills Upgrading in Microwave Assembly (SUMA), 10–30. MTD Training
5. Thornburg, D.L., Thall, E.S., Brous, J.: Manual of Materials for Microwave Tubes, 23–199. Electronic Technology Laboratory, Radio Corporation of America, Virginia (1961)
6. Weissler, G.L., Carlson, R.W.: Vacuum Physics and Technology, 181–321. Methods of Experimental Physics, Vol. 14, California
7. Klocke, F.: Recommended Machining Parameters for Copper and Copper Alloys, 5–55. German Copper Institute, Fraunhofer Institute for Production Technology, Germany (2010)
8. Ausschuss für wirtschaftliche Fertigung: "Arbeitsergebnisse – AWF-Betriebsblatt 158" [Committee for economic fabrication: „Work results – AWF working paper 158"] Beuth-Verlag, Berlin

9. Sadowy, M.: Untersuchungen zur Theorie der Fließspanbildung. III Schnittkräfte, Spanformen und Spanstauchung (Fortsetzung) [Research into the theory of continuous chip formation III Cutting forces, chip forms and compressions (Continuance) Maschinenmarkt **73**(10), S. 167–174 (1967)
10. Dionosio, R., Paoloni, C.: The basics of travelling wave tube amplifiers, European Space Agency, Lanchester University, The 46th European Microwave Conference (2016)
11. Espe, W.: Materials of High Vacuum Technology, Vol. 1. Pergamon, Oxford, Metals and Metalloids (1966)
12. Gilmour, A.S.: Microwave Tubes. Artech House Inc, Norwood (1986)
13. Sgobba, S.: "Materials for High Vacuum Technology": An Overview. CERN.
14. Li, Y., Liu, X.: Vacuum Science and Technology for Accelerator Vacuum Systems. Cornell University, Ithaca, NY
15. Krist, T.: Metallindustrie – Zerspantechnik – Verfahren, Werkzeuge, Einstelldaten Metalindustry - Machining technique – Process, tools, setting data] Hoppenstedt Technik Tabellen Verlag Darmstadt: 1993, 22. Auflage
16. Spenrath, N.M.: Technologische Aspekte zum Feinstdrehen von Kupferspiegeln (Technological aspects of micro-turning of copper reflectors). Dissertation RWTH Aachen (1991)

Development and Modeling of the Plasma-Assisted Sub-THz Backward-Wave Oscillator

Nikita M. Ryskin[1,2(✉)], Vladimir N. Titov[1,2], Roman A. Torgashov[1,2], Prerna Unadkat[3], Vishant[3], Anand Abhishek[3], Sahil Jain[3], and Niraj Kumar[3]

[1] Saratov Branch, Kotelnikov Institute of Radio Engineering and Electronics, Saratov, Russia
ryskinnm@gmail.com

[2] Saratov State University, Saratov, Russia

[3] CSIR-Central Electronics Engineering Research Institute, Pilani, India

Abstract. Development of high-power and portable sources of coherent electromagnetic radiation operating at sub-THz (0.1–0.3 THz) frequency band is important for many scientific and industrial applications. Among such sources, devices with pseudospark (PS) electron beam have attracted interest since they are capable of producing high power and do not require an external magnetic field to confine the electron beam. In this article, we present the results of development of the W-band backward-wave oscillator (BWO) driven by a self-focused PS sheet electron beam. The high-frequency interaction structure based on the staggered dual grating slow-wave structure (SDG SWS) is designed and simulated using CST Studio. The SWS provides wide bandwidth and high Pierce interaction impedance at the backward spatial harmonic. Then, interaction of the plasma focused high-current-density electron beam with electromagnetic wave in the SWS is simulated using the time-domain 3-D particle-in-cell CST Particle Studio simulator. The PS-based hollow-cathode electron source providing a 6.3-A, 30-kV sheet electron beam is developed and experimentally tested. Fabrication issues of the SWS by using computer-numerical-control (CNC) nano-machining are discussed. The interaction structure is fabricated and assembled.

Keywords: Pseudospark Discharge · Ion Focusing · Sheet Electron Beam · Backward-Wave Oscillator · W band

1 Introduction

Many applications require high-power, portable sources of sub-THz radiation. At frequencies above 0.1 THz, using vacuum-tube electron devices (VEDs) is the only possible solution to attain kilowatt power level. High-aspect-ratio sheet electron beams have been proposed since they can combine high current with reasonable current density. However, focusing of such beams is difficult and require high magnetic field ~ 1 T. Devices with ion-focused electron beams propagating in plasma allow operation without heavy and bulky magnetic focusing system. Therefore, hollow-cathode electron sources based on the pseudospark (PS) discharge have attracted a lot of interest [1]. Such sources are capable of producing high-current, high-energy electron beams.

S. K. Ghosh et al. (Eds.): VEDA 2024, SPPHY 424, pp. 271–277, 2025.
https://doi.org/10.1007/978-981-96-5174-0_39

In [2], we developed and experimentally tested a PS-based hollow-cathode source with sheet electron beam for high-power millimeter-band generation. This paper is aimed at the design and development of the W-band backward-wave oscillator (BWO) driven by a self-focused PS sheet electron beam.

The rest of the article is organized as follows. In Sect. 2, we present the results of design, development, and fabrication of the high-frequency interaction structure. In Sect. 3, hot-test operation of the BWO is simulated by 3-D particle-in-cell (PIC) code. Sect. 4 reports progress in the development of the PS-based hollow-cathode electron beam source.

2 SWS Design, Simulation, and Fabrication

The high-frequency beam-wave interaction structure is based on a staggered double-grating slow wave structure (SDG SWS). Such SWS has wide bandwidth, is easy to fabricate by modern microfabrication technologies such as computer-numerical-control (CNC) nano-machining, and is capable to adopt a high-aspect-ratio electron beam [3]. Therefore, it is often used in sub-THz VEDs.

The W-band SDG SWS was designed and simulated using CST Studio [4]. In Fig. 1, dimensional schematic of the SWS unit cell is presented.

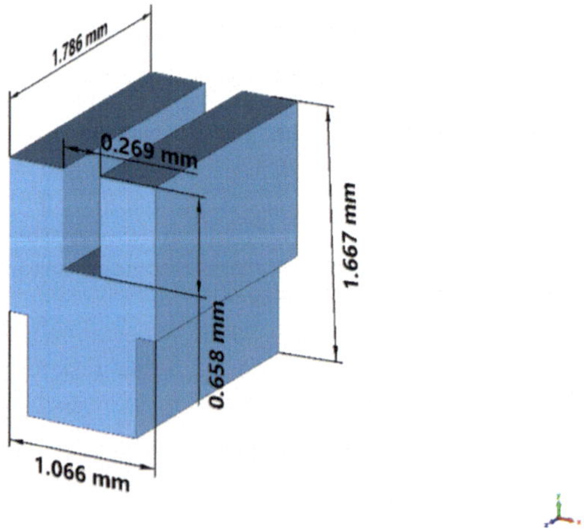

Fig. 1. Schematic of unit cell of the SDG SWS

Dispersion diagram of the designed SWS is presented in Fig. 2. The SWS has wide bandwidth of 40 GHz. As is typical for SDG SWSs [3, 5], it supports propagation of two modes, namely, symmetric and antisymmetric one, shown in Fig. 2 with red and blue color, respectively. At $0 < \varphi < 2\pi$ (φ is phase shift per SWS unit cell), the symmetric mode has anomalous dispersion, i.e., it is suitable for backward-wave operation. In addition, the symmetric mode has much higher Pierce interaction impedance than the antisymmetric one. Within the 90–100 GHz frequency band, interaction impedance of the symmetric mode is 10–20 Ohm, while for the antisymmetric mode it is 3–4 Ohm. Therefore, the electron beam interacts mainly with the backward harmonic, while its interaction with forward harmonic of the antisymmetric mode is negligible.

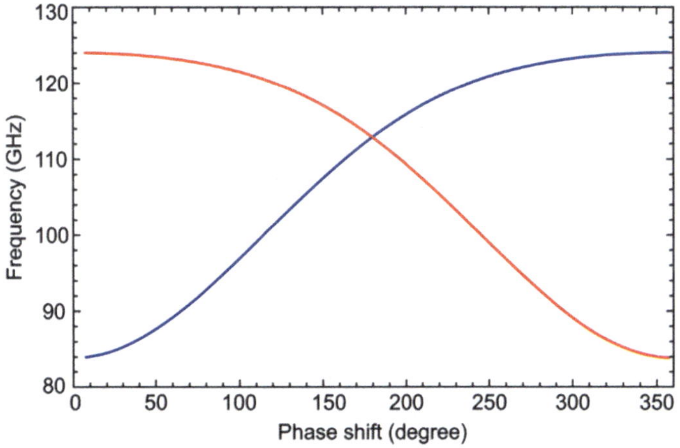

Fig. 2. Dispersion diagram of the SDG SWS

In [6], we reported successful fabrication and cold-test measurements of the SDG-SWS interaction structure integrated with beam adapter and spent-beam collector. The beam adapter serves for the shaping and collimation of a sheet electron beam generated from the PS-based hollow cathode. The structure was fabricated by CNC micromachining. The machine has a high-speed spindle installed with an RPM range of 6000–60000 RPM and a coolant system to lower the spindle's temperature during high-speed milling. The structure fabricated with the optimized machining parameters of the CNC micro-milling process was found to be uniform and has lesser variations in terms of dimension along the length and width of the interaction structure. The surface roughness is ~ 150 nm, which is significantly less than the skin depth at that frequency, and the overall dimensional variation is ~ 10 microns.

3 Simulation of the BWO Hot-Test Operation

Time-domain 3-D particle-in-cell (PIC) simulation of interaction of the high-current-density electron beam with high-frequency electromagnetic wave in the interaction structure was performed using CST Studio [4]. The 6.3-A, 30-kV electron beam was assumed in accordance with the experimental measurements (see Sect. 4). At this voltage, synchronism with the backward slow wave at nearly 85 GHz is attained. The interaction structure consists of 20 identical SDG cells with beam adapter at one end and a tapered transition to a rectangular waveguide at another end. The transition consists of 5 SDG cells with linearly decreasing vane height.

Although 3-D PIC modeling has been widely used for simulation and optimization of microwave devices with PS-based electron source, most of works actually considered an electron beam focused by a strong external magnetic field and propagating through the structure filled with the medium with dielectric permittivity $\varepsilon = 1 - \omega_p^2 / \omega^2$, where ω_p is the electron plasma frequency, see e.g. [7–9]. This simplified approach ignores processes of plasma ionization and ion focusing of the beam.

In this work, we use more physically based approach, which takes into account self-consisting process of plasma ionization by the electron beam. It uses a Monte Carlo Collision (MCC) model of ionization [10] implemented in CST Studio Suite code. The interaction structure is initially filled by a neutral monoatomic gas (Argon) background with a preset density, pressure and temperature. Ion-electron pairs are generated through the collision of an electron with a neutral background gas particle in accordance with the ionization cross section as a function of the electron kinetic energy.

In the simulations, we observed gradual formation of a plasma channel, which focused the electron beam. However, the results of simulation strongly depend on the gas pressure, as is shown in Fig. 3.

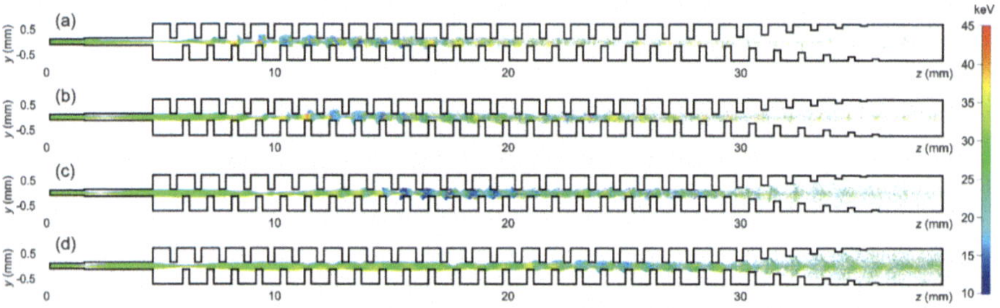

Fig. 3. Beam trajectories at different values of gas pressure: (a) 9 Pa at $t = 12$ ns, (b) 18 Pa at $t = 12$ ns; (c) 45 Pa at $t = 5.5$ ns, (d) 45 Pa at $t = 8$ ns.

With the increase of gas pressure, beam transport improves as the beam propagates further down the interaction region. At the same time, the increase of gas pressure shifts down the region of effective beam-wave interaction, which is indicated by movement of sufficiently decelerated electrons (shown in blue) along the SWS. Thus, even if the gas pressure is high enough for beam transport, there exists an optimal length of the interaction region, above which no further increase of interaction power is possible. Figure 3(c) and (d) demonstrate the nonstationary character of beam transmission and beam-wave interaction.

In Fig. 4, output signal waveforms calculated at different values of the gas pressure are presented. At $P = 9$ Pa, oscillation mode with nearly constant amplitude establishes after a transient process with duration of ~ 10 ns, as shown in Fig. 4(a). This value of amplitude corresponds to output power of approximately 10 kW. At $P = 18$ Pa, the output power increases to nearly 30 kW and the duration of transient process decreases (Fig. 4(b)).

When the gas pressure increases to $P = 45$ Pa, the oscillation mode becomes completely nonstationary (Fig. 4(c)) and no establishing of steady-state oscillation is observed. At $t \approx 5.5$ ns, the output power becomes as high as ~ 75 kW. At that moment

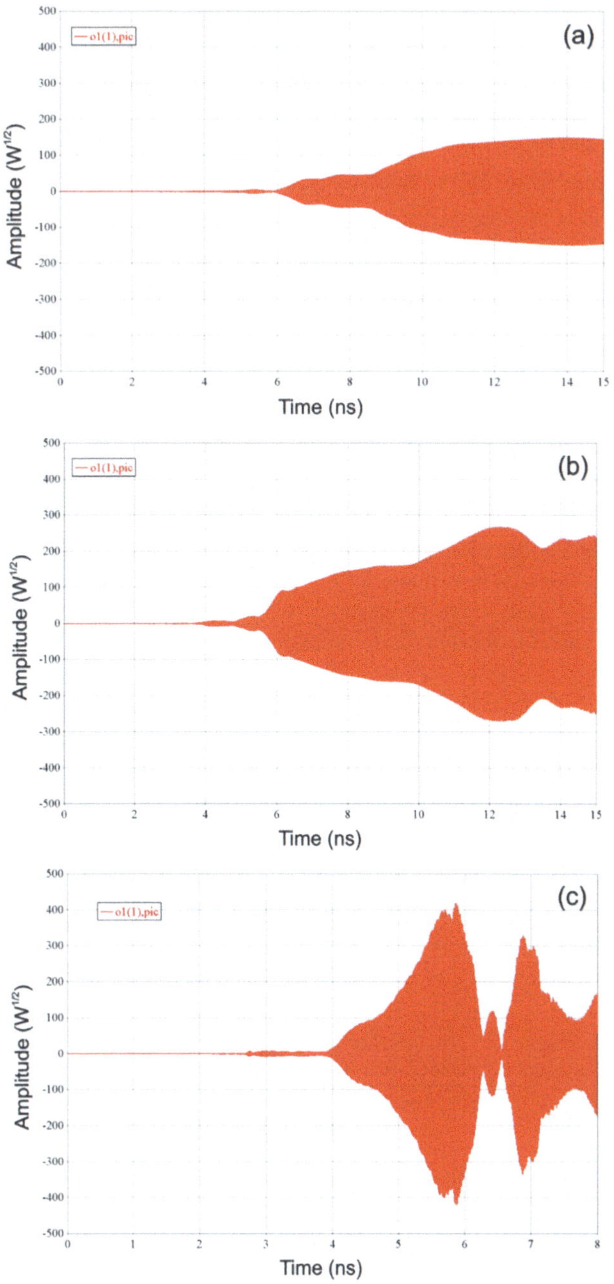

Fig. 4. Output signal waveforms as gas pressure $P = 9$ Pa (a), 18 Pa (b), and 45 Pa (c).

of time, there appear many strongly decelerated electrons, as is shown in Fig. 3(c). However, after that, the oscillation amplitude and power drop abruptly. Although at $t = 8.0$ ns the electron beam penetrates deeper into the interaction space (Fig. 3(d)), the power is much smaller than at $t = 5.5$ ns. This demonstrates the complex nature of the ion-focused electron beam interaction with the high-frequency field. Further investigation of self-consisted beam-plasma dynamics including the process of ion channel formation, beam focusing and transport is required to reveal the underlying physical phenomena and to optimize the beam-wave interaction.

4 PS-Based Electron Beam Source Development

PS-based electron beam sources are capable of generating high current density and high energetic electron beam [11–14]. A three-gap structure of PS-based electron beam source has been developed for application in the sub-THz BWO. It comprises mainly of a hollow cathode, two floating anodes, three insulating discs, and one grounded anode disc. The grounded anode disc has an aperture of 1.5 mm × 0.25 mm, which is suitable for the beam adapter of the beam-wave interaction structure. The entire structure is a flange-based design and it is filled with Argon gas. The electron beam source is tested for an applied gap voltage of 10–30 kV in self-breakdown mode and experiments at each voltage level are repeated five times. A typical V-I curve of the electron beam source is presented in Fig. 5. Due to the presence of two floating anodes in the gap region, the applied voltage collapses in successive breakdown mode.

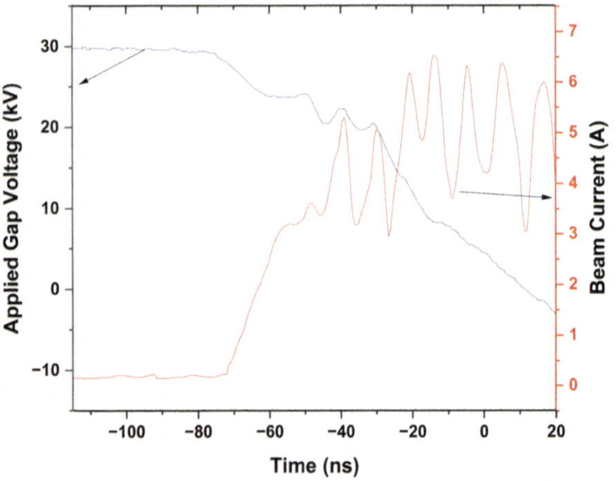

Fig. 5. Typical experimental voltage and current waveform of the electron beam at an applied gap voltage of 30 kV and gas pressure of 17 Pa.

Acknowledgment. This work was supported by the Russian Science Foundation (grant # 22-49-02017) and DST India (grant # DST/IC/RSF/2022/280).

References

1. Sarkar, M., Kumar, N.: Pseudospark discharge-based electron beam sources: a review. IEEE Trans. Plasma Sci. **52** (2024)
2. Kumar, N., Abhishek, A., Vishant, Singhal, K., Gurjar, N., Jain, S., Starodubov A.V., Ryskin, N.M.: Pseudospark-driven high-current miniaturized voltage-tunable sheet-electron-beam source. IEEE Trans. Electron Dev. **68**(12), 6482–6486 (2021)
3. Shin, Y.-M., Barnett, L.R., Luhmann, N.C.: Phase-shifted traveling-wave-tube circuit for ultrawideband high-power submillimeter-wave generation. IEEE Trans. Electron Dev. **56**(5), 706–712 (2009)
4. CST Studio Suite Electromagnetic Field Simulation Software, Dassault Systemes Deutschland GmbH, https://www.3ds.com/products-services/simulia/products/cst-studio-suite (last accessed on September 15, 2024)
5. Rozhnev, A.G., et al.: Studying characteristics of the slow-wave system of the traveling-wave tube with a sheet electron beam. Radiophys. Quantum Electron. **56**(8–9), 542–553 (2014)
6. Kumar, N., Vishant, Ranjan, O., Abhishek, A., Starodubov, A.V., Titov, V.N., Ryskin, N.M.: A technique for planar integration of components of W-band backward-wave oscillator to eliminate misalignment. IEEE Electron Device Lett. **44**(2), 313–316 (2023)
7. Yin, Y., He, W., Zhang, L., Yin, H., Robertson, C.W., Cross, A.W.: Simulation and experiments of a W-band extended interaction oscillator based on a pseudospark-sourced electron beam. IEEE Trans. Electron Dev. **63**(1), 512–516 (2016)
8. Xie, J., Zhang, L., Yin, H., He, W., Ronald, K., Phelps, A.D.R., Chen, X., Zhang, J., Alfadhl, Y., Yuan, X., Meng, L., Cross, A.W.: Study of a 0.35 THz extended interaction oscillator driven by a pseudospark-sourced sheet electron beam. IEEE Trans. Electron Dev. **67**(2), 652–658 (2020)
9. Torgashov, R.A., et al.: Investigations on high-current-density sheet-beam driven V-band slow wave oscillator with a hollow cathode. IEEE Trans. Plasma Sci. **49**(1), 263–268 (2021)
10. Birdsall, C.K.: Particle-in-cell charged-particle simulations, plus Monte Carlo collisions with neutral atoms. PIC-MCC. IEEE Trans. Plasma Sci. **19**(2), 65–85 (1991)
11. Unadkat, P., Jain, S., Kumar, N.: A novel multi-aperture r-θ configuration-based pseudospark-driven electron beam source. IEEE Electron Device Lett. **45**(8), 1520–1523 (2024)
12. Unadkat, P., Kumar, N.: Study of space charge potential for propagation of plasma-assisted electron beam. IEEE Trans. Plasma Sci. **52**(5), 1669–1675 (2024)
13. Abhishek, A., Kumar, N., Vishant, Prabhu, S., Mishra, A., Dixit, N., Khare, S.: Analysis of dynamic impedance for different phases of pseudospark-driven electron beam. IEEE Trans. Electron Dev. **70**(11), 5916–5920 (2023)
14. Abhishek, A., Kumar, N., Vishant, Meena, B.L., Jain, S., Prabhu, S., Mishra, A., Dixit, N., Khare, S.: Electrical characterization of plasma-assisted electron beam source with sub-mm sheet aperture. IEEE Trans. Electron Dev. **70**(2), 771–775 (2023)

Simulation of Dielectric Loaded Interaction Structure for Ka-Band Gyro-TWT Amplifier

Anju Kumari, Kamlesh, M. K. Alaria(✉), and S. K. Ghosh

CSIR- Central Electronics Engineering Research Institute (CEERI), Pilani, India
mka@ceeri.res.in

Abstract. Design and simulation of the loaded waveguide interaction part for millimeter wave Gyro-TWT amplifier has been reported. The cold parameters of the waveguide interaction structure like dispersion and interaction impedance have been computed by commercially available software CST Microwave studio. The bandwidth enhancement technique for a Gyro-TWT at high power and gain by loading the highly lossy rings in the waveguide has been presented in this paper. The design approach of dielectric loaded waveguide interaction structure has been described. A new type and advance dielectric lossy material (AlN-SiC) rings has been used to enhance the bandwidth characteristics of the Gyro-TWT amplifier. PIC simulation predicts that the peak power 80 kW at Ka-band in the dielectric loaded Gyro-TWT for 65 kV beam voltage and 10A beam current.

Keywords: Gyro traveling wave tube (Gyro-TWT) · Dispersion · lossy waveguide · Aluminium nitride Silicon carbide (AlN-SiC)

1 Introduction

Gyro-TWT amplifier is very important microwave sources for the generation of RF power radiation in the microwave to millimeter wave region of EM spectrum. High resolution radars and long distance communication systems are the major user of Gyro-TWT amplifiers [1]. In such amplifier, an interaction structure which is cylindrical in shape and has smooth wall, is used for the propagation of electromagnetic wave. The wave propagates with a speed greater than the velocity of light hence they are known as fast wave devices [2]. This results in coalescence of beam mode line and waveguide mode dispersion hyperbola over a narrow frequency range. In order to achieve wider bandwidths, special types of lossy structure are employed [3]. Gyro-TWT with a lossy interaction waveguide has a hyperbolic dispersion curve. Gyro-TWT must be operated far from the cut-off. The cylindrical waveguide is loaded with a lossy ceramic material to suppress backward-wave oscillations [4]. The design of lossy structure for Gyro-TWT requires the working of the proper RF field known as the operating mode, RF field profile, proper excitation of RF field estimated through start oscillation current, Eigen frequency, quality factor, competing modes, efficient RF power growth, etc. [5, 6]. The waveguide design is carried out by eign mode analysis of cold waveguide and beam wave interaction analysis for various interaction structure design compatible with the

S. K. Ghosh et al. (Eds.): VEDA 2024, SPPHY 424, pp. 278–282, 2025.
https://doi.org/10.1007/978-981-96-5174-0_40

design output parameters, bandwidth, output power, gain and efficiency etc. is achieved [7].

2 Design of Lossy Interaction Structure

The RF performance of Gyro-TWT depends mostly on its interaction structure. Therefore, dielectric loaded interaction region of Gyro-TWT is very important [8]. The design approach of lossy ceramic loaded interaction structure for Gyro-TWT has been discussed. The proposed interaction structure slows down the axial component of the phase velocity of the RF to facilitate its interaction with electron beam, which is in contrast with conventional tubes where the beam is cylindrical [9]. In Gyro-TWT a hollow metallic wave-guide with proper shape and size is employed. To achieve wide bandwidth in the Gyro-TWT is a major challenge. The design of unique interaction structures loaded by lossy rings [10]. The accurate design and analysis of the interaction structure would lead to the perfection in the characteristics of Gyro-TWT in terms of output power, gain, bandwidth and efficiency. Based on the analysis, the lossy ring radius, thickness and materials have been finalized [11]. A model of RF interaction structure for electromagnetic simulation is designed in CST Microwave studio. The design of loaded lossy RF interaction structure has been carried out. The characteristics, like dispersion diagram, the azimuthal interaction impedance and the attenuation constant of the interaction structure have been analyzed [12]. The effects of parameter variations on the different performance of the lossy ring-interaction structure have been performed by CST microwave studio [13]. The simulated model of lossy interaction structure for Gyro-TWT is shown in the Fig. 1. The techniques of periodic lossy loaded interaction waveguide performed the excellent results and stability. The dielectric material AlN-SiC is a good for lossy materials. The dispersion of AlN-SiC loaded rings interaction structure has been obtained. The operating mode TE_{01} is considered in the design, the electric field is maximum in the center of rings and waveguide [14]. Figure 2 shows the plot between propagation constant and frequency for operating modes. Figure 3 shows the interaction impedance curve of lossy ring loaded waveguide structure. The amplitude of peak value of the axial electric field has been used for interaction impedance and effect on the gain of the Gyro-TWT.

Initially, each electron with the energy 65 keV enters into the waveguide structure with input, the waveguide cavity in form of TE_{01} mode and transfers an amount of the energy to the electromagnetic wave in the structure. Figure 4 shows the TE_{01} mode in the lossy interaction structure. The analysis shows that the interaction structure get produce 90 kW output power, 48 dB gain for 65 kV and 10A electron beam. Figure 5 shows the output power performance of lossy loaded interaction structure achieved using CST.

3 Conclusion

The design and analysis of lossy waveguide structure for millimetre wave Gyro-TWT amplifier has been presented. Particle in Cell simulation of the Gyro-TWT generates maximum RF power of 85 kW for TE_{01} mode at 30 GHz with an efficiency of 25%. Dispersion and interaction impedance analysis of lossy ceramic ring waveguide structure of millimetre wave Gyro-TWT have also been obtained.

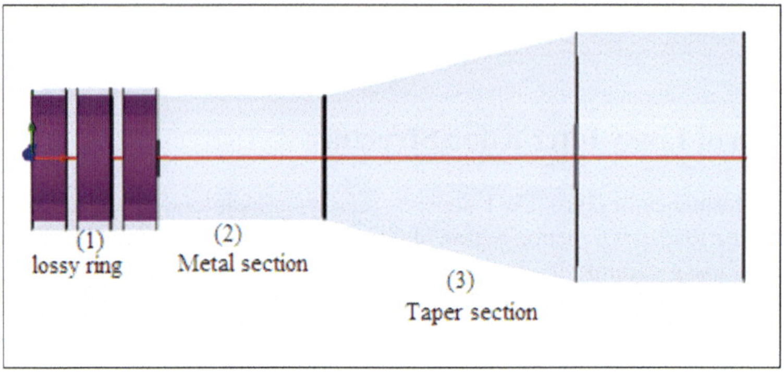

Fig. 1. Simulated model of loaded RF interaction structure for Gyro-TWT

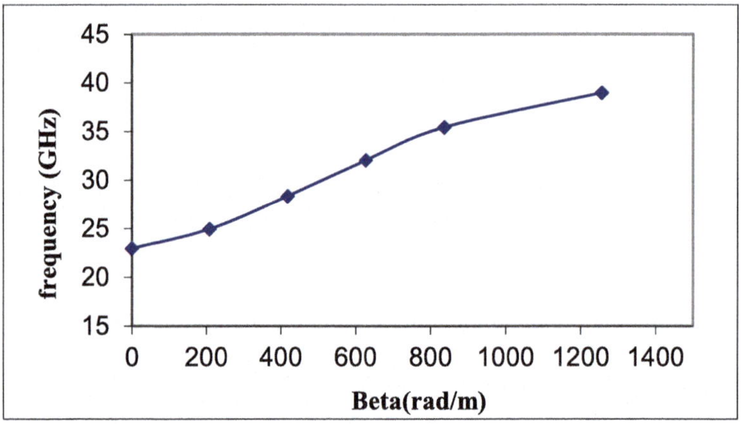

Fig. 2. Dispersion characteristics of AlN-SiC ceramic rings loaded structure

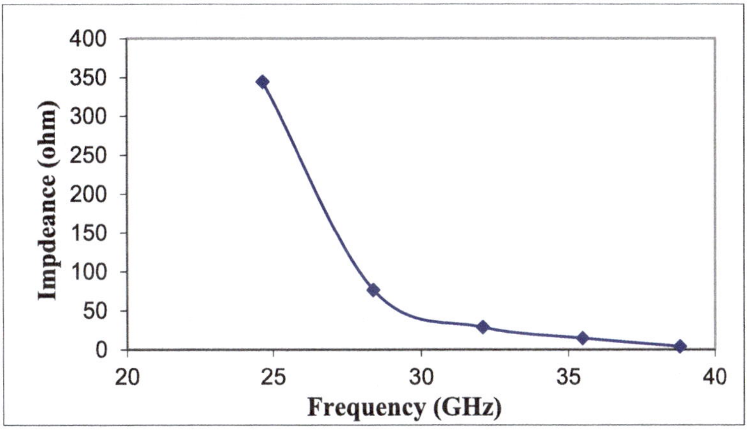

Fig. 3. Interaction impedance characteristics of loaded ceramic ring structure

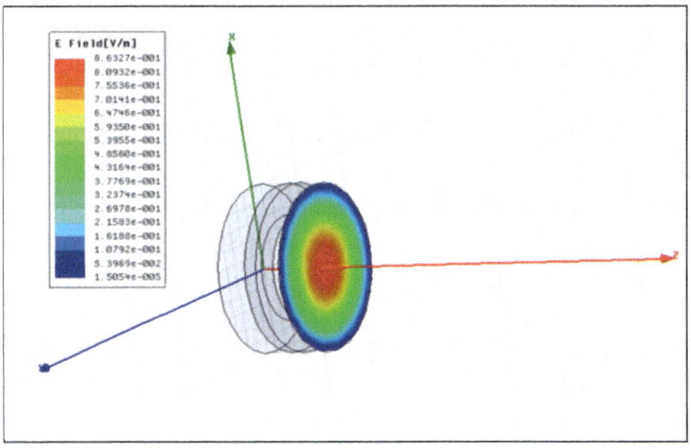

Fig. 4. Electric field pattern (TE_{01} mode) in the RF interaction structure

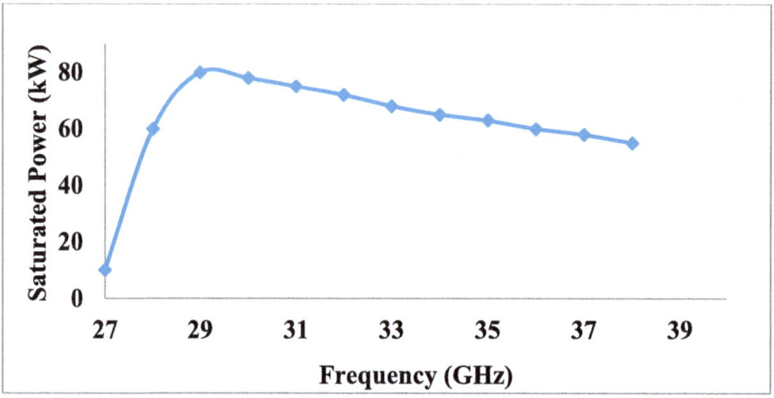

Fig. 5. Simulated output power response of lossy loaded interaction structure

References

1. Edgecombe, C.J.: Gyrotron Oscillators: Their Principles and Practice. Taylor & Francis Ltd. (1993)
2. Gilmour, Jr., A.S.: Klystrons, Traveling Wave Tubes, Magnetrons, Crossed-Field Amplifiers, and Gyrotrons. Artech House, Boston, MA, USA (2011)
3. Du, C.H., Liu, P.K.: A lossy dielectric-ring loaded waveguide with suppressed periodicity for Gyro-TWTs applications. IEEE Trans. Electron Dev. **56**, 2335–2342 (2009)
4. Alaria, M.K., Choyal, Y., Sinha, A.K.: Analysis of interaction structure for ka-band Gyro-TWT. In: IEEE International Conference on Plasma Science, Edinburgh, UK, pp. 11–15, 8–12 July (2012)
5. Singh, G., Chandra, S.M.S.R., Bhaskar, P.V., Jain, P.K., Basu, B.N.: Analysis of dispersion and interaction impedance characteristics of an azimuthally-periodic vane-loaded cylindrical waveguide for a Gyro-TWT. Int. J. Elect. **86**, 1463–1479 (1999)
6. Singh, G.: Analytical study of the interaction structure of vane-loaded Gyro-traveling wave tube amplifier. Prog. Electromag. Res. B **4**, 41–66 (2008)
7. Alaria, M. K., Ghosh, S. K.: Design of dielectric loaded interaction structure for Q-band Gyro-TWT. IEEE Trans. Plasma Sci. **50**(8), 2342–2347 (2022)
8. Tang, Y., Luo, Y., Xu, Y., Yan, R., Jiang, W., Zheng, Y.: Design of a novel dual-band Gyro-TWT. IEEE Trans. Elect. Dev. **61**, 3858–3863 (2014)
9. Du, C.H., Chang, T., Liu, P.K.: Design of a W-band Gyro-TWT amplifier with lossy ceramic-loaded circuit. IEEE Trans. Electron Dev. **60**, 2388–2394 (2013)
10. Alaria, M.K., Ghosh, S.K., Choyal, Y., Sinha, A.K., Jain, P.K.: Design and simulation of lossy interaction structure for Ka-band Gyro-TWT. IEEE Trans. Plasma Sci. **41**(8), 2264–2268 (2013)
11. Chu, K.R., Barnett, L.R., Lau, W.K., Chang, L.H., Kou, C.S.: Recent developments in millimeter wave Gyro-TWT research at NTHU. Int. Electron Dev. Meeting Tech. Dig., 699–702 (1990)
12. Garven, M., et al.: A gyrotron-traveling-wave tube amplifier experiment with a ceramic loaded interaction region. IEEE Trans. Plasma Sci. **30**, 885–893 (2002)
13. Leou, K.C., McDermott, D.B., Luhmann, N.C.: Large-signal characteristics of a wide-band dielectric-loaded Gyro-TWT amplifier. IEEE Trans. Plasma Sci. **24**, 718–726 (1996)
14. CST Microwave Studio: Computer Simulation Technology. Darmstadt, Germany (2019)

Three-Dimensional Simulation Study of Pseudospark Driven Electron Beam Source

Prerna Unadkat[1,2], Sahil Jain[1,2], and Niraj Kumar[1](✉)

[1] CSIR- Central Electronics Engineering Research Institute (CEERI), Pilani, India
prerna.ceeri22a@acsir.res.in

[2] Academy of Scientific and Innovative Research (AcSIR), Ghaziabad, India

Abstract. The generation of high-power THz using BWO requires an electron beam with high current density and high energy. PS discharge-driven electron beam source is a gas discharge-based electron beam source capable of producing high-current density, high-energy pulsed electron beam within a short duration of time. Also, the gas discharge-based e-beam source can be used conveniently with a plasma-assisted beam-wave interaction structure to develop a plasma-assisted THz BWO. A 3D simulation study of the electron beam source can assist in providing a better understanding of the effect of various parameters on the electron beam generation to produce the desirable beam. In this work, the simulation study of a Pseudospark electron beam source is carried out using the Particle-in-cell (PIC) method. For this study, a 3D simulation model is developed using the software VSim Multiphysics. The electrostatic field penetration inside the hollow cathode (HC) cavity, the particle-field dynamics, and the current profile of the generated e-beam at the anode are studied for the gap voltage 20 and 30 kV, and the HC gas pressure 1, 10, and 20 Pa.

Keywords: 3D particle-in-cell (PIC) simulation · VSim Multiphysics · Pseudospark discharge · high-current density e-beam · Backward wave oscillator

1 Introduction

A Backward wave oscillator (BWO) is among the promising candidates suitable for the generation of high-frequency radiations such as THz [1]. It can be used for the development of a compact high-power THz source. It mainly comprises an electron beam source for the emission of high-energy electron beam followed by the interaction structure for supporting the generation of the desirable frequency and finally, the spent beam collector for spent beam collection at the end of the BWO source [2–4]. The electron beam source must be capable of generating an electron beam with a high current density specifically in the case of THz BWO [5, 6]. This is because the beam has to be propagated through a miniaturized interaction structure with dimensions in the sub-mm range. The propagation of a high-current density beam also requires a focusing mechanism for guiding a focused electron beam through the drift space of the interaction structure. Application of external magnets in the sub-mm dimensions is challenging

S. K. Ghosh et al. (Eds.): VEDA 2024, SPPHY 424, pp. 283–287, 2025.
https://doi.org/10.1007/978-981-96-5174-0_41

and adds to the complexities and size of the device. As an alternative, plasma-assisted electron beam focusing can be utilized to overcome the mutual space charge repulsive forces in the beam and reduce the beam divergence [7]. A hollow cathode electron beam source based on the Pseudo-spark (PS) discharge mechanism can be feasibly used in combination with the plasma-assisted beam-wave interaction structure so that the gas discharge generated beam can be propagated at a long distance by the self-focusing fields [8]. Apart from being plasma-assisted, the PS-driven electron beam source is capable of generating high-current density and high-energy electron beam within a very short time duration [9]. This provides rapid current rise and generation of a highly intense electron beam desirable for the operation of a THz BWO [9]. Owing to the role of the electron beam in the generation of high power, it is crucial to perform the analysis of electron beam generation through the PS discharge electron gun. To achieve the goal, in this work an analysis of electron beam generation with mm size cross-section for THz BWO is carried out using a 3-dimensional simulation model. The analysis is carried out for anode voltage 20 and 30 kV and the neutral gas pressure 1, 10, and 20 Pa. Simulation studies are performed to investigate the role of electrode potential difference and the gas pressure on the electrostatic field penetration inside the hollow cathode (HC), electron beam generation, and the beam current density at the anode. The 3D simulation model of the PS-driven electron beam source is discussed in Sect. 2. The results of the analysis are presented and discussed in Sect. 3 and the summary of the work is presented in Sect. 4.

2 Simulation Model

The 3D simulation model of the PS-driven electron beam source developed using particle-in-cell (PIC) simulation software, VSim Multiphysics by Tech-X [10] is shown in Fig. 1.

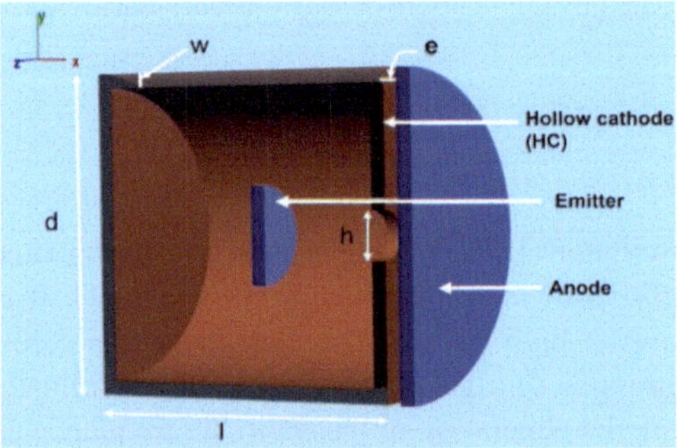

Fig. 1. 3D simulation model of PS-driven e-beam source developed in VSim Multiphysics

The 3D model comprises a metallic hollow cylinder that forms the hollow cathode (HC) of diameter (d) 66 mm and length (l) 66 mm. The wall thickness (w) of the HC cylinder is 3 mm. An aperture of diameter (h) 10 mm is created at a plain wall of the HC. An anode metal disc of diameter 10 mm and thickness 3 mm is placed facing the aperture containing wall of the HC. A gap distance (e) of 3 mm is maintained between the HC aperture wall and the anode. The HC cavity is filled with neutral Argon gas. An emitter in the shape of a disc is placed inside the HC cavity for the emission of seed electrons required for the initialization of the gas discharge [9]. The emitter and HC are kept at ground voltage and a high potential difference (in kV) is applied at the anode. In this work, 20 kV and 30 kV voltage are applied at the anode to study the effect of applied gap voltage on the gas breakdown. Also, the effect of pressure on the gas breakdown is studied using the 3D model at the background gas pressures 1, 10, and 20 Pa.

3 Results and Discussion

In a PS-driven electron beam source, the gas discharge takes place by the anode field penetration inside the HC cavity through the HC aperture [11]. The initial electrostatic field penetration accelerates the seed electrons from the emitter towards the aperture [12, 13]. The seed electrons while propagating towards the apertures can ionize the neutral Argon molecules present inside the HC cavity via impact ionizations leading to the production of Argon ions and secondary electrons. This leads to the generation of a plasma medium inside the HC cavity (Fig. 2) [11]. The plasma medium comprises copious electrons that can be extracted from the HC aperture for the generation of a high-current density electron beam [14].

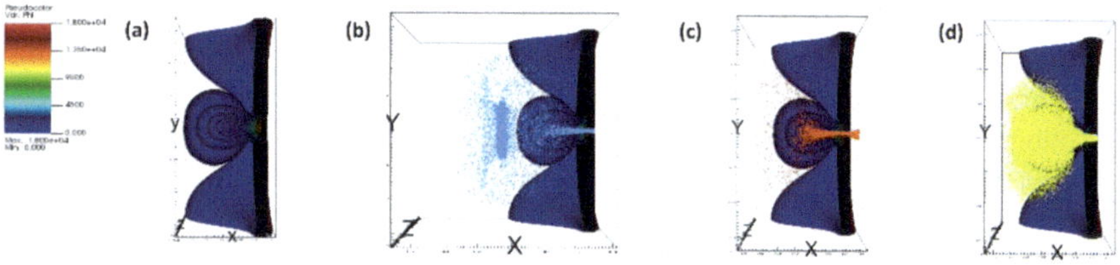

Fig. 2. 3D contour plot showing (a) the increased electrostatic field penetration, (b) the primary electrons, (c) the secondary electrons, and (d) the Argon ions formation inside the HC cavity

The extent of initial field penetration inside the HC cavity depends upon the voltage applied at the anode. Figure 3a shows the time-varying beam current profile of the electron beam generated through a constant cross-section, collected at the anode at different applied gap voltages. The gas pressure inside the HC cavity is kept constant at 10 Pa. From Fig. 3a, the high voltage of 30 kV results in faster breakdown of gas at ~ 50 ns whereas a comparatively lower potential of 20 kV results in sustained breakdown at ~ 70 ns. However, a higher beam current density of 28.66 A/cm^2 is obtained in the case of 20 kV compared to the 22.29 A/cm^2 obtained at 30 kV. This is because, keeping the aperture dimensions constant, a high field at the anode will lead to a high extent of field penetration inside the HC cavity. This, in turn, will result in faster gas breakdown and electron beam generation, emitted through the aperture. The faster breakdown at 30 kV prevents the formation of a high-density plasma inside the HC cavity which results in a reduced beam current density at the anode. Apart from the applied gap voltage, the pressure of gas also plays an important role in the beam current density at the anode.

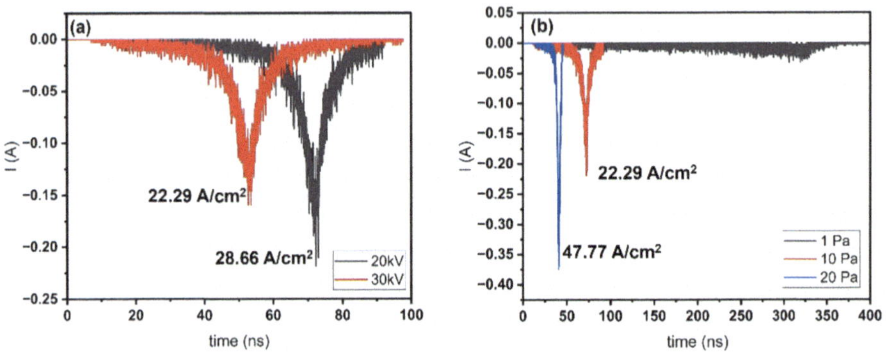

Fig. 3. Current curves showing the breakdown time and beam current density at different (a) anode voltages, and (b) background gas pressure

Figure 3b shows the time-varying beam current profile of the electron beam generated through a constant cross-section, collected at the anode at different background gas pressures. The anode voltage is kept constant at 20 kV. From Fig. 3a, b faster breakdown at ~ 50 ns is observed at 20 Pa with a beam current density of 47.77 A/cm^2. Comparatively, at 10 Pa the breakdown time is ~ 75 ns but the beam current is 22.29 A/cm^2. However, at 1 Pa the breakdown is sustained for a longer duration ~ 325 ns and results in a negligible beam current density. The density of gas inside the HC cavity is the function of the gas pressure. And gas density determines the mean free path for collision. At a high gas pressure of 20 Pa, the mean free path is very low resulting in a higher degree of ionization and production of a high-density plasma. This causes the faster gas breakdown and the generation of an electron beam with high current density at 20 Pa. The mean free path in the case of 10 Pa is comparatively high leading to a low degree of ionization and low-density plasma formation. This leads to a sustained breakdown with comparatively less current density electron beam generation. Whereas, in the case of 1 Pa the gas density inside is extremely low leading to a very high mean free path, and thus, it takes a longer duration for discharge to take place and at discharge the current density produced is negligible.

4 Summary

The 3D simulation study of the PS-driven electron beam source has been performed using VSim Multiphysics. The electrostatic field penetration leading to plasma formation and electron beam generation has been studied using the 3D model. It has been observed that a low gap voltage of 20 kV leads to a sustained gas breakdown with a high-current density beam generation at the anode as compared to the high voltage of 30 kV. Apart from that, the higher gas pressure of 20 Pa of the neutral gas inside the HC cavity led to a faster gas breakdown with the generation of high current density as compared to low-pressure values. The analysis will be useful in determining the operational parameters for generating a high-beam current density using a PS-driven electron beam source for use in plasma-assisted THz BWO.

References

1. Dhillon, S.S., et al.: The 2017 terahertz science and technology roadmap. J. Phys. D: Appl. Phys. **50**(4), Art. no. 4 (2017). https://doi.org/10.1088/1361-6463/50/4/043001
2. Booske, J.H., Kumbasar, A.H., Basten, M.A.: Phys. Rev. Lett. **71**, 3979 (1993)
3. Jain, A., Vishant, Kaur, M., Kumar, N.: Optimization of W-band interaction structure developed using three different micro-fabrication techniques. Rev. Sci. Instrum. **95**, 104702 (2024)
4. Jain, S., Vishant, Kaur, M., Kumar, N.: Parameterization of micro-milling process for fabrication of 0.22 THz interaction structure with improved surface roughness and dimensional tolerance. In: 2023 IEEE Microwaves, Antennas, and Propagation Conference (MAPCON), Ahmedabad, India, pp. 1–5 (2023). https://doi.org/10.1109/MAPCON58678.2023.10464027
5. Unadkat, P., Kumar, N.: Study of space charge potential for propagation of plasma-assisted electron beam. IEEE Trans. Plasma Sci., 1–7 (2024). https://doi.org/10.1109/TPS.2024.3394031
6. Kumar, N., et al.: Pseudospark-driven high-current miniaturized voltage-tunable sheet-electron-beam source. IEEE Trans. Electron Dev. **68**(12), 6482–6486 (2021). https://doi.org/10.1109/TED.2021.3119555
7. Carmel, Y., Shkvarunets, A., Nusinovich, G.S., Rodgers, J., Bliokh, Y. P., Goebel, D. M.: Electron beam dynamics in Pasotron microwave sources. Phys. Plasmas **10**(12), Art. no. 12 (2003). https://doi.org/10.1063/1.1624248
8. Zhang, L., Phelps, A.D.R., Ronald, K., Cross, A.W.: Simulations of the self-focused pseudospark-sourced electron beam in a background ion channel. IEEE Access **9**, 160938–160945 (2021). https://doi.org/10.1109/ACCESS.2021.3131488
9. Unadkat, P., Jain, S., Kumar, N.: A novel multi-aperture r-θ configuration-based pseudospark-driven electron beam source. IEEE Electron Device Lett. **45**(8), 1520–1523 (2024). https://doi.org/10.1109/LED.2024.3413147
10. Tech-X: VSim Reference Manual. Release 12.0.3. Tech-X Corporation, Boulder, CO (2022)
11. Beouf, J.-P., Pitchford, L.C.: Pseudospark discharges via computer simulation. IEEE Trans. Plasma Sci. **19**(2), 286–296 (1991). https://doi.org/10.1109/27.106826
12. Jain, S., Unadkat, P., Shukla, S., Singh, A.K., Ryskin, N.M., Kumar, N.: Analysis of breakdown characteristics of a pseudospark-driven electron beam source. IEEE Trans. Electron Dev. **71**(11), 7094–7098 (2024). https://doi.org/10.1109/TED.2024.3462684
13. Unadkat, P., Priya, P., Kumar, N.: Analysis of pseudospark-driven electron beam generation and its propagation inside high-power THz BWO. In: 2023 IEEE Microwaves, Antennas, and Propagation Conference (MAPCON), Ahmedabad, India, pp. 1–4 (2023). https://doi.org/10.1109/MAPCON58678.2023.10464023
14. Abhishek, A., et al.: Performance evaluation of three-gap pseudospark-driven miniaturized electron beam source. IEEE Trans. Electron Dev. **71**(9), 5691–5695 (2024). https://doi.org/10.1109/TED.2024.3429307

Analysis and Simulation of a Boost-Marx Based Negative High Voltage Pulse Generator

Aashish Ranjan[1,2(✉)] and Anand Abhishek[1,2]

[1] CSIR-Central Electronics Engineering Research Institute, Pilani, India
aashishranjan1497@gmail.com
[2] Academy of Scientific and Innovative Research, Ghaziabad, India

Abstract. This paper presents a design, simulation, and analysis of an advanced negative high voltage nanosecond pulse generator (NSPG) utilizing a boost-based multi-stage Marx generator with solid-state SiC-MOSFET switches. The proposed topology offers significant advantages over conventional spark-gap-based Marx generators, including higher pulse repetition frequencies, reduced size, and improved efficiency. By integrating series-connected SiC MOSFETs and an interconnected boost stage, the NSPG can generate adjustable high-voltage pulses with minimized rise and fall times. MATLAB-based simulation results demonstrate the NSPG's ability to produce pulses in the range of −25 kV to −28 kV at a pulse repetition frequency (PRF) of 20 kHz. Additionally, the results highlight important parameters such as output pulse duration, rise and fall time along with its synchronization with switching signal. This advanced NSPG system has potential applications in various fields, such as microwave generation, plasma research, and industrial processes.

Keywords: Pulse power · High voltage · Marx generator

1 Introduction

Nanosecond HV pulse generators are utilized in various applications such as plasma [1, 2], biological application [3], trigger supply [4], microwave [5], ozone generation [6] and many more. Conventional pulse generators are limited by spark-gap switches, which have long recovery times that constrain their operation to low pulse repetition frequencies (PRF) [1]. This hampers performance in fast pulse applications and introduces inefficiencies due to the physical size and maintenance needs of the switches. Further, with the advent of solid-state switching devices like silicon carbide MOSFETs, which offer sufficient breakdown voltage and fast dynamic response with minimal recovery time, spark-gap switches can now be effectively replaced in many pulse generator circuit topologies such as Marx generator [2–6]. The Marx generator developed several decades ago, remains a vital topology for producing high-voltage, short-duration pulses due to its simple design and versatility. Traditionally, Marx generators use spark gaps for voltage multiplication. While spark gaps are simple and reliable for high voltage ultra-short duration pulses, however, they have drawbacks such as electrode erosion,

S. K. Ghosh et al. (Eds.): VEDA 2024, SPPHY 424, pp. 288–295, 2025.
https://doi.org/10.1007/978-981-96-5174-0_42

inconsistent triggering, and limitations in high-repetition-rate applications due to recovery time. Hence nowadays advanced solid-state devices which are almost solving the issues of spark gaps are selected by researchers to obtain pulses with high precision at a high repetition rate. Beyond pulse generators, solid-state switching devices are increasingly used in various applications, including renewable energy systems, microgrids, and high voltage generation, these applications previously relied primarily on spark-gap switches [7–10].

Modern advancements in semiconductor technologies have significantly enhanced the Marx generator's capabilities. Devices such as Insulated Gate Bipolar Transistors (IGBTs) and Metal-Oxide-Semiconductor Field-Effect Transistors (MOSFETs) have largely replaced spark gaps in many applications. Semiconductors offer precise control over pulse parameters, including polarity, rise and fall times, duration, and repetition rates, making them ideal for applications requiring consistent and reliable pulse characteristics. Additionally, advanced materials like Silicon Carbide (SiC) and Gallium Nitride (GaN) can provide higher breakdown voltages and faster switching speeds. These improvements enable higher repetition rates, better efficiency, and a more compact and robust design, expanding the generator's use in fields like defence, biomedical devices, and high-power microwave systems. The lower breakdown voltage of semiconductor switches can limit their use in high-voltage applications compared to spark gaps. However, this limitation can be addressed by connecting semiconductor switches in series.

For high-voltage pulse applications [11–15], conventional Marx generators require a large number of stages, with each stage having a switch, which increases the number of switching devices. Traditionally, semiconductor switches cannot handle high voltage at a single stage, so multiple Marx stages with lower breakdown voltage levels are connected to achieve a high-voltage pulse [11]. However, this article addresses these limitations by utilizing a boost circuit and series-connected MOSFET switches in the proposed pulse generator topology. This design features higher voltage at individual stages, reducing the number of stages, and uses multiple series-connected switching devices to manage high voltage at each stage with lower voltage drop.

This paper describes a boost Marx-based nanosecond pulse generator that integrates a boost circuit with a Marx circuit, to minimize the number of stages and component counts. Section 2 details the design topology and its operational principles. Section 3 presents the simulation outcomes, while Sect. 4 offers a summary of the findings along with recommendations.

2 Analysis and Implementation of Boost Marx-Based Nanosecond Pulse Generator

The proposed boost Marx-based nanosecond pulse generator with multiple series-connected switching devices offers several advantages, including reduced voltage drop, less synchronization issues, and enhanced controllability over control and triggering events.

The schematic circuit diagram of the modular boost Marx-based generator is shown in Fig. 1. Each stage or module contains a triggering switch, two diodes, and one capacitor.

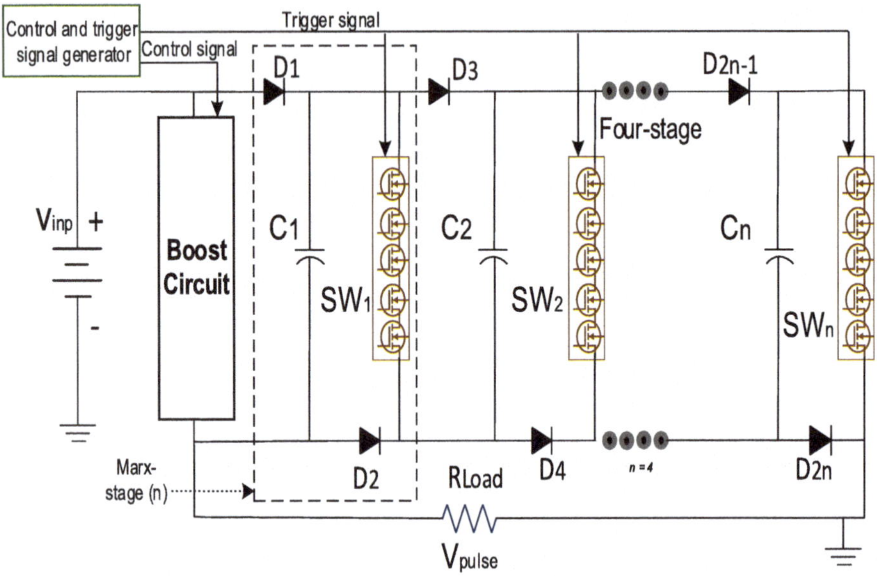

Fig. 1. Schematic diagram of boost-interlinked Marx nanosecond pulse generator.

The DC voltage source (V_{inp}) inputs to the boost circuit, which is interconnected with the Marx generator. The trigger switches are labelled SW_1, SW_2, …, SW_n, and the diodes are labelled D_1, D_2, …, D_{2n}, the stage capacitors of equal value are labelled as C_1, C_2., C_n and R_{Load} represents the resistive load. The proposed boost-based Marx generator operates on the principle of capacitor charging and discharging to produce high-voltage, short-duration pulses. Initially, the DC supply charges an inductor of the boost circuit for a fixed time interval. The first stage of the Marx generator is linked directly to the boost circuit, which includes a control switch (not shown in Fig. 1). Additional stages can be added or removed based on voltage requirements. Pulse voltage and duration are controlled by adjusting the switching period of the control and trigger switches (S_1, S_2,…, S_n), each trigger switch containing five series-connected MOSFETs to enhance voltage-withstanding capabilities. Multiple 1–2 kV SiC-MOSFETs in series are used to achieve higher voltage-withstanding capabilities, although this configuration can introduce synchronization challenges. The boost circuit initially stores energy in the inductor (L_{Boost}) when the trigger switches are off. Subsequently, all switches conduct for brief intervals, causing the Marx stage capacitors' voltage (V_{dc}) to discharge in series, generating high-voltage, short-duration pulses. The switches' dynamic behaviour during turn on and turn off period for nanosecond duration significantly affects the rise and fall time of output pulses. The track inductance and other parasitics also contribute delay and distortion of output pulses. The output pulse depends on the number of capacitors connected in series during the discharge period and is estimated as:

$$V_{out} = N \times V_{dc} \tag{1}$$

where N is the number of capacitors, however, the pulse width mainly depends upon the turn-on period of the trigger switch.

3 Simulation Results and System Analysis

The design and simulation of the boost Marx-based generator were carried out using MATLAB Simulink, where the system was configured to produce variable pulse amplitudes and durations. This was achieved through a variable-step size configuration with a sample time of less than 25 ns, ensuring high precision and control. The simulation results provide detailed insights into the generator's performance, including key metrics such as pulse duration, amplitude, and the rise and fall times of the pulses. The simulation results demonstrate the system's ability to generate extremely short, high-voltage pulses with precise synchronization to the switching signals, crucial for applications requiring high temporal accuracy. The flexibility of the design allows for both single and multiple pulse generation across a range of pulse durations and voltage amplitudes, confirming its effectiveness for applications that require adjustable high-voltage outputs. The MATLAB-based results validate the pulse generator's capability to produce consistent, short-duration pulses, highlighting its applicability in fields such as pulsed power systems, microwave generation, and plasma research.

3.1 System Configuration and Simulation Results

In this section, the simulation results of the proposed boost Marx-based pulse generator are evaluated, showing pulse amplitudes ranging from −25 kV to around −28 kV and pulse durations between 100 ns and 500 ns (Table 1).

Table. 1. Simulation parameter for boost Marx-based nanosecond pulse generator.

Parameters	Value
V_{inp}	400 V
V_{pulse}	Approx. (−25 to −28 kV)
L_{Boost}	5 mH
$C_1 = C_2 = \ldots C_n$	3.3 uF
R_{load}	1 MΩ
Switching frequency	20 kHz
Pulse duration	Approx. (100–500 ns)

To generate a 500 ns pulse at a repetition rate of 20 kHz, the trigger pulse must fall within 500 ns, and the control switch should stay on for more than 500 ns to ensure sufficient energy storage for pulse formation, as indicated by the ~ 500 ns trigger pulse shown in Fig. 2. Where the switching pulse repeats itself after every 50µs to generate a repetitive short duration pulses.

The corresponding output pulse for the 500 ns switching signal is depicted in Fig. 3. This figure demonstrates that the generated pulse has an amplitude of approximately −28 kV and a duration of around 500 ns. The generation pattern of this output pulse is

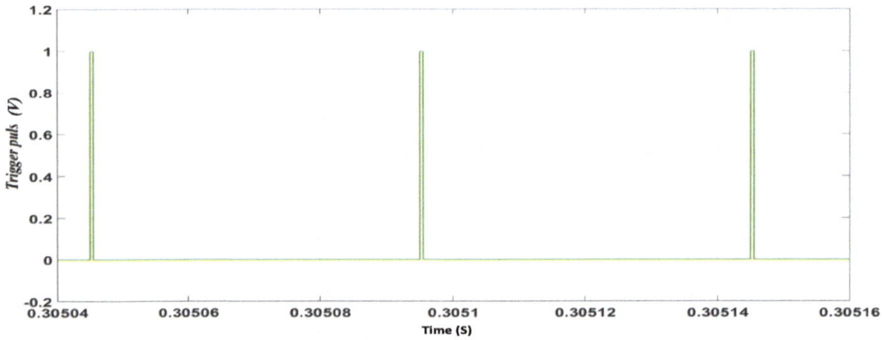

Fig. 2. Switching waveform for trigger signal.

closely synchronized with the trigger pulse, highlighting the precise control in the pulse generation process.

In contrast, Fig. 4 provides a detailed illustration of the pulse width, which is also approximately 500 ns. Additionally, the recorded rise time and fall time for this pulse are measured to be 19.01 ns and 19.41 ns, respectively.

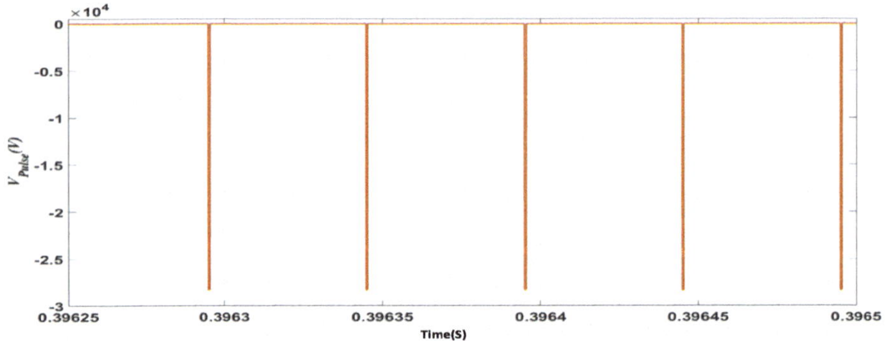

Fig. 3. 20 kHz repetitive pulse of amplitude approx. −28 kV and pulse duration of ~ 500 ns.

These measurements indicate the rapid transition characteristics of the pulse, reflecting the efficiency of the switching mechanism used in the system. The proposed pulse generator is designed to generate variable pulses without the need to alter the number of stages (n). Instead, it achieves variable pulse amplitudes by varying the turn-on timing of the control switches. By adjusting the turn-on period of the trigger pulse, the output pulse duration can also be modified. This method, however, introduces a difference in the turn-on timing of the trigger signal for different pulse voltage levels. Consequently, by varying the turn-on duration of both the control switch and the trigger switch, the pulse generator can produce pulses with an amplitude of approximately −25 kV, a pulse duration of around 100 ns, and a pulse repetition rate of 20 kHz, as shown in Fig. 5. A detailed analysis of the 100 ns pulse is presented in Fig. 6. This figure illustrates a single pulse with a rise time of 19.3 ns and a fall time of 19.24 ns, based on results captured

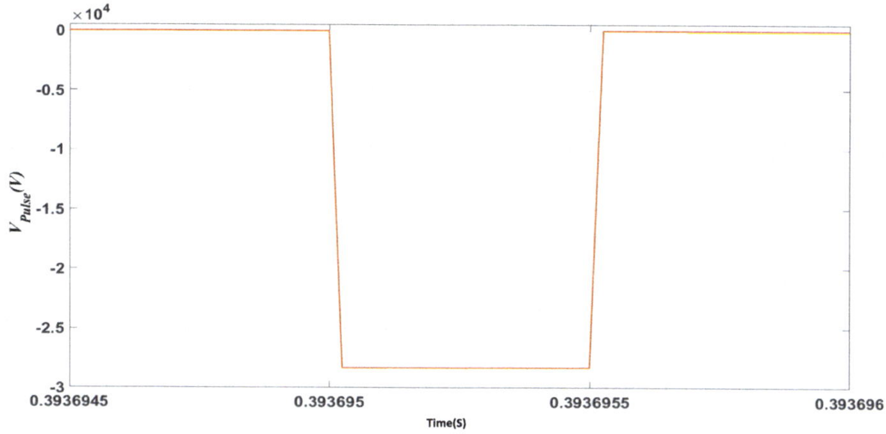

Fig. 4. ~500 ns pulse width single output pulse waveform of amplitude approx. −28 kV.

from MATLAB Simulink simulations. The precise control over both the rise and fall times demonstrates the effectiveness of the pulse generator in producing high-fidelity pulses for various applications.

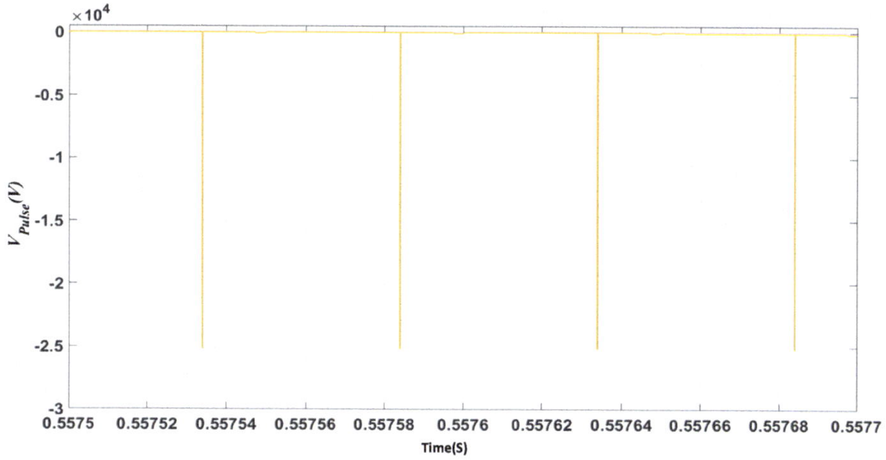

Fig. 5. Pulse generator ~ 100 ns, approx. −25 kV amplitude repetitive pulse wave of pulse repetition frequency of 20 kHz.

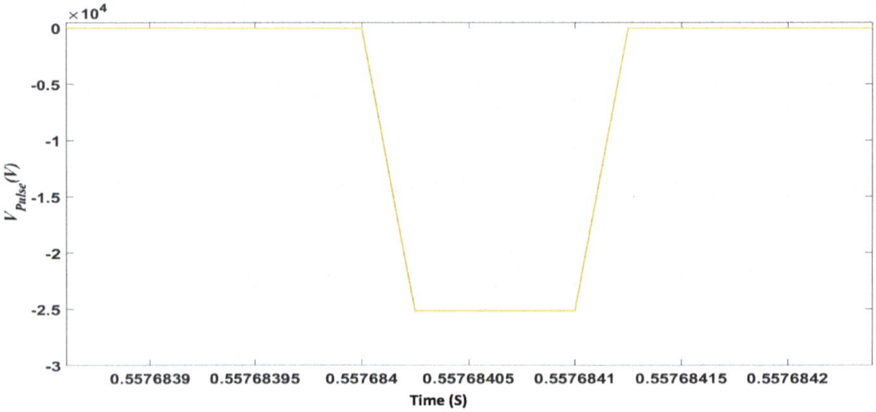

Fig. 6. ~100 ns pulse duration single pulse waveform of amplitude approx. −25 kV.

4 Conclusion

The advanced boost Marx-based nanosecond generator has been developed and simulated for generating negative high-voltage nanosecond pulses. Unlike traditional Marx generators, this design incorporates a boost circuit and an additional control switch, enabling voltage boosting and reducing the number of required stages. Fewer stages minimize synchronization challenges among trigger switches. The control switch's varying turn-on time allows for variable pulse amplitudes between approximately −25 and −28 kV. Pulse amplitude is influenced by the number of stages and the boosting factor. Controlling pulse width from 100 to 500 ns requires precise synchronization of the duty cycle and on-time of control and trigger switches. To generate such precise nanosecond pulse duration at a 20 kHz repetition rate the MOSFETs switching devices require proper synchronization and efficient switching. To manage high voltage and reduce stress across switches, the design uses five series-connected MOSFETs for both control and trigger switches. This Marx generator can generate a wide range of voltage amplitudes and pulse widths by adjusting stages and switching patterns.

Acknowledgement. The authors express their sincere gratitude to the Director and the members of High-Power Microwave Systems Group (HPMSG), CSIR-Central Electronics Engineering Research Institute, Pilani, Rajasthan, India, for their valuable suggestions in the research work.

References

1. Matharoo, G.S., Walsh, J.L.: Design and characterization of a solid-state Marx generator for plasma-activated water generation. IEEE Trans. Plasma Sci. **51**(10), 3070–3079 (2023). https://doi.org/10.1109/TPS.2023.3315874
2. Abhishek, A., Kumar, N., Pal, U.N., Singh, B., Akbar, S.A.: Implementation of trigger unit for generation of high-current-density electron beam. IEEE Trans. Electron Dev. **68**(7), 3582–3587 (2021). https://doi.org/10.1109/TED.2021.3079290

3. Yao, C., Zhang, X., Guo, F., Dong, S., Mi, Y., Sun, C.: FPGA-controlled all-solid-state nanosecond pulse generator for biological applications. IEEE Trans. Plasma Sci. **40**(10), 2366–2372 (2012). https://doi.org/10.1109/TPS.2012.2188908
4. Sack, M., Müller, G.: Modular trigger generator for an overvoltage-triggered Marx generator. IEEE Trans. Plasma Sci. **40**(10), 2618–2624 (2012). https://doi.org/10.1109/TPS.2012.2196292
5. Pinguet, S., Dupéroux, J.-P., Delmote, P., Bieth, F., Bischoff, R.: Short-pulse Marx generator for high-power microwave applications. IEEE Trans. Plasma Sci. **41**(10), 2754–2757 (2013). https://doi.org/10.1109/TPS.2013.2277981
6. Huiskamp, T., van Oorschot, J.J., Pereira, M.T., Redondo, L.M.: Ozone generation with a flexible solid-state Marx generator. In: 2018 IEEE International Power Modulator and High Voltage Conference (IPMHVC), Jackson, WY, USA, pp. 147–150 (2018). https://doi.org/10.1109/IPMHVC.2018.8936658
7. Ranjan, A., Abhishek, A., Kumar, N., Verma, B.K.: Evaluation of phase-shift control for a high voltage DC supply using multi-level Cockcroft-Walton multiplier. In: 2023 IEEE International Conference on Power Electronics, Smart Grid, and Renewable Energy (PESGRE), Trivandrum, India, pp. 1–6 (2023). https://doi.org/10.1109/PESGRE58662.2023.10404757
8. Abhishek, A., Ranjan, A., Singh, B., Akbar, S.A.: Performance evaluation of a 500 W bidirectional converter for DC microgrid. In: 2022 IEEE International Conference on Power Electronics, Smart Grid, and Renewable Energy (PESGRE), Trivandrum, India, pp. 1–6 (2022). https://doi.org/10.1109/PESGRE52268.2022.9715753
9. Ranjan, A., et al.: IEEE 17th India Council International Conference (INDICON). New Delhi, India **2020**, 1–7 (2020). https://doi.org/10.1109/INDICON49873.2020.9342318
10. Abhishek, A., Ranjan, A., Devassy, S., Verma, B.K., Ram, S.K., Dhakar, A.K.: Review of hierarchical control strategies for DC microgrid. IET Renew. Power Generation **14**(10), 1631–1640 (2020). https://doi.org/10.1049/iet-rpg.2019.1136
11. Ranjan, A., Abhishek, A., Kumar, N., Verma, B.K.: Review of nanosecond pulse generator with high-power switching devices. IEEE Trans. Elect. Dev. **71**(9), 5165–5176 (2024). https://doi.org/10.1109/TED.2024.3429300
12. Abhishek, A., Akbar, S.A.: Linear and nonlinear control strategy for full bridge buck converter. In: 2015 International Conference on Electrical, Electronics, Signals, Communication and Optimization (EESCO), Visakhapatnam, India, pp. 1–6 (2015). https://doi.org/10.1109/EESCO.2015.7254007
13. Abhishek, A., Kumar, N., Vishant, Singh, B.: Design of pulse power supply for high current density electron beam source. In: 2022 23rd International Vacuum Electronics Conference (IVEC), Monterey, CA, USA, pp. 17–18 (2022). https://doi.org/10.1109/IVEC53421.2022.10292428
14. Abhishek, A., et al.: Performance evaluation of three-gap pseudospark-driven miniaturized electron beam source. IEEE Trans. Electron Dev. **71**(9), 5691–5695 (2024). https://doi.org/10.1109/TED.2024.3429307
15. Ranjan, A., Abhishek, A., Kumar, N., Ryskin, N.M., Titov, V.N., Torgashov, R.A.: Simulation and analysis of solid-state switch-based high voltage pulse power supply using Marx generator. In: 2024 Photonics & Electromagnetics Research Symposium (PIERS), Chengdu, China, pp. 1–8 (2024). https://doi.org/10.1109/PIERS62282.2024.10618835

Investigation of X-band Metamaterial Based Backward Wave Oscillator

Ambikesh Kumar, Rajnish Kumar, and Manpuran mahto(✉)

National Institute of Technology Patna, Patna, India
{ambikeshk.ph23.ec,rajnishk.phd19.ec,mmahto}@nitp.ac.in

Abstract. An X-band metamaterial slow-wave structure (MTM SWS) has been suggested in which leverages the reverse Cherenkov radiation property of metamaterials to function as a backward wave oscillator (BWO). The proposed MSWS helps in device miniaturization and intense electron bunching. An X-band MTM BWO with complementary electric split ring resonators (CeSRRs) integrated inside a circular waveguide makes up the electron-beam structure. The cold test simulation is performed to determine dispersion and coupling impedance properties. The transmission and reflection coefficients of the proposed MSWS are also evaluated.

Keywords: Metamaterial · CeSRR · high power microwave (HPM) · Transmission properties

1 Introduction

Firstly, left-handed material (LHM) introduced by Veselago in 1968. Metamaterial is defined based on its properties which are not found or exited in the natural materials [1]. That means metamaterial is a unique property that overcomes normal material. The metamaterial is obtained by changing the properties of normal material like relative permittivity and relative permeability and the changes in the property allows the metamaterial to behave in more controllable manner. The important properties of the metamaterials are: (i) negative group velocity and refractive index, (ii) reverse Doppler Effect and Cherenkov radiation [2]. Applications of the metamaterial is widely used in microwave, millimeter wave, terahertz wave, cloaking, phase shifter, and antenna. In last few years, metamaterials are being used in various electron beam devices for system miniaturization and performance improvement [3]. In addition, metamaterial assisted slow wave structure (MTM SWS) offers high average interaction impedance [4].

The introduction of split-ring resonator (SRR) arrangements allowed for the achievement of negative effective permeability as well as permittivity to fulfill the requirement of metamaterial properties. It has also been demonstrated that the complementary electric-SRR (CeSRR) exhibit the similar properties [5]. Furthermore, the MTM SWS is designed to operate below the waveguide's cutoff frequency. Since, the wave is interacting with the electrons in the longitudinal direction; parallel CeSSR plates are used for the transverse magnetic (TM) mode. Different mode configurations can be obtained by various

S. K. Ghosh et al. (Eds.): VEDA 2024, SPPHY 424, pp. 296–301, 2025.
https://doi.org/10.1007/978-981-96-5174-0_43

geometrical arrangement [6]. Since, it works on the phenomenon of negative dispersion; power flows in the antiparallel direction to the motion of electron beams.

In 2002, Marques and Esteban et al. reported in an empty waveguide, electromagnetic waves flow through quasi-TE or TM modes below the cut-off frequency [7]. This characteristic is important for the compactness of the waveguide, as the RF interaction structure consist of an arrangement of split ring resonators (SRRs) that are arranged periodically within a square waveguide. Still, it is unsuitable for HPM devices because of the drawback of dielectric failure. Duan et al. [8] solved the problem of dielectric breakdown for SRR by designing a planner CSRR MTM loaded in a square waveguide. Wang et al. [9] designed an all-metal MTM SWS that operates at a frequency of 2.454 GHz. This design is well-suited for use in concise high-power vacuum electron devices (VEDs). Kumari et al. [10] design CeSRR in the groove placed on the inner ring outer circumference and observe that frequency and average interaction impedance increase, after that s-band CeSRR scaled to x –band of operating frequency is 9.2 GHz. This proposed paper comprises an x-band unit cell with a resonant frequency is increases by decreasing the radius of the CeSRR unit cell [8–10] (Fig. 1).

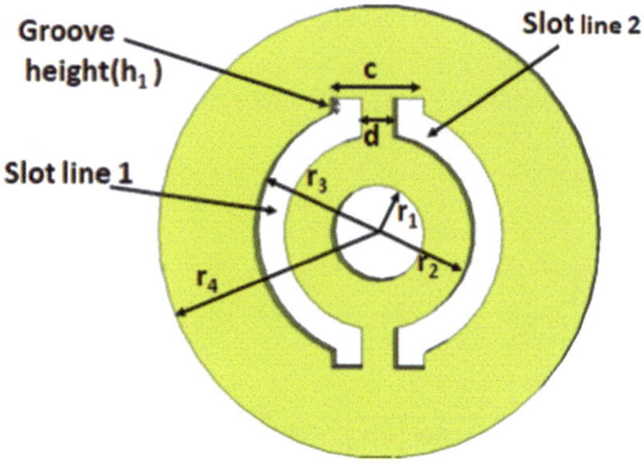

Fig. 1. Proposed CeSRR unit cell operating in X-band

2 Design

In this structure, X-band MTM SWS that is loaded into circular waveguide indicating the versatility and applicability of the design across different frequency ranges. In the article, we have investigated the metamaterial inspired SWS for high power microwave (HPM) operating in X-band frequency. This designed architecture is modelled in "CST studio suite" furthermore we have examined its transmission characteristics. An X-band MTM BWO with CeSRR integrated within a circular waveguide that makes the electron-beam structure. Overall, this approach was taken to validate the proposed concept, demonstrating the effectiveness of the modified CeSRR in enhancing the performance of metamaterial-based slow wave structures. The findings support the potential for improved efficiency and miniaturization in high-frequency applications [11]. The physical measurements of the conventional CeSRR are displayed in Table 1.

The mean interaction impedance for the suggested X-band CeSRR is calculated by using post-processing tools in the CST Eigenmode solver. The results indicate that the modified design achieves a higher average interaction impedance, which is beneficial for device efficiency [10, 11]. Interaction impedance is an important factor in judging any vacuum electron devices for high power output. If the interaction impedance is high at the dominant mode and lower at the higher order modes the devices have better performance. Interaction impedance is determined by the

$$K_{cn} = |E_{zn}|^2/2\beta_n^2 p_w$$

Here, βn indicates the nth spatial harmonic of the phase constant, E_{zn} indicates the axial element of the electric field magnitude, additionally Pw indicates the transmission power along the axial direction. In this paper at dominant mode attenuation is minimum so provides better stability at this operating frequency. Attenuation is described as $\alpha =$ Ploss / 2P, where P represents the total power flowing along axial direction of the electric field, and Ploss denotes the power loss.

Table 1. Parameters of CeSRR unit cell

Parameters	Values (mm)	Parameters	Values (mm)
r_4	7	c	3
r_3	4	d	1
r_2	3	P	6
r_1	1.5	T	1

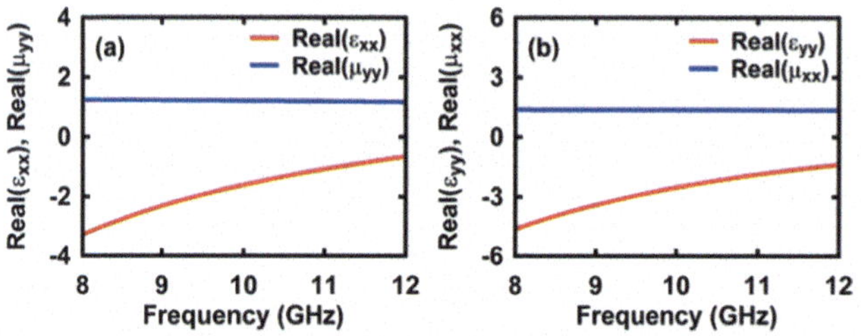

Fig. 2. Retrieved constitutive parameter (a) ε_{xx} and μ_{yy} (b) ε_{yy} and μ_{xx}.

3 X-Band Mtm Sws Model and Results

The 'CST frequency domain' solver is used for extracting the permittivity and permeability from the parametric extraction method Fig. 2(a) and (b) shows the effective.

ε and μ corresponding to the polarized electric field alongside the x and y axis, while Fig. 3 depicts the surface current distribution at 11.2 GHz. The high frequency

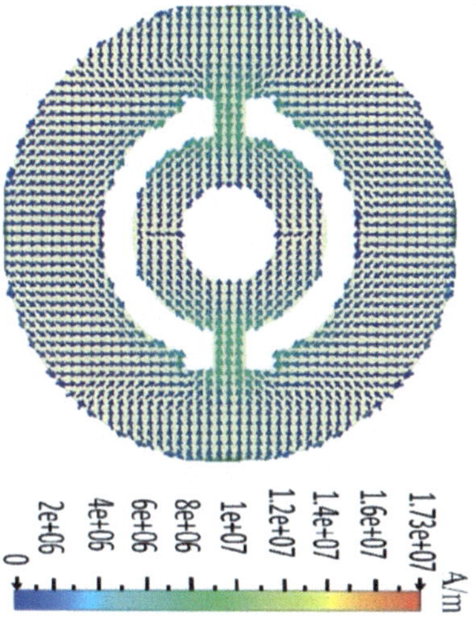

Fig. 3. Surface current distribution at 11.2 GHz.

characteristics (i.e., dispersion and average interaction) of the proposed modelled are presented in Fig. 4, utilizing the Eigenmode solver of CST. Figure 4(a) illustrates the X-band dispersion curve of the metamaterial slow-wave structure (MTM SWS). The dispersion diagram exhibits negative dispersion for the n = 0, that indicates the curve appears as a "backward wave." The X-band CeSRR-based on MTM SWS offers upper average interaction impedance compared to its S-band counterpart, owing to geometrical modifications that enhance the length of the effective path for the high-frequency currents. It demonstrates that the gadget is in the forward wave regime for the zeroth spatial harmonics. The average interaction impedance of the X-band MTM SWS for mode 1, should be more than 700 Ω.

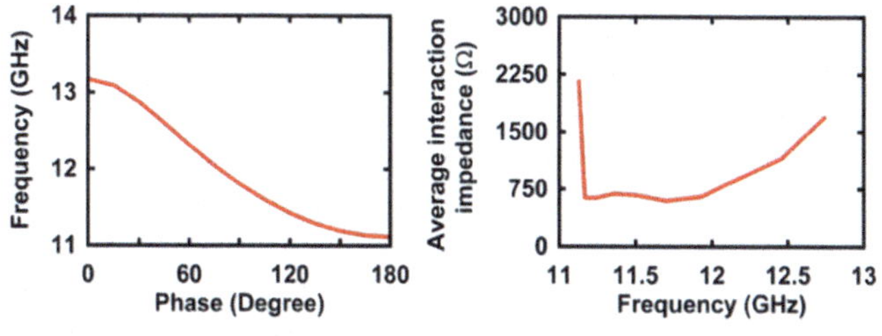

Fig. 4. (a) Dispersion characteristics and (b) Interaction impedance curve

Transmission properties of the recommended structure is determined by periodically loading CeSRR in a circular waveguide along with the couplers and coupling probes as shown in Fig. 5 The waveguide coupler features an inner radius of 13 mm and a length of 43 mm. The length of the circular waveguide segment, which includes the MTM SWS, is 69 mm, with an inner radius of 8 mm. This MTM SWS consists of an array of complementary electric split ring resonators (CeSRRs) whose outer radius matches the radius of the waveguide. Ten cavities are formed by the 11 unit cells and the 5 mm gap between two CeSRRs in order to create SWS. The each side of flanges are 3 mm. The X-band MTM SWS (D) has a total length of 161 mm. Standard flanges equipped with SMA connector used as the coupling probes. The radii of inner and outer conductor of the SMA connector are 0.65 mm and 2.05 mm respectively.

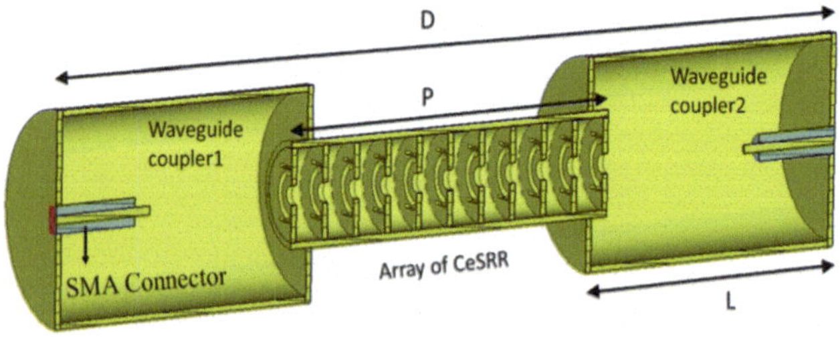

Fig. 5. Simulation of proposed structure with waveguide couplers and Coupling probe.

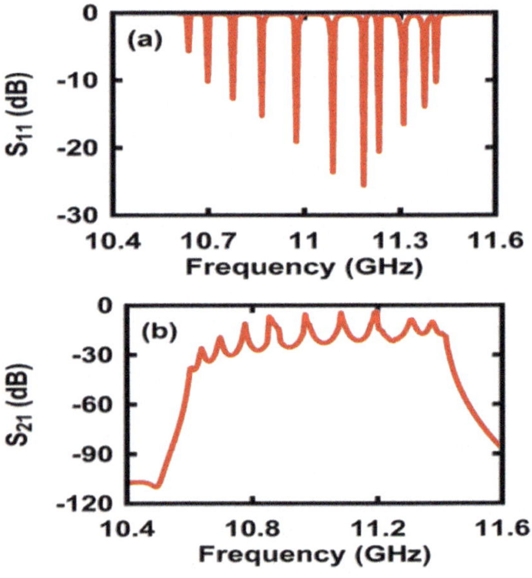

Fig. 6. (a) reflection coefficient (S_{11}), (b) transmission coefficient of proposed structure.

The simulated reflection and transmission of the suggested structure is represented in Fig. 6(a) and (b). From fig. we can observe that the resonances are because there are ripples within the frequency range from 10.6 to 11.4 GHz. These ripples depend upon the unit cell of CeSRRs used in the given structure. It can be seen that resonance frequency shown in Fig. 6(a) having a resonance depth of −25 dB at 11.2 GHz. The ripple in Fig. 6(b) demonstrates the pass-band of the structure in the frequency range of 10.6–11.4 GHz.

4 Conclusion

In this paper, X-band CeSRR metamaterial has been proposed and designed. It is named because of it is operating at resonance frequency i.e., 11.2 GHz. The interaction impedance of the X-band MTM SWS, is more than 700 Ω. We have loaded the X-band complementary electric split ring resonator (CeSRR) within a circular waveguide to create an X-band MTM SWS, and the study shows that the reflection coefficient (S11) of this structure goes to depth of −25 dB at 11.2 GHz. This proposed X-band CeSRR will be useful to make the device miniaturized. The proposed structure finds its application in linear accelerators and high-power microwave sources.

References

1. Veselago, V.G.: The electrodynamics of media with simultaneously negative values of permittivity and magnetic permeability. Sov. Phys. I-Uspekhi **47**, 509–514 (1968)
2. McGregor, I., Hock, K.M.: Complementary split-ring resonator-based accelerating structure. J. Instrum. **8**(5), 05009 (2013)
3. Rowe, T., Booske, J.H., Behdad, N.: Metamaterial-enhanced resistive wall amplifiers: theory and particle-in-cell simulations. IEEE Trans. Plasma Sci. **43**(7), 2123–2131 (2015)
4. Xu, H., et al.: Effective-medium models and experiments for extraordinary transmission in metamaterial-loaded waveguides. Appl. Phys. Lett. **92**(4), 667 (2008)
5. Wang, Y., et al.: S-band high-efficiency metamaterial microwave sources. IEEE Trans. Electron Devices **63**(9), 3747–3752 (2016)
6. Guha, R., Bandyopadhyay, A.K., Varshney, A.K., Datta, S.K., Basu, B.N.: Investigations into helix slow-wave structure assisted by double-negative metamaterial. IEEE Trans. Electron Devices **65**(11), 5082–5088 (2018)
7. Wang, X., Duan, Z., Wang, F., Li, S., Jiang, S., Jin, L., Wang, Z., Wei, Y., Gong, Y.: High-frequency characteristics of a metamaterial slow wave structure. In: IEEE International Vacuum Electronics Conference (IVEC), pp. 345–346, June (2018)
8. Wang, Y., et al.: All-metal metamaterial slow-wave structure for high power sources with high efficiency. Appl. Phys. Lett. **107**(15), 153502 (2015)
9. Wang, X., et al.: Novel S-band metamaterial extended interaction Klystron. IEEE Electron Device Lett. **41**(10), 1580–1583 (2020)
10. Kumari, S., Jain, P.K., Mahto, M.: Study of a modified complementary electric split ring resonator-based metamaterial slow wave structure. IEEE Trans. Plasma Sci., August 11 (2023)
11. Marqués, R., Martel, J., Mesa, F., Medina, F.: Left-handed-media simulation and transmission of EM waves in subwavelength split-ring resonator-loaded metallic waveguides. Phys. Rev. Lett. **89**(18), Art. no. 183901 (2002)

Development of RF Window for S-band Klystron

A. S. Dhavale[1(✉)], T. Arvind[2], Mentes Jose[3], S. Shrotriya[4], Rama Rao[4], M. Pande[4], P. C. Saroj[1], R. I. Bakhtsingh[1], and M. Mascarenhas[5]

[1] Accelerator and Pulse Power Division, Bhabha Atomic Research Centre, Trombay, Mumbai 400085, India
ajd@barc.gov.in

[2] Centre for Design and Manufacturing, Bhabha Atomic Research Centre, Trombay, Mumbai 400085, India

[3] Ion Accelerator Development Division, Bhabha Atomic Research Centre, Trombay, Mumbai 400085, India

[4] Accelerator Controls Division, Bhabha Atomic Research Centre, Trombay, Mumbai 400085, India

[5] Beam Technology Development Group, Bhabha Atomic Research Centre, Trombay, Mumbai 400085, India

Abstract. BARC has initiated a program on the indigenous development a Klystron that can deliver 6 MW (peak)/24 kW (average) power. Ceramic window is one of the vital sub-systems that need special attention due to its susceptibility to RF, thermal and multipacting breakdown. The paper describes the design of ceramic window. It consists of a cylindrical waveguide with Alumina disk operating in TE11 mode that is terminated in a std. WR284 rectangular waveguide. Window parameters like ceramic material, disk thickness, diameter, cylindrical waveguide length, diameter and length of WR284 waveguide is optimized using CST Microwave studio. Alumina material is used as ceramic due to its high dielectric constant (k = 9.9) and low loss tangent (tan δ ~ 0.0001). A maximum bandwidth of ~8 MHz is obtained at central frequency of 2857 MHz at a return loss of 21 dB after optimization. Maximum tangential and normal electric field of 463 V/m and 328 V/m is observed in the center of the Alumina disk. Maximum magnetic field of 1.1 A/m is observed at a radius of 32.86 mm. Maximum power loss density in the overall structure is ~16.86 W/m^3. Thermal analysis is carried out using ANSYS. A cooling jacket is designed; water flow rate is estimated. Multipacting analysis is carried out using CST particle studio at different power levels at different initial positions of the primary electron. A transition is designed between output cavity of klystron and the RF window to minimize reflection losses.

Keywords: Klystron · RF window · Alumina

1 Introduction

Particle accelerators are widely used in the various sectors from basic research to industry, medical, food preservation to isotope production. BARC is involved in the development of high energy, high current RF accelerators for the different applications. Majority

S. K. Ghosh et al. (Eds.): VEDA 2024, SPPHY 424, pp. 302–309, 2025.
https://doi.org/10.1007/978-981-96-5174-0_44

of the components and important sub-systems like RF cavity, couplers, magnets, beam diagnostics equipment are indigenized and are successfully developed and are used in the Accelerators except for the RF power source. To overcome this drawback, a program on the development of indigenous S-band Klystron is initiated in collaboration with CEERI, Pilani. It is aimed to develop a Klystron that can deliver 6 MW (peak)/24 kW (average) power. Ceramic window is one of the vital sub-systems that need special attention due to its susceptibility to RF, thermal and multipacting breakdown. The paper describes the RF design of ceramic window and transition between output cavity of klystron and the ceramic window.

2 Design of RF Window

2.1 RF Design

A RF window is designed for the S-band Klystron. Alumina material is used as the ceramic due to high dielectric constant (k = 9.9) and low loss tangent (tan $\delta \sim 0.0001$) [1, 2]. The 3D view of RF window simulated in CST microwave studio is in Fig. 1.

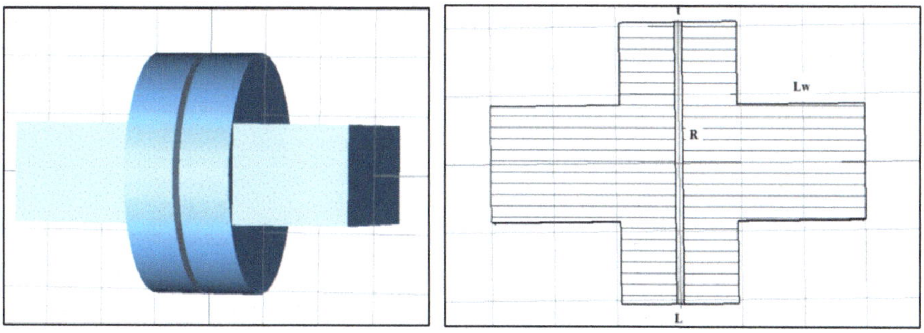

Fig. 1. 3D model and section view of RF window.

Geometrical parameters like cavity radius R, cavity length L, ceramic thickness t, waveguide length Lw is optimized to maximize bandwidth and to reduce the return loss at the resonant frequency. Figure 2 shows the change in resonant frequency with R; as R reduces frequency increases. Figure 3 shows the effect of ceramic thickness on resonant frequency; higher is the thickness lower is the frequency. Figure 4 shows the effect of cavity length, L on resonant frequency. Increase in the length leads to increase in the frequency. Figure 5 shows that increasing waveguide length frequency can be lowered.

After optimization of all the parameters, bandwidth of ~8 MHz is achieved about the central frequency of 2857 MHz at a return loss of −21 dB (See Fig. 6). Return loss at 2857 MHz is < -90 dB. To estimate the power loss in the window analysis is done in the frequency domain. A TE11 mode is generated in the window. Electric and magnetic field distribution is shown in the Fig. 7. Maximum tangential and normal electric field of 463 V/m and 328 V/m is observed in the center of the Alumina disk. Maximum magnetic field of 1.1 A/m is observed at a radius of 32.86 mm. Maximum power loss density in the overall structure is ~16.86 W/m^3.

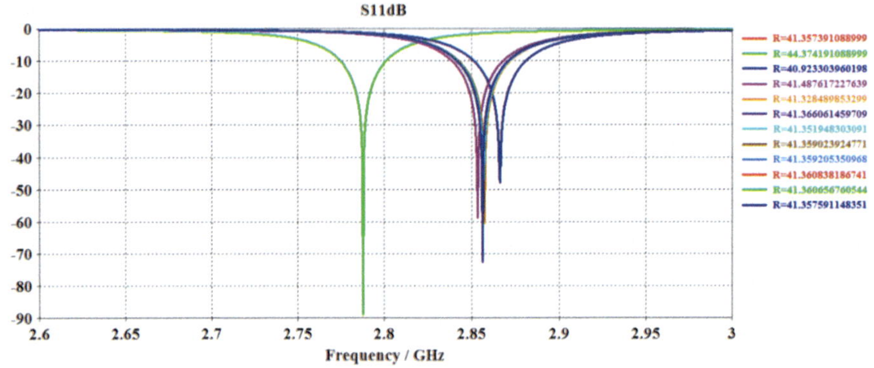

Fig. 2. Effect of cavity radius, R.

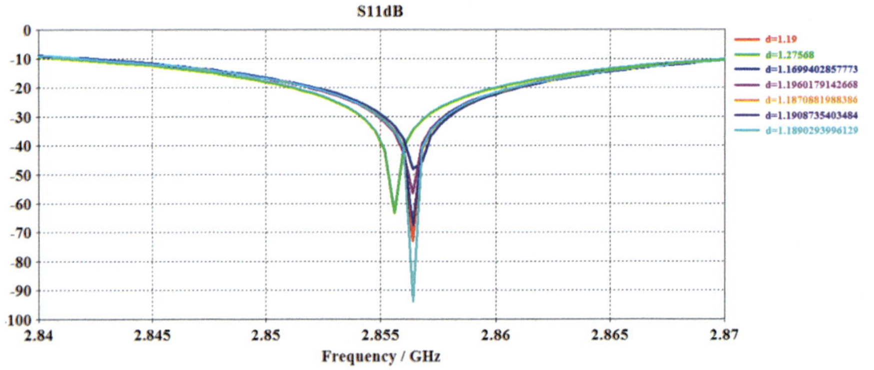

Fig. 3. Effect of ceramic disk thickness, t.

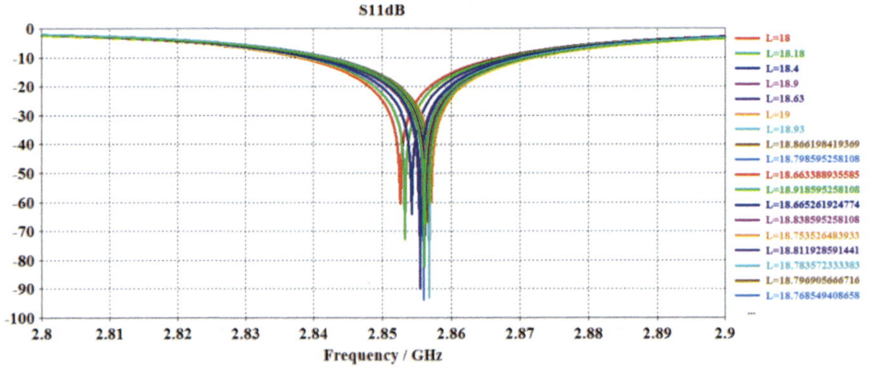

Fig. 4. Effect of cavity length, L.

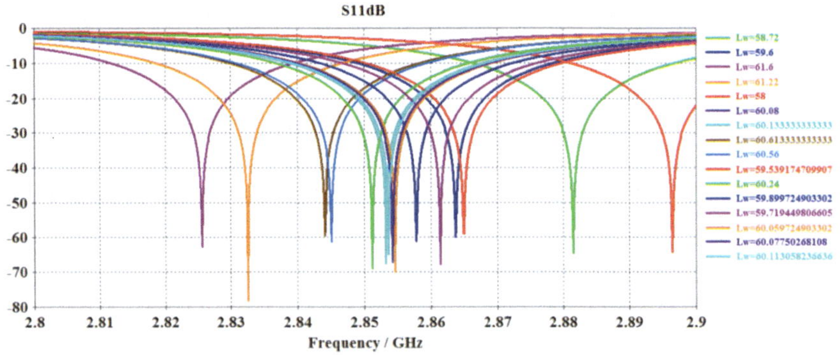

Fig. 5. Effect of waveguide length, Lw.

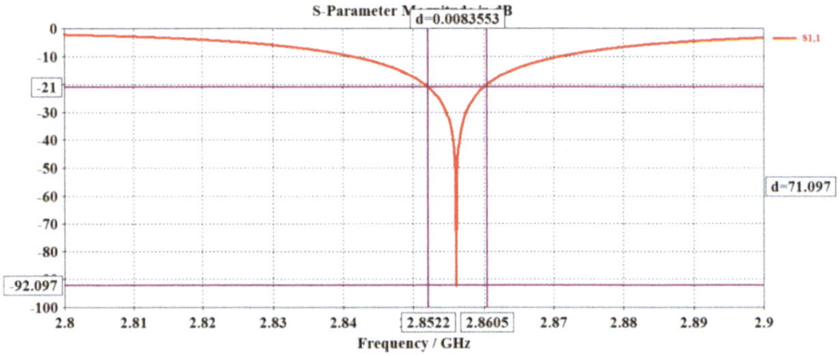

Fig. 6. Bandwidth of RF window.

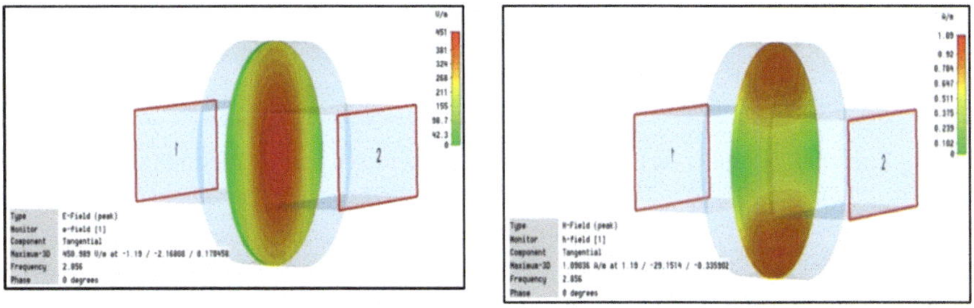

Fig. 7. Electric and magnetic field distribution in the ceramic disk.

2.2 Thermal Analysis

Thermal analysis of the RF window assembly is carried out using ANSYS. The maximum temperature of 32.5 ^{0}C is observed in the centre of the ceramic disk. Overall temperature over full window assembly varied from 28.2 to 32.5 °C. A cooling jacket (see Fig. 8) is designed. After providing water cooling (flow rate ~ 8 lit/min) maximum temperature on the ceramic disk reduced to 24.6 °C while overall temperature in the assembly varied between 20 to 24.6 °C as shown in Fig. 9.

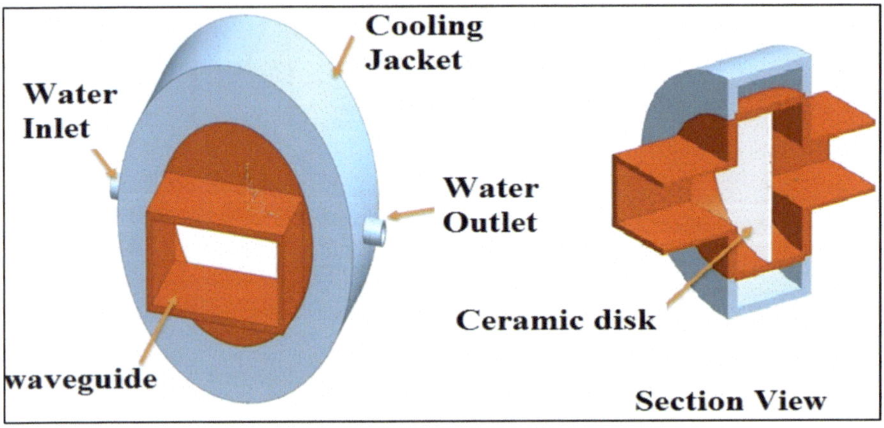

Fig. 8. Cooling Jacket for RF window.

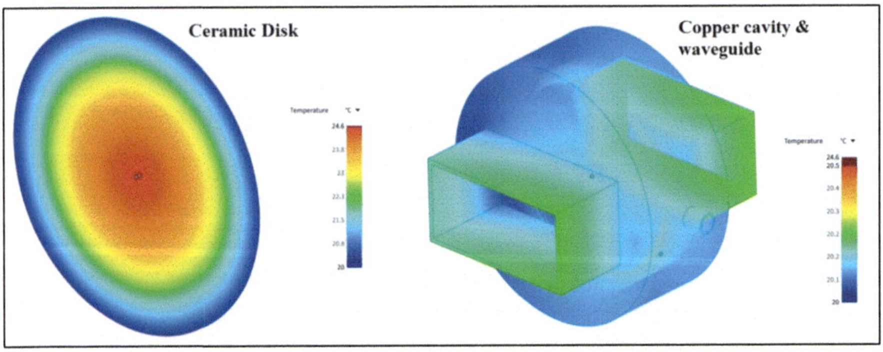

Fig. 9. Temperature profile on the ceramic disk, copper cavity and waveguide after cooling.

2.3 Multipacting Analysis

Performance of RF window is greatly affected by multipacting, a resonant buildup of electrons in the RF volume at the operating power levels. Multipacting depends on the geometry, RF power level and the secondary emission yield (SEY) of the surface. To estimate multipacting, a PIC (Particle-In-Cell) solver within the CST module is employed.

SEY depends on the material, surface conditions [3–5], surface impurities [6], angle of incidence and energy of primary electron. Vaughan emission model is used to model SEY of Copper and Alumina [7, 8]. Figure 10 illustrates SEY as a function of incident electron energy. Electromagnetic fields calculated by EM solver are imported into the PIC

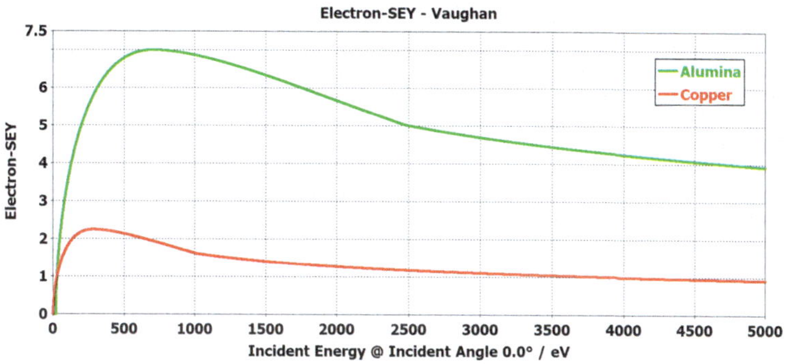

Fig. 10. SEY as a function of incident energy in eV.

solver. About 1000 seed electrons with random energy are introduced in the structure. The PIC solver tracks these electrons; secondary electrons are generated when the primary electron hits the surface. The number of electrons remaining after 50 ns cycles (equivalent to 142 RF cycles) serves as a good indicator of the probability of multipacting. Field levels are scaled to simulate various power levels, with simulations extending up to 6 MW. It is observed that the likelihood of multipacting increases above 350 kW, peaks beyond 1 MW, and declines above 3 MW. Figure 11 shows typical particle growth with time at 402 kW power for 50 ns. Further simulations covering 100 ns (284 RF cycles) reveal that, at 1.2 MW, the probability of multipacting is highest. However, the two-order increase in particle growth suggests this might be a soft barrier that can be mitigated through RF conditioning. Figure 12 shows the particle growth observed after 100 ns at various power levels up to 6 MW.

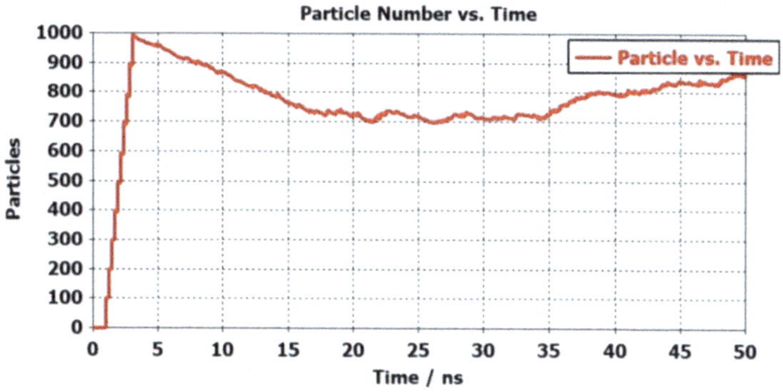

Fig. 11. Particle growth at 402 kW RF power with time

2.4 Transition Between Output Cavity of Klystron and RF Window

A power is coupled from the output cavity to the external waveguide. Ceramic window is used to provide a vacuum barrier between UHV cavity region from the atmosphere.

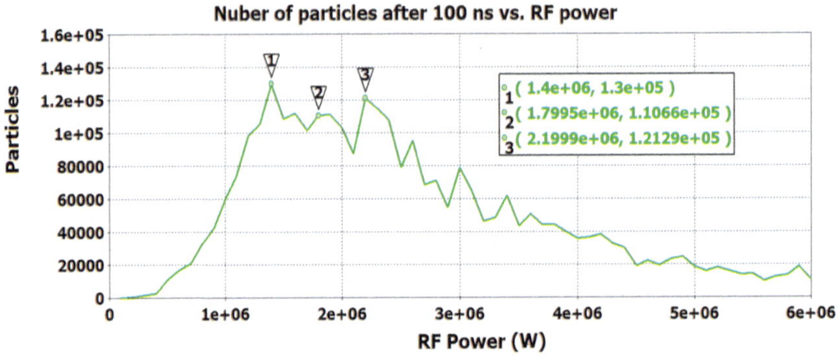

Fig. 12. Particle growth at different RF power levels; soft multipacting barriers at 1.2 MW, 1.8 MW and 2.2 MW are marked.

For the maximum power transfer, quality factor of the cavity shall be matched with the quality factor Q_{ext} of RF window assembly. Figure 13 shows the sectional view of klystron output cavity with the RF window. The output cavity dimensions given in ref. [9] are used for the simulation. A rectangular coupling aperture is provided between cavity and RF window waveguide section. Quality factor, Q_0 of the cavity is ~7900. A rectangular aperture of size 18.2 mm × 16 mm gives the Q_{ext} ~ 7700. Further optimization is in progress.

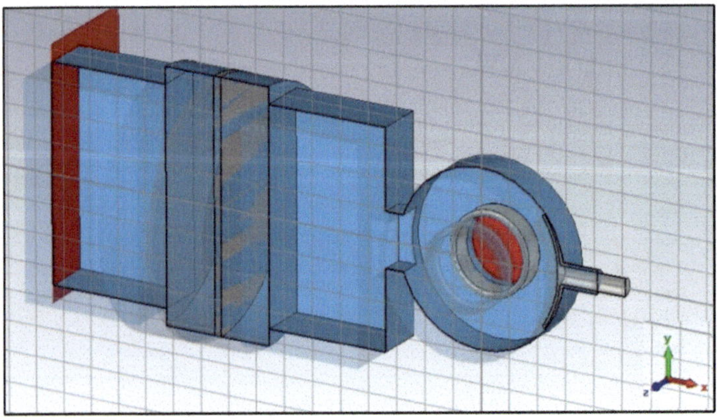

Fig. 13. Klystron output cavity with the RF window.

3 Conclusion

In conclusion, RF, thermal design and multipacting analysis of RF window is completed. Feasibility of design will be checked by fabricating prototype.

Acknowledgement. Authors would like to thank Dr. G.Joshi, Head, ACnD, BARC and Dr. Rajesh Kumar, Head, IADD, BARC for their support. Authors would also like to thank CSIR-CEERI, Pilani Klystron development team for technical input.

References

1. Neubauer, M.: Rugged ceramic window for RF applications – phase 1, Nov. 2012, DE-FG02-08ER85171
2. Thielk, S.U., Agustsson, R., Carriere, P., Kutsaev, S.V.: Evaluation of low-loss alumina material for high-power RF windows, 14th International Particle Accelerator Conference. J. Phys.: Conf. Ser. **2687**(2024), 082044. IOP Publishing https://doi.org/10.1088/1742-6596/2687/8/082044
3. Meng, X., Xu, Y., Lian, Z., Wang, J., Dan, W., Qi, K.: Secondary roughness effect of surface microstructures on secondary electron emission and multipactor threshold for PTFE-filled and PI-filled single ridge waveguides. J. Phys. D: Appl. Phys. **57** (2024)
4. Chunjiang, B., Tiancun, H., He, Y., Guanghui, M., Rui, W., Na, Z., Wanzhao, C.: Analysis of secondary electron emission using fractal method. Chin. Phys. B **30** (2021)
5. Gibaru, Q., Inguimbert, C., Belhaj, M., Toulouse, F., Raine, D.: Lambert, Monte Carlo Simulation Of 3d Surface Morphologies For Secondary Electron Emission Reduction (2021)
6. Larciprete, R., Grosso, D.R., Commisso, M., Flammini, R., Cimino, R.: The Chemical Origin of SEY at Technical Surfaces (2013)
7. Vaughan, J.R.M.: A new formula for secondary emission yield. IEEE Trans. Electr. Dev. **36**, 1963–1967 (1989)
8. Vaughan, R.M.: Secondary emission formulas. IEEE Trans. Electr. Dev. **40**, 830 (1993)
9. Design and Development of 6 MW Peak, 24 kW Average Power S-band Klystron, Updated Know-How Document, June 2020, CSIR-CEERI Pilani Report

Thermal Simulation of L-Band Tunable Pulse Magnetron

Shivendra Maurya[1,2(✉)], N. Kanagaraj[1], and Rajendra Kumar Verma[1,2]

[1] CSIR-Central Electronics Engineering Research Institute, Pilani 333 031, India
smaurya@ceeri.res.in
[2] AcSIR-Academy of Scientific and Innovative Research, Ghaziabad 201 002, India

Abstract. In this paper we present the thermal simulation results of a 300 kW L-band tunable pulsed magnetron using thermo-electric module of commercial software ANSYS Multiphysics. The resonant structure of this magnetron is a double ring hole and slot type structure with oxide coated cathode. The heater voltage and current are in the range of 9–10 V and 15–16 A respectively. The electrical load in term of heater voltage has been applied between both terminals of the heater. It is predicted by simulation that the total power dissipation in the magnetron anode block can be easily managed by the forced air cooling, and rise in temperature will not affect the frequency of oscillation of magnetron.

Keywords: Interaction structure · Pulsed magnetron · Thermal analysis · Oxide coated cathode · Particle-in-Cell (PIC) simulation

1 Introduction

The magnetron is a highly efficient, high power and low-cost device for generating microwaves from kW to GW-levels at centimeter wavelengths [1]. These magnetrons may be either CW or pulsed magnetrons. One of the major applications of CW magnetron at 2.45 GHz frequency is for microwave heating applications. Other frequency of interest for industrial heating and UHF communication application is 915 MHz. Magnetrons with CW power from 50 to 100 kW at 915 MHz are commercially available and are being used for different heating applications. The pulse magnetron at 915 MHz have limited application mainly in UHF communication and Radars. Earlier thermal design of magnetrons was little known due to non-availability of commercial thermal design tools. But now days 3 D commercial software are available and efficient thermal design of magnetrons are possible. The most favorable cathode for this particular type of magnetron is indirectly heated triple carbonate (Ba-Sr-Ca) CO3 oxide coated cathode. Heater made of tungsten wire with 3% rhenium is used to heat the emitting surface of the cathode to the required temperature (~850–950 °C). The thermal simulation has been carried out in three stages i.e. Thermal simulation of helical heater, Thermal simulation of heater-cathode assembly, and thermal simulation of complete magnetron including heat generated by heater as well as electron impact on anode surfaces. The paper has been divided in four sections including introduction Sect. 1. Section 2–4 presents the thermal

S. K. Ghosh et al. (Eds.): VEDA 2024, SPPHY 424, pp. 310–314, 2025.
https://doi.org/10.1007/978-981-96-5174-0_45

simulation of the heater, cathode and complete magnetron followed by conclusion in Sect. 5.

2 Thermal Simulation of Helical Heater

ANSYS is one of the best tools for the thermal design and analysis of vacuum electronic devices and is being successfully used by many authors [2–4]. The helical heater of this magnetron is made of tungsten-rhenium alloy. The rhenium is added to improve the ductility to the heater [5]. The optimized heater diameter and number of turns has been taken as 0.94 mm and 24 respectively. The required input voltage (9–10 V) has been applied between the heater terminals to create resistive heating of the heater. Figure 1 shows the variation of temperature on heater surface for a heater voltage of 10 V. The maximum temperature achieved is 1358^0C which is sufficient to heat the emitting surfaces of the cathode to the required temperature 850–950 °C [6].

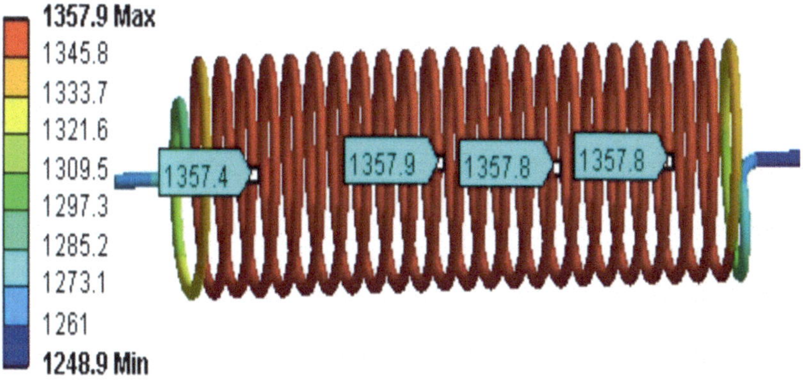

Fig. 1 Temperature distribution on heater surfaces.

3 Thermal Simulation of Helical Heater and Cathode

The proposed cathode for L-band magnetron under discussion is an indirectly heated triple oxide coated (carbonates of Ba, Sr, and Ca) cathode. So, the main role of the hater is to heat the emitting surface of the cathode to a temperature so the electrons can be emitted from the cathode surface having a work function of 1.2–15 eV [7]. Also design of heater should be such that it heats the cathode surface uniformly. The cathode sleeve and endcaps are made by Nickel-222. The required input voltage of 10 V has been applied between the two terminals of the cathode and surfaces to surface boundary condition has been applied between heater and cathode. The heater is heated by resistive heating and cathode by the thermal radiation. The optimum heater voltage i.e. heater power can be obtained by dip test of the cathode [1]. Figure 2 shows the variation of temperature on heater and emitting surfaces of the cathode. It is observed that the temperature of cathode varies by 0.3% along the emitting surface. As this cathode operates in the space charge limiting region and this temperature variation does not affect the performance of the cathode. It is also observed that the heater temperature has been increased by ~4.5% as compared to that of heater only case [8]. This increase is due to surface to surface radiation boundary between outer surface of heater and inner surface of cathode.

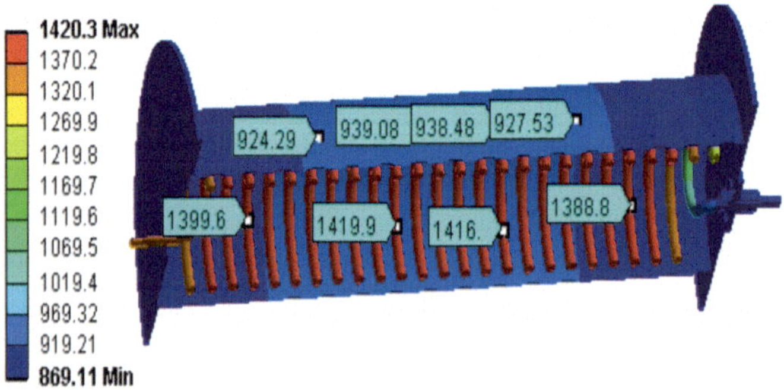

Fig. 2 Temperature distribution on heater and cathode surfaces.

4 Thermal Simulation of Magnetron with Cathode

The complete geometry of the eight vane 300 kW L-band double ring strapped pulse magnetron modelled in solidworks is as shown in Fig. 3. Later on, this model is imported in ANSYS 3D code for thermal simulations [9]. The major parts and corresponding materials are anode block (oxygen free high conductivity copper), heater and cathode support rod (tungsten), heater wire (W 3%R) and heater cathode insulator ceramic. The heater and cathode support rod support the cathode so that cathode is always concentric with the magnetron anode block. They also act as two terminals of heater. The cooling fins have been created for effective cooling and better thermal management of anode block. There are basically two mechanism by which the magnetron gets heated (a) by heater power, and (b) by electron bombardment on anode i.e. the electrons who don't lose their energy effectively, strikes on anode and generate heat energy. These two sources of heating have been modeled in ANSYS. The heater power has been applied directly in terms of voltage between heater and cathode terminals. For this magnetron the 1200 W of power is generated due to beam collision on anode vanes, which has been estimated from CST PIC simulation. This power has been applied as heater flux on inner surfaces of anode. The total inner anode surface area excluding eight slots is found to be 3480 $\times 10^{-6}$ m^2, hence the heat flux to be applied is 1200/3480 $\times 10^{-6} = 3.45 \times 10^6$ W/m^2. The surface to surface radiation boundary has been applied between heater and cathode (inner wall) and cathode (outer wall) and inner surface of anode. The outer surface of the magnetron is cooled by forced air. This cooling has been applied on all the outer surfaces of magnetron taking the heat convection coefficient of air as 500 W/m^{2}K [2].

The temperature distribution on different parts of magnetron is as shown in Fig. 4. The maximum temperature is found to be 1532 °C and minimum temperature is at the

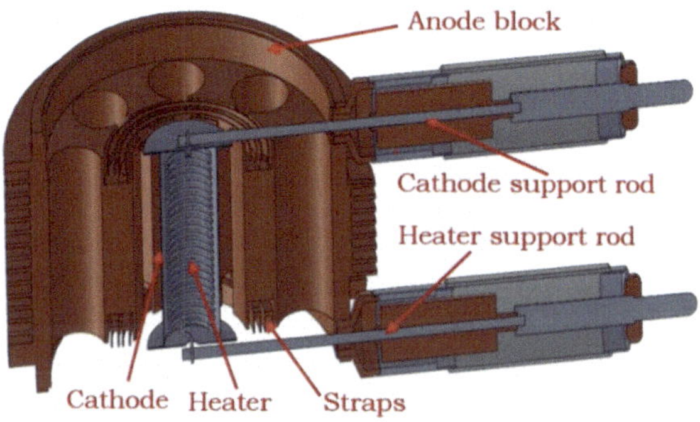

Fig. 3 3D model of complete magnetron.

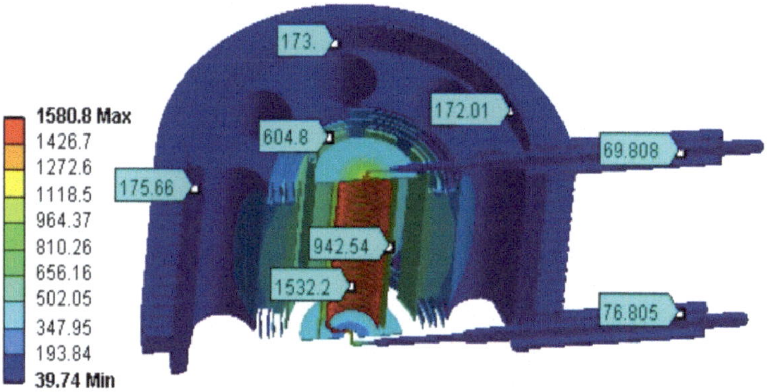

Fig. 4 Temperature distribution on different parts / surfaces of the magnetron.

heater/cathode support rods. The temperature on middle of emitting surface of cathode is found to be 942 °C, which is sufficient for electron emission in oxide coated cathodes. The temperature of the anode inner surfaces is found to be 604 °C, which is quite bellow the melting point of copper and it is expected that will not change the dimension as well as the frequency of the magnetron.

5 Conclusion

Thermal analysis of the magnetron has been also carried out in three stages i.e. heater only, heater-cathode assembly, and complete magnetron tubes. It is predicted that the power loss in the magnetron can be easily managed by the forced air cooling, and rise in temperature will not affect the frequency of oscillation of magnetron.

References

1. Collins, G.B.: Microwave Magnetrons. McGraw-Hill New York (1948)
2. Matlabjon et al.: Thermal analysis of a strapped magnetron. IEEE Trans. Electr. Dev. **58**(8) (2011)
3. Bhat, K.S., Sreedevi, K., Ravi, M.: Thermal analysis of electron gun for travelling wave tubes. Appl. Surf. Sci. **253**(2), 679–682 (2006)
4. Sharma, R.K., Bera, A., Srivastava, V.: Thermal and structural analysis of electron gun assembly for a C-band 60 W space TWT. Int. J. Microw. Opt. Technol. **4**(5), 309–314 (2009)
5. Ramalingam, M.L.: Investigation of sintered tungsten, rhenium additive alloys for high temperature space application. Ph.D. Dissertation, Arizona state university, May 1986
6. Jagdish Kumar et al.: Design considerations and experimental estimation of temperature coefficient of resistance of W-Re heater for high power pulse magnetrons. J. Electromagnetic Wave Appl. 1353926 (2017). https://doi.org/10.1080/09205071
7. Bergman, T.L., Lavine, A.S., Incropera, F.P., Dewitt, D.P.: Introduction to Heat Transfer, 6th edn. Wiley, Hoboken, NJ, USA (2011)
8. Vyas, S. K.: Thermal design of megawatt power-level X-band coaxial magnetron. IEEE Trans. Electr. Dev. **69** (2022)
9. ANSYS, Inc., Canonsburg, PA. http://www.ansys.com

Cold and Beam-Wave Interaction Analysis for Ka Band 500 W Helix Travelling Wave Tube for Earth Station

Jitender Singh(✉), Aashka Oza, Ambrish Ghadiya, Priyanka Natani, Ramagiri Santhosh Kumar, and Shilpi Soni

Space Applications Centre, ISRO, Ahmedabad, India
jitender@sac.isro.gov.in

Abstract. Cold and beam-wave interaction analysis of Periodic Permanent Magnet (PPM) focused Ka-band (27–31 GHz) 500 W ground TWT for satellite uplink application having saturated gain >50dB, electronic efficiency >17% and overall efficiency >50% are presented in this paper. This paper deals with the design and simulation procedure of TWT components mainly focused on electron gun, slow wave structure and multi-stage depressed collector.

Keywords: Helix TWT · SWS · Depressed collector · Dual-Anode electron gun

1 Introduction

The Helix TWT (traveling wave tube) is an important part of the modern electronic communication systems. Helix TWTs are most widely used amplifiers in applications where wide bandwidth, high efficiency and light-weight is desired [1].

A CW (continuous wave) TWT delivering saturated RF output power of 500 W operating in the frequency band 27–31 GHz designed and simulated for satellite uplink application in earth stations. While the TWT can deliver 500 W saturated output power, its usual operation will be at 350 W output power in the input back-off.

A dual anode electron gun designed to provide beam current of 200 mA and beam radius of 0.2 mm at cathode voltage of 14.5 kV. To focus the high current density electron beam, confined flow focusing technique is used. PPM parameters are estimated using analytic equations and optimized in CST [2] particle tracking solver to achieve 100% DC beam transmission. Helix tape and support rods parameters were selected based upon dispersion and interaction impedance requirement. To suppress the BWO (Backward wave oscillation) with sufficient margin, a dual section sever is employed. Helix pitch profile and attenuator loss coating profile simulated and optimized using large signal modeling code. Multi-stage depressed collector is employed to recover the power from the spent beam to achieve overall tube efficiency greater than 50%.

S. K. Ghosh et al. (Eds.): VEDA 2024, SPPHY 424, pp. 315–321, 2025.
https://doi.org/10.1007/978-981-96-5174-0_46

2 Design Procedure

Helix TWT consists of three main sub-assemblies namely electron gun, slow-wave-structure and multistage-depressed collector. In this section, design and simulation procedure of the individual assemblies and the integrated particle simulation procedure is described.

2.1 Slow Wave Structure (SWS)

Helix SWS consists of a tape-helix in the center and three rectangular dielectric support rods around it for mechanical support and thermal conduction path [3]. Tape and dielectric rod parameters are selected to have required interaction impedance and dispersion characteristics for sufficient beam-wave interaction over wide bandwidth along with the good heat draining capability to conduct the heat generated due to RF losses or interception current. A single pitch is modeled in CST Microwave studio [4], as shown in Fig. 1, for high frequency dispersion characteristics and interaction impedance calculation. T-shaped dielectric rods used to obtain high interaction impedance. Optimized single cell parameters are listed in Table 1.

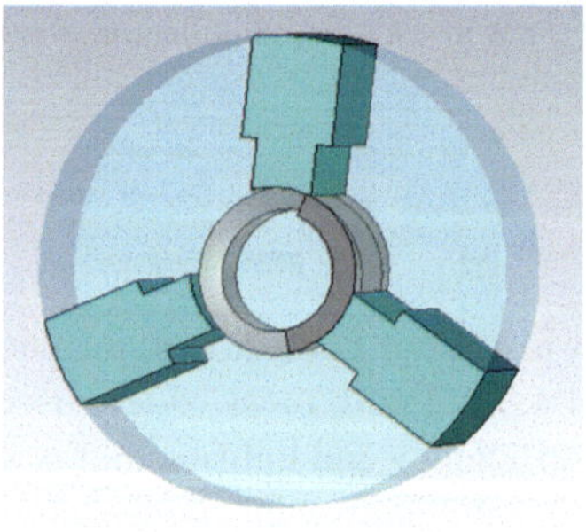

Fig. 1. Single cell helical SWS

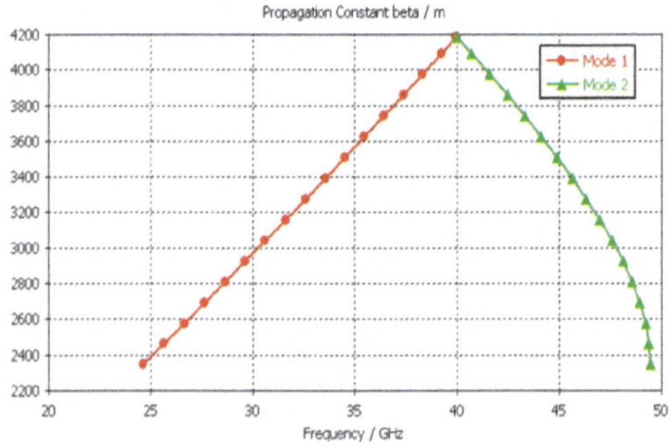

Fig. 2. Dispersion characteristics Pi-point frequency

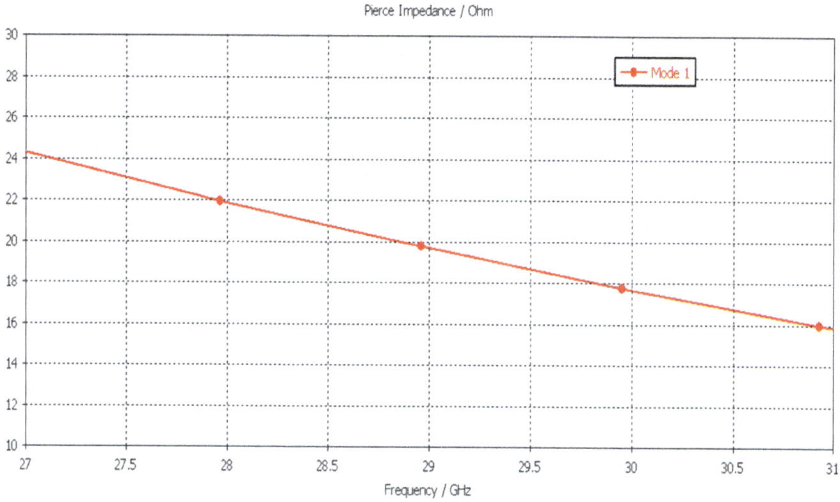

Fig. 3. Interaction impedance variation over frequency

Table 1. Helix SWS parameters

Parameters	Normalized value
Pitch	1
Helix ID	1.09
Barrel ID	4.25
Helix tape width	0.4
Helix tape thickness	0.2

For a given electron beam current and operating frequency, backward wave is suppressed by keeping the length of the uniform pitch section below the critical length at the operating frequency required to excite backward wave [5]. Figure 2 shows the fundamental mode and first negative space harmonic mode dispersion diagram. The intersection of both the curves indicates the pi-point frequency.

Propagation constant (β) and Pierce impedance (K) at various pitches calculated for 1-D large signal simulation of the SWS. Phase velocity variation as a fraction of speed of light, c is 0.216–0.21 and interaction impedance variation (Fig. 3) is 24–16 Ω across the frequency band 27–31 GHz. SWS parameters: β, K, helix dimensions and beam parameters: beam voltage, beam current are the inputs to the large signal code for simulating beam-wave interaction. A three section SWS or two sever section design is employed so that growth of backward wave can be suppressed to prevent the BWO. Attenuator loss coating value calculated to provide attenuation to the backward wave with sufficient margins. Input section pitch is optimized to have maximum gain, mid output section for maximum bunching and in the end, a negative step taper for maximum electronic efficiency. Schematic of the final optimized helix pitch profile is shown with normalized pitch values in Fig. 4.

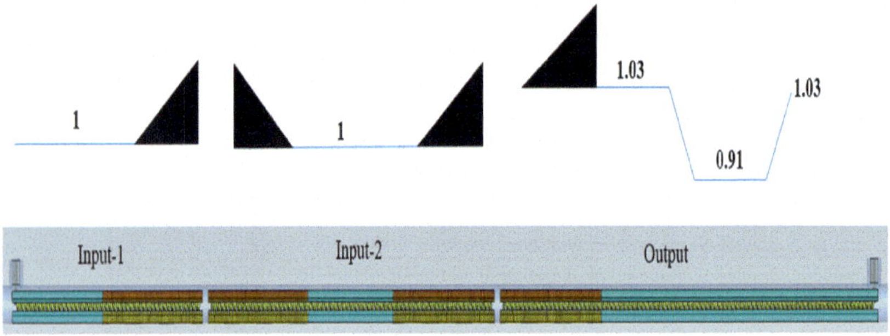

Fig. 4. Schematic diagram of Helix SWS normalized pitch profile

PIC (Particle-in-cell) simulation is carried out to ensure the RF performance and beam focusing using PPM (periodic permanent magnet) structure. For CST simulation conductivity of the helix material considered of Copper (5.8×10^7 S/m) and the loss tangent of the dielectric material is taken as 0.0005.

2.2 Electron Gun

A dual-anode electron gun [6] is designed to generate a well-focused electron beam of desired parameters (14.5 kV, 200 mA). Gun assembly comprises cathode, focusing electrode and ground anode as well as two anodes: Control anode for cathode current regulation and Ion barrier anode for improved cathode reliability. Electron gun geometry modeled in CST particle tracking solver with appropriate boundary conditions to simulate the gun performance. Cathode flux linkage optimized to improve beam laminarity to minimize beam scalloping as it traverses through the SWS tunnel [7]. Figure 5 shows CST tracking solver simulated particle trajectories.

2.3 Multistage Depressed Collector

By recovering the energy from the spent beam, heat dissipation in the collector reduces resulting in increase in overall tube efficiency. Using the spent beam energy diagram (Fig. 6) across the band, four-stage collector electrode potentials are calculated. Depressed collector resulted in overall tube efficiency greater than 50%. Collector electrode geometry is designed to maximize the power recovery along with minimum electron back-streaming [8]. Secondary electron emission (SEE) properties of copper are considered for the collector material in the simulation.

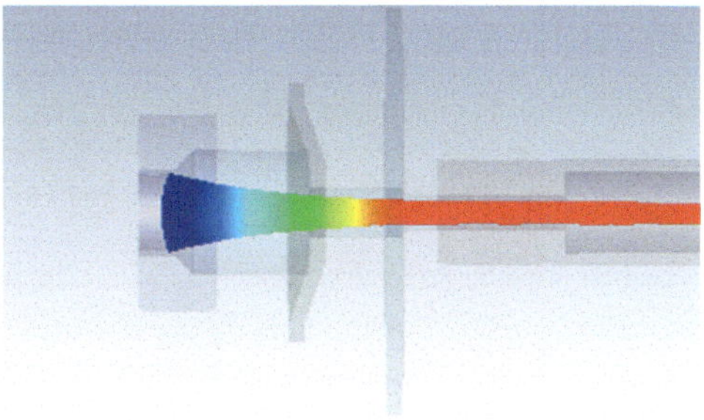

Fig. 5. Simulated particle trajectories

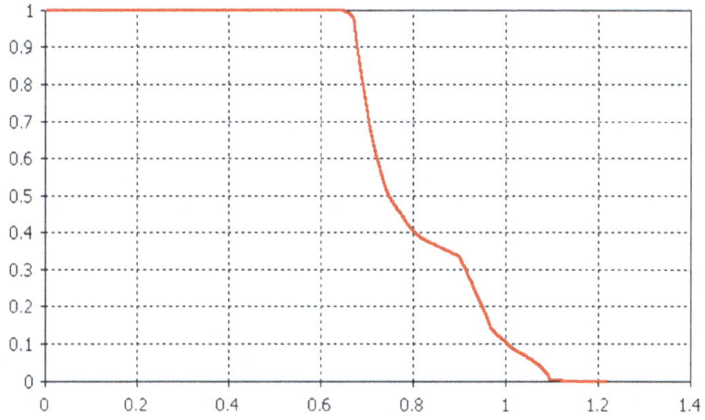

Fig. 6. Spent beam energy diagram

3 Results and Discussion

Integrated TWT simulation carried out by utilizing mutual interaction of various solvers such as electrostatic solver for collector electrodes potential profile, magneto-static solver for PPM focusing field and PIC solver for beam-wave interaction simulation. Table 2 shows the final simulated performance of Ka band (27–31 GHz) TWT. Figure 7 shows the CST simulation result of the amplified output signal. Saturated RF output power in excess of 500 W achieved over the frequency band with 100% beam transmission. Confined flow PPM focusing is used for efficient beam focusing so that the electron beam is not sensitive to perturbations. Figure 8 shows the output signal frequency spectrum in RF saturation condition. At saturation, second harmonic level is −40 dBc. Both the input and output couplers are designed using coaxial transformers to match helix impedance across frequency band, further couplers use a door-knob structure to convert coaxial port to standard WR34 waveguide.

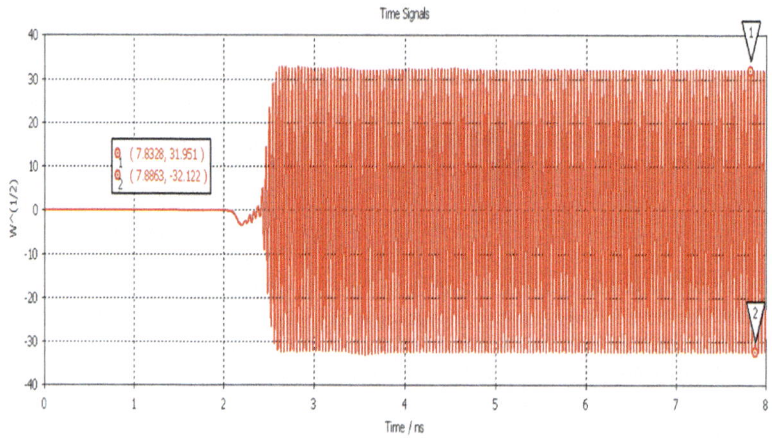

Fig. 7. CST PIC solver simulated saturated RF output power

Table 2. Ka band TWT simulated performance

Parameters	Value
Frequency band	27–31 GHz
Sat. output power	500 W min
Gain @ sat	50 dB
Second harmonic	−40 dBc
Electronic efficiency	>17%
Collector efficiency	>80%
TWT efficiency	>50%

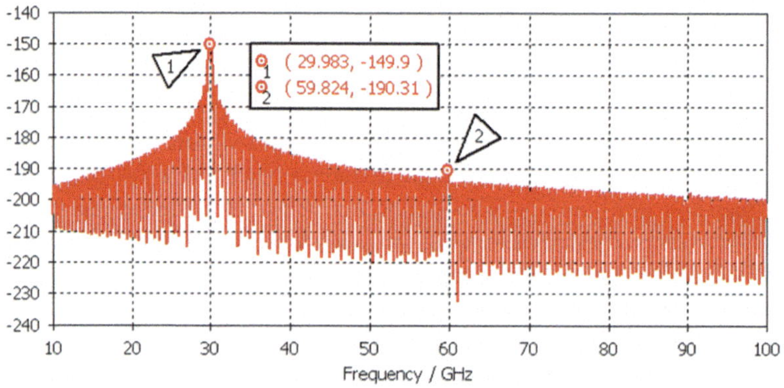

Fig. 8. Output signal frequency spectrum

4 Conclusion

In this paper, a Ka-band 500 W ground TWT is designed according to the detailed design procedure mentioned in Sect. 2. Initial design of TWT was done using large signal analytical code and further the design is verified and optimized with the help of CST particle studio. High efficiency of the tube was ensured by using velocity tapering in the output section and four-stage-depressed collector to minimize heat dissipation. Multi-sever design was employed to prevent BWO conditions. Simulated RF performance of the TWT is encouraging.

Acknowledgments. The authors would like to thank Director, SAC and Deputy Director HSTA, SAC for their constant support and encouragement.

References

1. Chong, C.K., Menninger, W.L.: Latest advancements in high-power millimeter-wave helix TWTs. IEEE Trans. Plasma Sci. **38**(6), 1227–1238 (2010). https://doi.org/10.1109/TPS.2010.2041940
2. Safi, D., Birtel, P., Meyne, S., Jacob, A.F.: A traveling-wave tube simulation approach with CST particle studio. IEEE Trans. Electron Devices **65**(6), 2257–2263 (2018)
3. Gilmour, A.S.: Principles of traveling wave tubes. Artech House Radar Library. Artech House, Boston (1994)
4. Computer Simulation Tool CST. www.cst.com
5. Feng, Y., Feng, J.: Design and simulation of multi-severs high frequency structure of millimeter wave helix TWT. In: 2015 8th UK, Europe, China Millimeter Waves and THz Technology Workshop (UCMMT), Cardiff, UK, pp. 1–4 (2015). https://doi.org/10.1109/UCMMT.2015.7460592
6. Ghadiya, A., Kumar, R.S., Soni, S., Singh, D.K.: Design and development of thermally stable dual anode electron gun for space applications. In: National Conference on Emerging Trends in Vacuum Electronic Devices & Applications, 19–21 January 2023
7. Singh, J., Oza, A., Ghadiya, A., Kumar, R.S., Soni, S.: Optimization of Beam Shaver Section for Helix TWT. In: 2023 IEEE Microwaves, Antennas, and Propagation Conference (MAPCON), Ahmedabad, India, pp. 1–4 (2023). https://doi.org/10.1109/MAPCON58678.2023.10464054
8. Deng, W.-K., et al.: Multiobjective design optimization and experimental verification of a high-performance 200-W Ku-Band CW helix TWT. IEEE Trans. Electron Devices **70**(7), 3878–3884 (2023). https://doi.org/10.1109/TED.2023.3279305

Design and Simulation of High Efficiency Ku Band 200 W Helix TWT

Aashka Oza(✉), Jitender Singh, Ambrish Ghadiya, Priyanka Natani, Ramagiri Santhosh Kumar, and Shilpi Soni

Space Applications Centre, ISRO, Ahmedabad, India
aashkaoza@sac.isro.gov.in

Abstract. This paper presents design details of Ku band 200W TWT with minimum saturated gain of 55 dB, electronic efficiency >30% and overall efficiency >60% in the frequency band of 10.7–11.7 GHz. This paper includes simulation results of TWT subassemblies as well as overall TWT performance. To achieve high electronic efficiency and interaction impedance, helix SWS parameters were optimized. The large signal modeling and simulations for SWS were carried out using large signal analytical code and Computer Simulation Technology (CST) particle studio [1]. Input section used a positive linear pitch taper for improving margins with respect to backward wave oscillation. Output section used a negative step taper for high electronic efficiency [2]. The Periodic Permanent Magnet (PPM) structure was optimized for optimum beam transmission [3]. Multi-stage depressed collector was designed to recover maximum spent beam power along with minimum electron backstreaming. Input coupler is a coaxial type coupler due to low power handling and high bandwidth requirement. Since the output coupler is required to withstand high power, a waveguide based coupler design (doorknob structure) was adopted [4, 5].

Keywords: High efficiency Helix TWT · PPM · Depressed collector · Dual-anode electron gun

1 Introduction

In space applications, the High Power Amplifier (HPA) is expected to be power efficient, wide-band and highly reliable. At higher frequencies, Helix Traveling Wave Tube Amplifier (TWTA) is more suitable than Solid State Power Amplifier (SSPA) as an HPA to meet payload requirements. A Ku band Helix TWT (10.7–11.7 GHz) has been designed and simulated within the context of communication payloads. The performance specifications are given in Table 1.

The electron gun is a pierce gun based design with beam voltage of 6.8kV and current of 102 mA. Dual anode configuration [6] is adopted wherein the Control Anode (CA) will be used for controlling the beam current throughout the operational life of TWT and the Ion Barrier Anode (IBA) is used for repelling any positive charges, generated in TWT structure beyond the electron gun, away from the cathode. A two section helix

S. K. Ghosh et al. (Eds.): VEDA 2024, SPPHY 424, pp. 322–329, 2025.
https://doi.org/10.1007/978-981-96-5174-0_47

Slow Wave Structure (SWS) is designed. The initial pitch profile is estimated using large signal code and subsequently the SWS is optimized in a 3D electromagnetic field simulator, CST (Computer Simulation Technology) [7], for RF performance. The RF input is transmitted into the SWS through input coupler and received at output coupler. The coupler design is influenced by bandwidth, size and power handling requirements. Both the input and output couplers are coaxial transformers. However, the input coupler is further transformed to SMA while the output coupler is transformed to WR 75 waveguide by the means of a door knob structure. The coupler designs are carried out in CST-MWS (CST Microwave Studio). For efficient collection of spent beam, characterized by minimum collector dissipation and minimum backstreaming of electrons from collector, a four stage depressed collector is considered.

Table 1. Ku band 200W TWT specifications.

Parameters	Value
Frequency band (GHz)	10.7–11.7
Saturated output power (W)	>200
Saturated gain (dB)	>55
Overall efficiency (%)	>60

2 Design and Simulation

This section describes the design and simulation procedure for individual assemblies of slow wave structure, electron gun and couplers. Integrated 3D EM particle simulation for slow wave structure and multistage depressed collector is also described.

2.1 Slow Wave Structure

To enhance the output power of TWT, high interaction impedance is required in SWS. The helix SWS comprises a helix and three T-shaped dielectric rods for mechanical and thermal support to the helix [8]. The key helix design parameters which help determine the extent of beam-wave interaction are helix tape size, helix pitch and helix mean radius. Table 2 summarizes the parameters chosen to optimize the dispersion characteristics of the SWS. The dispersion characteristics (Beta and K) are obtained from simulation of single pitch long section of SWS in CST. Beta and K at 11.2 GHz for various pitches are shown in Figs. 1 and 2.

The calculated propagation constants and interaction impedances for various pitches are input into large signal code to optimize the initial pitch profile. To suppress backward wave with sufficient margins, an appropriate amount of attenuator loss coating is assumed and the input helix is tapered. While the input section is optimized for maximum gain, the output helix is optimized for maximum efficiency and power. The pitch profile and

Table 2. Normalized values.

Parameters	Value
Pitch	1
Helix ID	1.81
Barrel ID	6
Helix tape width	0.54
Helix tape thickness	0.27

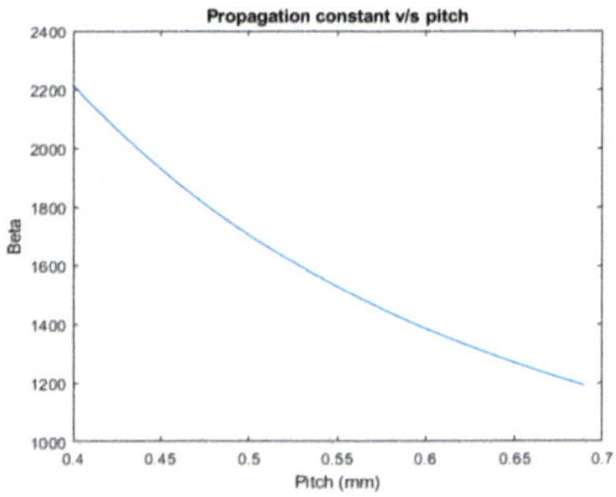

Fig. 1. Propagation constant.

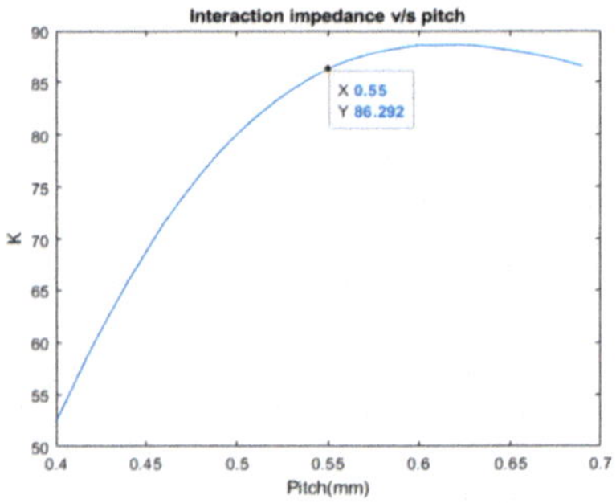

Fig. 2. Interaction impedance.

the resulting gain and power characteristics are shown in Figs. 3, 4 and 5. The blue, red and yellow characteristics are for 10.7 GHz, 11.2 GHz and 11.7 GHz respectively.

The Slow Wave structure with the initial estimates obtained from the large signal code are modeled and simulated in CST. Using CST PIC (Particle-in-cell) solver, further optimization of the helix SWS is carried out to obtain output power greater than 225W. In the CST model, the helix is simulated as copper plated and attenuator rods are simulated with a triangular rod loss coating with total loss of 10dB per rod, distributed along 14mm length.

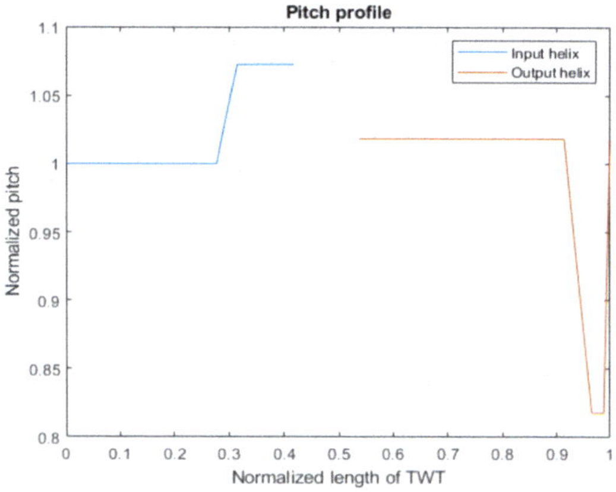

Fig. 3. Normalized helix pitch profile.

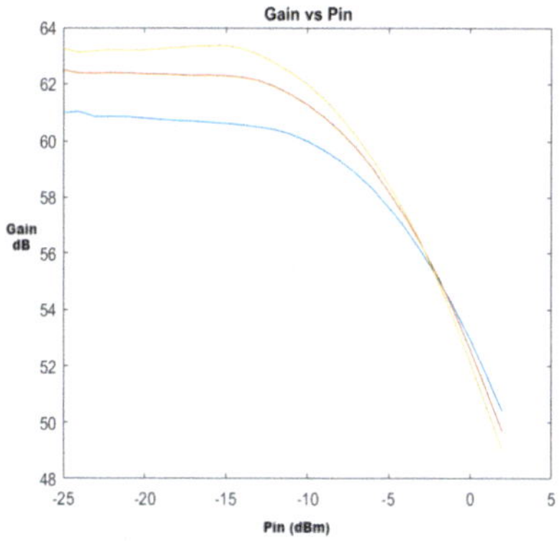

Fig. 4. Gain characteristics.

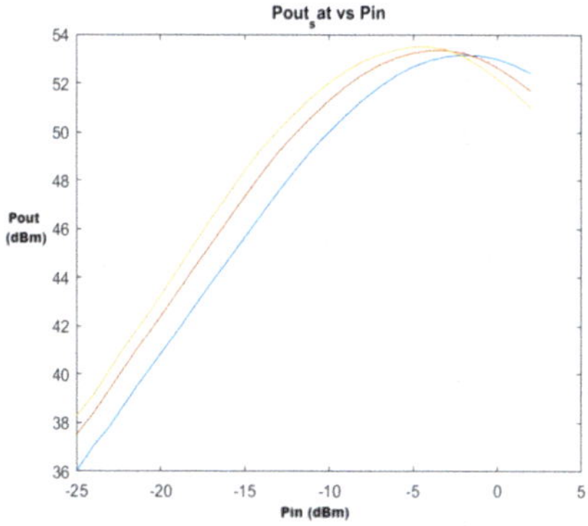

Fig. 5. Power characteristics.

The electron beam is focused using Periodic Permanent Magnet (PPM) structure to minimize scalloping and hence helix interception current. The PPM structure's magnetic field profile is shown in Fig. 6. The coupler design is motivated by minimum reflection and minimum transmission loss. The input coupler, a coaxial design, is optimized for impedance matching in the desired frequency range of 10.7 GHz to 11.7 GHz. Similarly, the output coupler adopts a door knob structure design. The simulated return loss results of couplers are shown in Fig. 7(a) and Fig. 7(b).

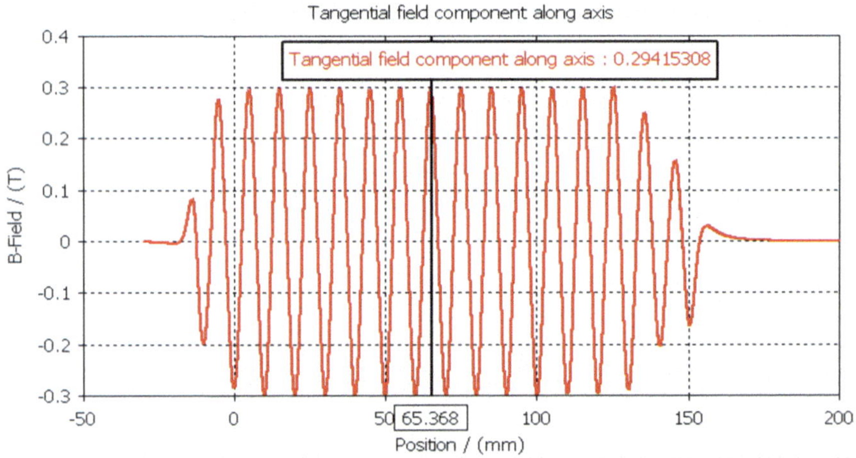

Fig. 6. The periodic permanent magnet profile employed for beam focusing with peak magnetic field of 2950 Gauss and period of 10 mm.

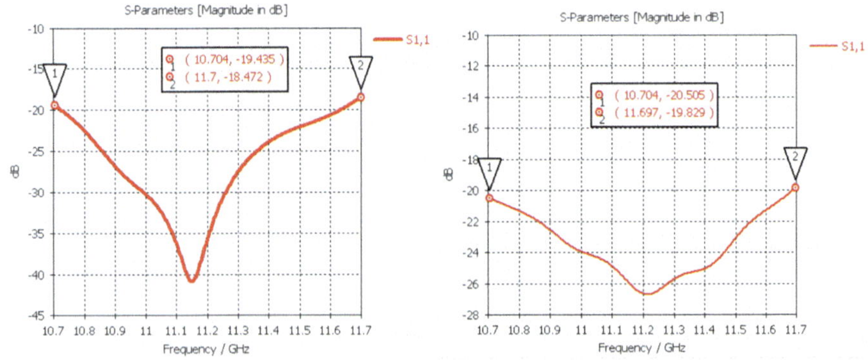

Fig. 7. (a) Input return loss with SMA connector (Marker 1: 10.7 GHz:RL = −19.43 dB, Marker 2: 11.7 GHz:RL = −18.47 dB), (b) Output return loss with waveguide type coupler (Marker 1: 10.7 GHz:RL = −20.5 dB, Marker 2: 11.7 GHz:RL = −19.83 dB).

2.2 Electron Gun

The electron gun design is planned on a Pierce type model with dual anode configuration. It consists of M-type dispenser cathode, beam focusing electrode, control anode and ion barrier anode. Optimization of inter electrode distances, electrode shapes and potentials is carried out for obtaining desired beam parameters in CST TRK (Track) solver. Table 3 lists the final optimized electron gun parameters.

Table 3. Electron gun parameters.

Parameter	Desired value
Beam current (mA)	102
Cathode voltage (kV)	−6.8
Beam focusing electrode voltage (kV)	−6.81
Control anode voltage (kV)	−2.69
Ion barrier anode voltage (V)	200

2.3 Multistage Depressed Collector

The overall TWT efficiency is improved by maximizing spent beam recovery in the collector. The collector voltages are designed using the spent beam energy data from SWS simulation. A multistage depressed collector configuration is taken up for minimizing collector dissipation and electron backstreaming. Secondary Electron Emission (SEE) of electrodes is also considered in simulation. The integrated simulation of SWS and collector is carried out in CST PIC solver. The simulation output includes the RF power output (shown in Fig. 8), helix interception current and dissipation in all collector stages as shown in Table 4. Several iterations are performed to optimize the above parameters for minimum collector dissipation across frequencies. Worst case collector dissipation is 114.7 W at 11.7 GHz. In the DC case, the collector dissipation is 67.8W. The overall achieved performance is summarized in Table 5.

Table 4. Integrated simulation output.

Frequency (GHz)	Output power (W)	Interception current (mA)	Collector 1 dissipation (W)	Collector 2 dissipation (W)	Collector 3 dissipation (W)	Collector 4 dissipation (W)
10.7	232	0.57	31.26	39.96	19.43	16.8
11.2	228	0.58	34.33	38.16	18.26	19.67
11.7	225	0.8	35.37	35.13	17.76	23.81
DC	–	0.02	0.88	0.28	18.63	48

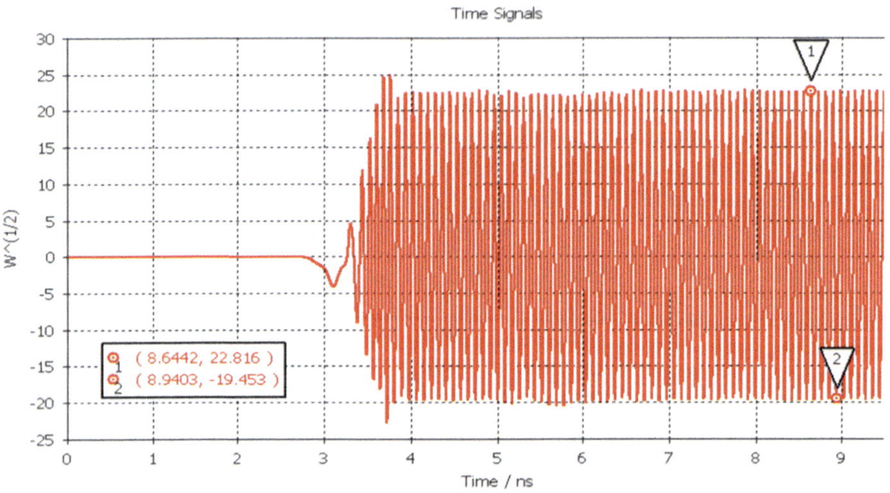

Fig. 8. Amplified RF output signal.

Table 5. Ku band 200 W TWT performance.

Parameter	Desired Value
Frequency band (GHz)	10.7 - 11.7
Saturated output power (W)	>200
Saturated gain (dB)	>55
Total phase shift (deg)	<60
Electronic efficiency (%)	>30
Collector dissipation (W)	115
Collector efficiency (%)	>72
TWT efficiency (%)	>60

3 Conclusion

Ku band 200W TWT is designed for space applications as per the details shown in the preceding section. Helix parameters were carefully chosen for achieving optimum interaction impedance and improving efficiency of the TWT. A two section helix TWT is designed using large signal code. Further optimization of various assemblies of TWT is done in CST solvers. Additional improvement in efficiency is expected by using graphite coated collector electrodes [9].

Acknowledgments. The Authors are grateful to Space Applications Center, Ahmedabad for enabling them to work on microwave power amplifiers for space application. Authors would like to thank Deputy Director HSTA, and Director SAC for their invaluable guidance, encouragement and support.

References

1. Safi, D., Birtel, P., Meyne, S., Jacob, A.F.: A traveling-wave tube simulation approach with CST particle studio. IEEE Trans. Electr. Dev. **65**(6), 2257–2263 (2018)
2. Deng, W.-K., et al.: Multiobjective design optimization and experimental verification of a high-performance 200-W Ku-band CW helix TWT. IEEE Trans. Electr. Dev. **70**(7), 3878–3884 (2023). https://doi.org/10.1109/TED.2023.3279305
3. Singh, J., Oza, A., Ghadiya, A., Kumar, R.S., Soni, S.: Optimization of beam shaver section for helix TWT. In: 2023 IEEE Microwaves, Antennas, and Propagation Conference (MAPCON), Ahmedabad, India, pp. 1–4 (2023). https://doi.org/10.1109/MAPCON58678.2023.10464054
4. Alaria, M.K., Ghosh, S.K.: Design of coaxial and waveguide couplers for helix TWT. In: Clerk Maxwell, J. (ed.) A Treatise on Electricity and Magnetism, vol. 2, pp. 68–73 (1892)
5. Oza, A., Singh, J., Ghadiya, A., Kumar, R.S., Soni, S.: Design of couplers incorporating fabrication limitations for helix TWT. In: 2023 IEEE Microwaves, Antennas, and Propagation Conference (MAPCON), Ahmedabad, India, pp. 1–4 (2023). https://doi.org/10.1109/MAPCON58678.2023.10463936
6. Ghadiya, A., Kumar, R.S., Soni, S., Singh, D.K.: Design and development of thermally stable dual anode electron gun for space applications. In: National Conference on Emerging Trends in Vacuum Electronic Devices & Applications, 19–21 January 2023
7. Computer Simulation Tool CST. www.cst.com
8. Gilmour, A.S.: Principles of traveling wave tubes. Artech House Radar Library. Artech House, Boston (1994)
9. Srivastava, V., Sharma, A.K., Bera, A., Sharma, R.K., Joshi, S.N.: Characterisation of multistage graphite collector for high efficiency TWT. In: National Conference on Microwaves, Antenna, Propagation and Remote Sensing, ICMARS-05, Dec. 20–22 (2005)
10. Carter, R.G.: Microwave and RF Vacuum Electronic Power Sources. Cambridge University Press

Slow Wave Structure for D Band Traveling Wave Tube Amplifier for High-Power FMCW Radar

C. Surya Prasath and Richards Joe Stanislaus(✉)

Vellore Institute of Technology, Vellore 600127, India
{suryaprasath.c,richards.stanislaus}@vit.ac.in

Abstract. Traveling Wave Tube Amplifiers (TWTAs) are essential for high-power RF amplification in applications such as satellite communications, radar systems, and electronic warfare. Key to enhancing the output power of TWTs in the millimeter wave frequency range is the increase of interaction current capacity and interaction impedance. A design novelty involves tilting the vertical section length by a phase angle of 30 degrees, which extends the interaction region within the Slow Wave Structure (SWS) and maximizes power output suitable for radar applications. In the D-band (130–150 GHz), preliminary analysis has shown an achievable bandwidth of 15% in modified Folded waveguide SWS. Utilizing a round electron beam configuration is crucial, as it ensures maximum electric field distribution at the centre, thereby enhancing the gain of the SWS. Within a square beam tunnel, this system employs a beam current of 70 mA with a beam velocity of 0.303 times the velocity of light. An axial magnetic field of 0.3T is applied to confine the electron beam within the beam tunnel. In the particle in cell simulations, with an input power of 62.5 mW, this design yields a minimum gain of 25.5 dB across the band and reaches a maximum gain of 28.9 dB at 141 GHz.

Keywords: Modified folded waveguide · Interaction impedance · Particle in cell simulation · slow wave structure · Dispersion analysis

1 Introduction

The most common vacuum electronic devices that amplify electromagnetic waves include Klystron amplifiers, Magnetrons, crossed-field amplifiers, and Traveling wave tube amplifiers (TWTAs) [1–5]. Since vacuum devices offer relatively larger output power than solid-state sources, a considerable amount of research has been done recently on vacuum sources expanding into the millimeter wave range. Due to the significant demands placed on high-power radiation sources operating at 140 GHz for their use in high-speed communications, homeland security, Radar, and biomedical imaging applications [6, 7]. Research on D-band frequency source TWTAs became crucial for domestic and international scientists [8–10]. Large areas were covered by the Radar wireless system by using multiple front-end amplifiers powered by TWTs. One of the objectives of the proposed TWT design strategy was low cost and adequate performance. This enabled the development of a low-cost power amplifier suitable for the Frequency

S. K. Ghosh et al. (Eds.): VEDA 2024, SPPHY 424, pp. 330–335, 2025.
https://doi.org/10.1007/978-981-96-5174-0_48

Modulated Continues wave Radar (FMCW) communication [11, 12]. As a result, we designed a Slow wave structure for the D-band TWTA source that combined affordability, output power, and lightweight components for communication systems operating in the millimeter wave range [13–15]. Folded waveguide slow-wave structures (FWG SWS) offer few advantages over other conventional SWS designs, such as helix and coupling cavities. A serpentine-shaped rectangular waveguide with a copper construction, flat dispersion, minimum high-frequency loss, high power capacity, simple processing, and high interaction impedance, the FWG SWS Traveling Wave Tube amplifier was a cutting-edge vacuum tube [16, 17].

In this paper, the authors present a modified folded waveguide [18] SWS (MFW-SWS) for the D-band millimeter wave frequency. The vertical length of the SWS was tilted at a particular angle to form the modified design in the Folded waveguide SWS, which results in higher interaction impedance compared to traditional folded waveguide structures. The dispersion analysis yielded a 12% operational bandwidth in the D-band range with a constant phase velocity of 0.303 [19, 20]. Particle-in-cell simulations were carried out in the completed SWS with a beam voltage of 24.6 kV in the operational bandwidth. A beam current of 70 mA was used to generate a circular pencil beam in the square beam tunnel with high interaction impedance in the SWS. Finally, particle cell simulation in the SWS produced a maximum gain of 28.9 dB [21, 22].

This article is structured into five sections. Section II discusses the SWS structure design using ANSYS Electronic desktop. Section III explains the eigenmode analysis for achieving the flat dispersion characteristics in the D-band. Section IV presents the particle cell simulation (PIC) and gain analysis. Section V presents the conclusion and future work.

2 SWS Unit Cell Design

Table 1. Preliminary value for D band SWS unit cell.

Variable	Description	Values
ax	Waveguide width	1.65
sh	Slanting height	0.96
wb	Waveguide breadth	0.85
ir	Inner bent radius	0.25
btx	Beam tunnel width	0.85
bty	Beam tunnel height	0.55

In this work, a modified folded waveguide SWS with a slanting height (sh) is considered instead of the vertical height of a conventional folded waveguide. This new design increases the region of interaction between the RF signal and electron beam throughout the SWS. In the design of the SWS unit cell with variable parameters - slanting height,

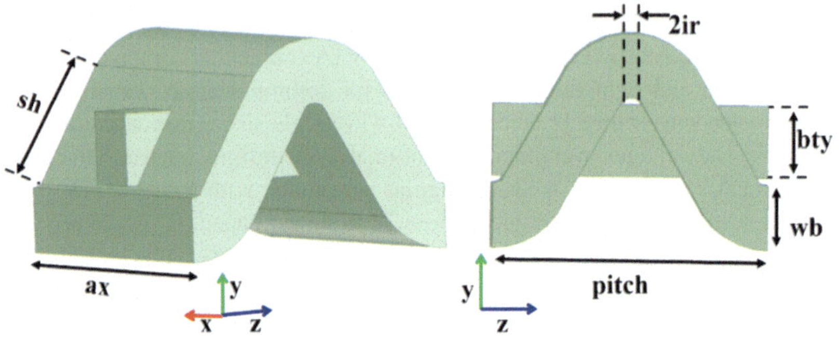

Fig. 1. Unit cell of MFW SWS

waveguide width, waveguide breadth, inner radius, and beam tunnel cross-section, the preliminary values considered are listed in Table 1. The unit cell of the D-band MFW SWS is modeled in ANSYS Electronics desktop (HFSS) as depicted in Fig. 1.

3 Dispersion Analysis

The listed parameters in Table 1 are modified to obtain flat dispersion characteristics across the operational D band range. The optimized values of the MFWG SWS unit cell attained from eigenmode analysis, are mentioned in Table 2. Figure 2 displays the optimized dispersion curve plotted with a line representing a constant phase velocity of 0.303 times the speed of light. The flat dispersion characteristics result in obtaining a 12% operation bandwidth (129–152 GHz) of the SWS being.

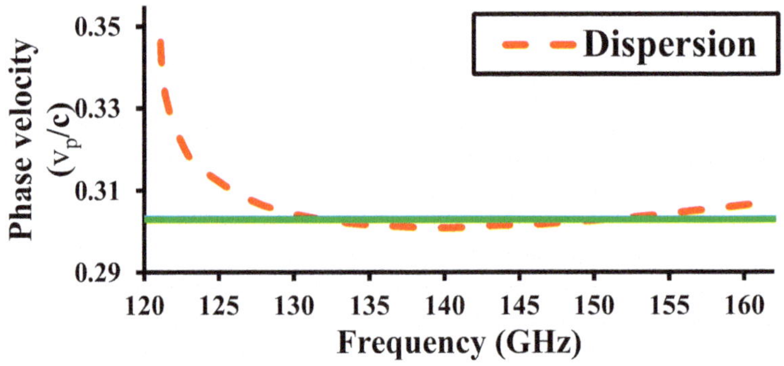

Fig. 2. Dispersion characteristics for MFW SWS unit cell

Table 2. Normalized value of D band SWS unit cell with one pitch distance

Variable	Description	Values (mm)
ax	Waveguide width	1.49
sh	Slanting height	0.58
wb	Waveguide breadth	0.19
ir	Inner bent radius	0.03
btx	Beam tunnel width	0.34
bty	Beam tunnel height	0.34

4 Beam-Wave Analysis

The particle in cell (PIC) simulations of the D-band Slow Wave Structure (SWS) circuit are performed in Dassault System's CST Studio Suite. This circuit consists of two sections without an attenuator, with a total of 80 periods of unit cells of the MFW SWS arranged periodically along the z-axis. To produce a circular electron beam with a beam velocity of 0.303 times speed of light, a beam voltage of 24.6 kV is employed. With a focusing constant magnetic field of 0.3 T, the beam is contained within the SWS with a current density of 216 A/cm^2. An input RF power of 62.5 mW is applied to the input port of the input section. Due to the continued beam-wave interactions across the high interaction impedance region within the SWS, an amplified RF output power with a gain of 28.9 dB is successfully collected at the output port. The PIC simulation is conducted across the entire operational bandwidth, and the resulting gain (dB) versus frequency (GHz) is displayed in Fig. 3. The operational bandwidth obtained from the dispersion characteristic results in a minimum gain of 25 dB at 129 GHz. A gain of over 26 dB is achieved across the operational D bang frequency range of 131–149 GHz with a deviation of less than 3 dB as observed in the gain plot in Fig. 3.

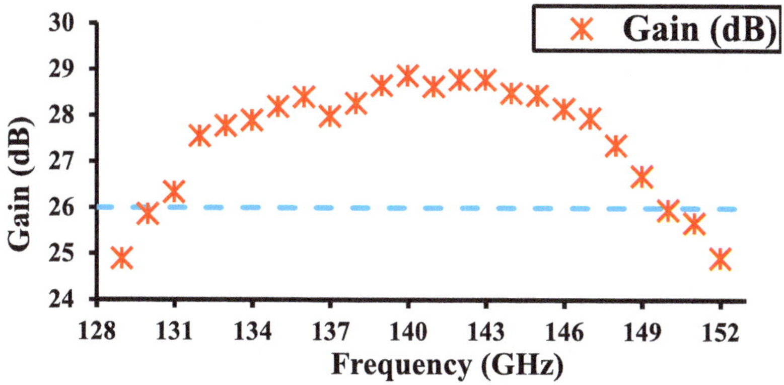

Fig. 3. Gain analysis of D band SWS

5 Conclusion

This research presents the design of the Modified Folded Waveguide (MFW) Slow-Wave Structure (SWS) for D-band traveling wave tubes (TWTs), with significantly enhanced interaction between the electron beam and RF signal. Through dispersion analysis, the structure is optimized to achieve a constant phase velocity of 0.303 times the speed of light. The optimized design demonstrates a maximum gain of 28.9 dB with an output power of 47.3 W and a minimum gain of 26 dB within the operational bandwidth of 131–149 GHz. The improved MFW SWS design not only meets the performance requirements for D-band applications but also ensures broader bandwidth and higher gain. The next step involves manufacturing and testing the SWS to validate the simulation results and affirm its practical performance in real-world applications.

References

1. Chattopadhyay, G.: Technology, capabilities, and performance of low power terahertz sources. IEEE Trans. THz Sci. Technol. **1**(1), 33–53 (2011)
2. Naveen Babu, G., Stanislaus, R.J.: Propagation of electromagnetic waves guided by perfectly conducting model of a tape helix supported by dielectric rods. IET Microwaves, Antennas & Propagation **10**(6), 676–685 (2016)
3. Ives, R.L.: Microfabrication of high-frequency vacuum electron devices. IEEE Trans. Plasma Sci. **32**(1), 1277–1291 (2004)
4. Rodgers, J., Firestone, T., Granatstein, V., Dronov, V., Antonsen, T., Ott, E.: Study and applications of wideband oscillations in high power pulsed traveling-wave tubes. In: Proceedings of the IEEE International Conference on Plasma Science (ICOPS), p. 250 (2005)
5. Komandin, G.A., Chuchupal, S.V., Lebedev, S.P., Goncharov, Y.G., Korolev, A.F., Porodinkov, O.E., Volkov, A.A.: BWO generators for terahertz dielectric measurements. IEEE Trans. THz Sci. Technol. **3**(4), 440–444 (2013)
6. Feng, J., Ren, D., Li, H., Tang, Y., Xing, J.: Study of high frequency folded waveguides BWO with MEMS technology. Teraherlz Sci. Technol. **4**(4), 164–180 (2011)
7. Jinjun, F., Electronics, I.M.V.: Vacuum Eleclronics **2**, 1–7 (2010)
8. Wenqiang, L., Yi, J., Quanfeng, Z., Peng, H., Linlin, H., Lei, Y., Hongbin, C.: Development of D-band continuous-wave folded waveguide traveling-wave tube. In: 2015 IEEE International Vacuum Electronics Conference (IVEC), pp. 1–3. IEEE
9. Wenqiang, L., Yi, J., linlin, H.: Design and experiment research for 0.14 THz folded waveguide traveling wave tubes. J. Terahertz Sci. Electr. Inf. Technol. **12**(3), 334–338 (2014)
10. Stanislaus, J., Gnanamoorthi, N.B.: Large-signal field analysis of a linear beam travelling wave tube amplifier for the anisotropically conducting tape helix slow-wave structure supported by dielectric rods. J. Electromagnetic Waves Appl. **32**(4), 439–470 (2018)
11. Montesano Martínez, D.: Development of low-cost frequency-modulated continuous-wave radars at S and W band. Doctoral dissertation, Telecomunicacion (2019)
12. Fan, Y., Bao, J., Aljumaily, M.S., Li, H.: Communications via frequency-modulated continuous-wave radar in millimeter wave band. In: 2019 IEEE Global Communications Conference (GLOBECOM), USA, pp. 1–7 (2019)
13. Babu, G.N., Stanislaus, R.J., Joshi, S.: Wave propagation characteristics in anisotropically conducting dielectric loaded tape helix slow wave structures. In: IEEE International Vacuum Electronics Conference (2014)
14. Hai-Rong, Y., Jin, X., Ling-Na, Y., Yu-Bing, G., Yan-Yu, W.: A wave beam interaction theory for folded-waveguide traveling wave tubes. Acta Phys. Sinica **61**(24), 244106 (2012)

15. Stanislaus, R.J., Babu, G.N.: Electron transit-time and exit velocity in linear beam travelling wave amplifier for a dielectric rod supported tape-helix slow wave structure, e-FERMAT Communications **2**(21), 1–53 (2017)
16. Tucek, J.C., Basten, M.A., Gallagher, D.A., Kreischer, K.E.: Operation of a compact 1.03 THz power amplifier. In: Proceedings of the 17th IVEC, Monterey, CA, USA, pp. 1–2 (2016)
17. Surya Prasath, C., Ray, I., Sneka, J., Matheshwara, A.S., Stanislaus, R.J.: Preliminary analysis of a folded waveguide slow-wave structure for 160 GHz traveling wave tube amplifier. In: 2021 14th UK-Europe-China Workshop on Millimetre-Waves and Terahertz Technologies (UCMMT), pp. 1–3. IEEE (2021)
18. Stanislaus, R.J.: Wide band high gain modified-folded-wave-guide slow wave structure for millimeter wave traveling wave tube amplifier in 6G and beyond back-haul communications. Physica Scripta (2024). https://doi.org/10.1088/1402-4896/ad88b9
19. Basu, R., Billa, L.R., Letizia, R., Paoloni, C.: Design of sub-THz traveling wave tubes for high data rate long range wireless links. Semicond. Sci. Technol. **33**(12), 124009 (2018)
20. Stanislaus, R.J., Bera, A., Sharma, R.K.: Design and analysis of a wideband staggered double-vane slow-wave structure for W-band amplifier. IEEE Trans. Plasma Sci. **49**(1), 251–257 (2020)
21. Yin, H., Xu, J., Yue, L., Gong, Y., Wei, Y.: A method to calculate output power for sheet-beam traveling-wave amplifiers. IEEE Trans. Electr. Dev. **59**(12), 3630–3634 (2012)
22. Stanislaus, R.J., Bera, A., Sharma, R.K.: Investigation of a sheet beam RF structure with Bragg reflector for W band amplifier. In: 2019 International Vacuum Electronics Conference (IVEC), pp. 1–2. IEEE (2019)

Optimization of Couplers with H-Plane Transition of Helix Traveling-Wave Tubes

Raktim Guha[1,2(✉)] and Sanjay Kumar Ghosh[1,2]

[1] Academy of Scientific and Innovative Research (AcSIR), CSIR-HRDG, Ghaziabad 201002, India
raktim.guha01@gmail.com
[2] Vaccum Electron Devices Group, CSIR-CEERI, Pilani, Rajasthan 333031, India

Abstract. A straightforward and efficient split-model optimization technique, utilizing both analytical and simulation methods termed as Hybrid method, was developed to optimize the coupling system of a helix traveling-wave tube (TWT). This innovative approach addresses the complex challenge of designing RF power coupling into a non-uniform helix pitch segment supported by attenuator-coated support rods within the output section of the slow-wave structure of the TWT. This design task becomes especially intricate when employing an H-plane-end launcher transition rather than more conventional E-plane transitions, such as probe or door-knob types, which are typically used to minimize the coupling system's size while improving S-parameters. By employing this split-model optimization technique, the simulation or computation time was reduced by over 95% compared to conventional optimization methods, all while maintaining the desired response in the voltage standing wave ratio (VSWR) within 1:1.4. The optimized VSWR obtained using the proposed split-model technique and the conventional simulation technique are agreed within 6–10% between them.

Keywords: Coaxial couplers · rectangular waveguide transition · impedance matching · helix slow-wave structure (SWS) · traveling wave tube (TWT)

1 Introduction

Helix traveling-wave tubes (TWTs) are widely used in communication, radar, and electronic warfare, with careful design of the RF power coupling system essential to meet requirements for wide bandwidth and high power. This design must account for factors like attenuator-coated dielectric helix support rods (homogeneous or non-homogeneous of conventional materials or innovative metamaterial), helix pitch (uniform or non-uniform), and metal envelope properties of the helix TWT [1–5]. Traditional analytical methods approximate the characteristic impedance of the helix using equivalent circuit analysis [6], but these simplified approaches often produce VSWR values that

S. K. Ghosh et al. (Eds.): VEDA 2024, SPPHY 424, pp. 336–341, 2025.
https://doi.org/10.1007/978-981-96-5174-0_49

diverge significantly from experimental results, prompting a shift toward 3-D simulation tools like CST and HFSS [7, 8]. However, commercial software can require extensive computation, especially as mesh sizes are refined for accuracy [1–5].

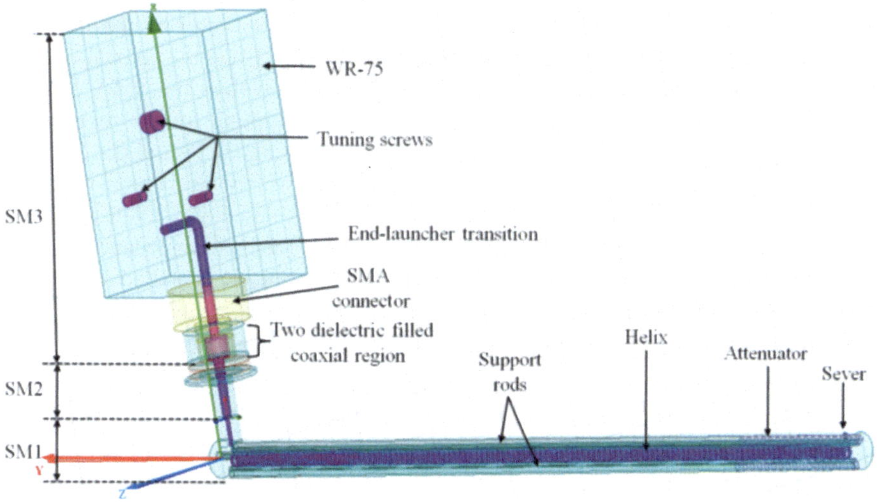

Fig. 1. The model of helix TWT along with output section with H-plane end-launcher transition followed by SMA connector.

To address these issues, a split-model (SM) optimization technique was introduced in [2], though it involves designing and simulating numerous helix models and couplers with various port and load impedances (e.g., 50, 55, 60 Ω), then combining the results to reach the desired VSWR—a time-intensive process. The improved SM technique presented here reduces optimization time by over 95% by eliminating redundant simulation pairs. This process is applied to only the output section of a two-section helix SWS with a distinct output non-uniform helix pitch profile, connected by SMA and H-plane end-launcher transitions, respectively. Additional refinements, including a piece-wise model for attenuator-coated dielectric support rods [9] and adjustments for coupler-to-helix transition discontinuity, further enhance accuracy. Section 2 outlines the SM technique, followed by results in Sect. 3, and conclusions in Sect. 4.

2 Proposed Split-Model Optimization Technique

In conventional optimization modelling, elements like the SWS, coupler, window, and connector/waveguide are combined into a single model, requiring extensive mesh cells and prolonged iteration time for accuracy [1, 3]. By separating the SWS from the entire coupling system and replacing it with its equivalent reflection coefficient during optimization, simulation time can be significantly reduced without sacrificing precision. In this proposed model, the SWS and transition sections are removed from the integrated model and substituted with their respective input impedances.

The helix with the coupling system (Fig. 1) is thus optimized through three split-models: (i) Split-Model 1 (SM1) includes the helix SWS with attenuator-coated dielectric rods, sever, and coaxial transition; (ii) Split-Model 2 (SM2) covers the coaxial coupler with the ceramic window; and (iii) Split-Model 3 (SM3) involves the SMA connector at the input and the SMA-to-H-plane end-launcher coupling for the rectangular waveguide at the output. SM1 is designed and simulated using HFSS and obtained the optimized VSWR (Fig. 2(a)) and hence input impedance which will be considered as load impedance for SM2. SM2 is modelled and optimized, based on VSWR response, using transmission line equations [10] in MATLAB. Now, the input impedance of SM2 is considered as the load impedance of SM3 and similarly, SM3 is modelled as SM2. In addition, SM2 and SM3 are modelled separately in HFSS by dividing the frequency band into multiple sample points and obtaining input impedance values from an Excel sheet, which can be applied in HFSS using post-processing. Through iterative adjustments, desired VSWR responses for SM2 and SM3 are achieved (see Fig. 2 in Sect. 3).

3 Results and Discussion

The effectiveness of the proposed simplified SM optimization method has been investigated for both the input coupler with SMA connector and the output coupler with SMA connector and connector to H-plane rectangular waveguide transition in X-band. The values of VSWR, obtained by this proposed approach of both the input and output couplers have been compared with those obtained by the conventional optimization method. Moreover, the proposed method reduces the optimization time by more than 95% as compared to the time required by the conventional method, the VSWR and the optimization time shown in Fig. 2 and Table 1, respectively.

The optimized values of VSWR of SM2 agree closely with those obtained from using the conventional approach (Fig. 2(b)). Thus, concerning the input section one can obtain the optimized values of VSWR of SM3 (Fig. 2(c)). Furthermore, the disagreement, more so at lower frequencies, also owes to the assumption of the excitation of only the dominant TE_{10} mode in the H-plane end-launcher transition of the rectangular waveguide in the output section [11]. The impedance matching can be improved by inserting multiple tuning screws [12] (Fig. 1). Further, by adjusting the radius and height of the tuning screws inside the rectangular waveguide one can achieve a nearly flat VSWR response within the desired frequency band (Fig. 2(d)). Also, the computational time of the proposed method is very low and can be estimated in a normal desktop computer. Thus, through this study on the input and output coupling system of a helix TWT, it is established that the proposed simplified split-model optimization method yields better and flat VSWR of the system over the band of interest and at the same time it takes very less computational time.

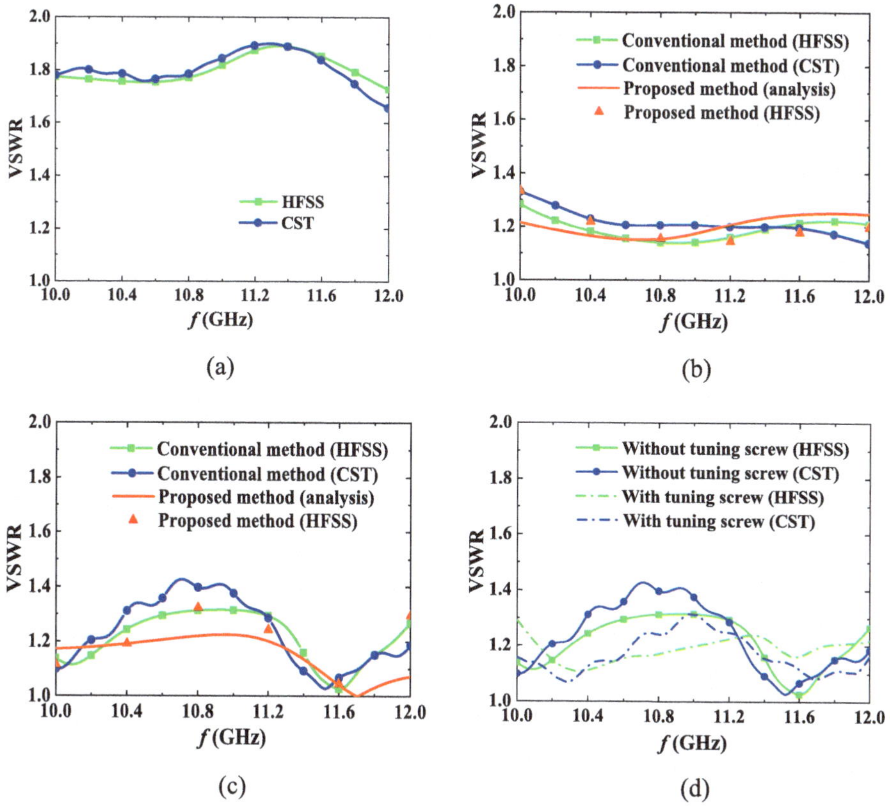

Fig. 2. Comparison of VSWR versus frequency responses: (a) SM1—obtained using HFSS and CST; (b) SM2 (same as input SM2)— and (c) SM3 (rectangular waveguide WR75, obtained using conventional and proposed optimization method of output coupler and H-plane coupled coaxial to rectangular waveguide transition (without any tuning screw); (d) considering tuning screws for further impedance matching.

Table 1. Comparison of the conventional method and the proposed simplified split-model optimization method.

Parameters	Time taken by each simulation		Total number of mesh cells	
	CST	HFSS	CST	HFSS
SM1 Conventional method	~ 3 h 50min	~ 1 h 05 min	13313300	215867
(SM1 + SM2) Proposed method	~ 5 h 10min	~ 1 h. 20 min	17578034	289336
(SM2 only) Conventional method	–	~ 1min	–	22540
(SM1 + SM2 + SM3) Proposed method	~ 6 h 40min	~ 1 h 45 min	25286250	325019
(SM3 only)	–	~ 1min50 sec	–	41042

4 Conclusion

A simplified split-model optimization method has been proposed for the optimization and design of the coupling system of a helix TWT. The proposed method is faster than the other methods in vogue in yielding results, though not at the cost of accuracy. However, there is scope to improve the analysis by taking into account losses in the transmission line model in the analysis. The future scope also includes the validation of the VSWR of the coupling system predicted by the proposed optimization method against experimental measurement.

Acknowledgements. The authors are thankful to the Director, CSIR-CEERI, Pilani, Rajasthan, India for allowing us to publish the work. Raktim Guha is thankful to the Council of Scientific and Industrial Research (CSIR) for granting the Senior Research Fellowship to him.

Disclosure of Interests. No potential conflict of interest was reported by the author(s).

References

1. Agarwal, A.K., Raina, S., Kumar, L.: A novel approach for simulation of coaxial coupler for helix TWTs using HFSS. In: 4th IEEE Int. Vacuum Electron. Conf. Proc., pp. 58–59. Seoul, Korea (South) (2001)
2. Ghosh, T.K., Carter, R.G., Challis, A.J., Rushbrook, K.G., Bowler, D.: Optimization of coaxial couplers. IEEE Trans. Electron Devices **54**(7), 1753–1769 (2007)
3. Rao, K.V., Naidu, V.B., Rao, P.R.R., Datta, S.K.: Simulation of RF coupler of a multi-section TWT with matched sever-loss. In: Proc. IEEE Int. Vac. Electron. Conf. (IVEC), pp. 455–456. Rome (2009)
4. Guha, R., Bandyopadhyay, A.K., Varshney, A.K., Datta, S.K., Basu, B.N.: Investigations into helix slow-wave structure assisted by double-negative metamaterial. IEEE Trans. Electron Devices **65**(11), 5082–5088 (2018)

5. Guha, R., Ghosh, S.K.: Dispersion control and size enlargement of helical slow-wave structure by double-positive metamaterial assistance. IEEE Trans. Electron Devices **69**(2), 771–776 (2022)
6. Jain, P.K., Basu, B.N.: Electromagnetic wave propagation through helical structures. In: Singh, O.N., Lakhtakia, A. (eds.) Electromagnetic Fields in Unconventional Materials. Wiley, Hoboken, NJ (2000)
7. CST AG, Darmstadt, Germany. CST STUDIO SUITE. Accessed: 2018. [Online]. Available: www.cst.com
8. HFSS V18.0: Ansoft Corporation. www.ansoft.com
9. Naidu, V.B., et al.: Three-dimensional electromagnetic analysis of attenuator-coated helix support rods of a traveling-wave tube. IEEE Trans. Electron Devices **56**(5), 945–950 (2009)
10. Rizzi, P.A.: Microwave Engineering: Passive Circuits. Prentice Hall International, New Delhi (1988)
11. Yang, D.Y.: Design and fabrication of an end-launched rectangular waveguide adapter fed by a coaxial loop. J. Inf. Commun. Converg. Eng. **10**(2), 103–107 (2012)
12. Gesche, R., Löchel, N.: Two cylindrical obstacles in a rectangular waveguide-resonances and filter applications. IEEE Trans. Microw. Theory Tech. **37**(6), 962–968 (1989)

Latest Advancements in Snow Depth Measurement Using Electromagnetic Signals: A Review

Soju J. Alexander[1,2](✉) and Chandrabhan Patel[1,2]

[1] CSIR-Central Building Research Institute (CSIR-CBRI), Roorkee, India
soju.alexander@cbri.res.in
[2] Academy of Scientific and Innovative Research (AcSIR), Ghaziabad, India

Abstract. India, particularly the northern regions like the Himalayas, is prone to frequent avalanches due to heavy snowfall, steep terrain, and climatic conditions. In recent years, India has witnessed several deadly avalanches, such as the October 2022 Draupadi ka Danda avalanche in Uttarakhand, which claimed the lives of 27 climbers, and the February 2021 glacier burst and avalanche in Chamoli, which led to over 200 casualties. Snow depth measurement is critical for predicting avalanches in mountainous and polar regions. Recent advancements in electromagnetic wave technologies, such as radar and microwave sensing, have revolutionized the accuracy and scalability of snow depth monitoring.

This paper provides a comprehensive review of state-of-the-art techniques used in snow depth measurement, focusing on systems utilizing electromagnetic waves across the microwave, millimeter-wave, and radar spectra. Key technologies discussed include Frequency-Modulated Continuous Wave (FMCW) radar operating at 60 GHz, which offers high-resolution snowpack characterization, particularly for localized measurements [1]. Synthetic Aperture Radar (SAR) systems utilizing C-band frequencies, in conjunction with deep learning models, have effectively estimated snow depth in complex terrains such as forested areas, offering hope for improved avalanche prediction [2]. Airborne multi-channel ultra-wideband (UWB) FMCW radar systems demonstrate the capability for large-scale, high-precision snow depth monitoring, particularly in remote regions, providing a promising solution for challenging terrains [3]. Passive microwave radiometry, integrated with machine learning algorithms like random forest model, further enhances snow classification and depth prediction on a broader scale, paving the way for more accurate and scalable snow depth monitoring [4].

Uncertainty in snow depth measurement remains a challenge, particularly in complex environments. However, ongoing research into radiative transfer models and signal processing techniques shows promise in reducing errors [5]. This review synthesizes findings from diverse approaches, highlighting the strengths, limitations, and future directions for improving snow depth measurement accuracy. By integrating these technologies and methods, future snow depth measurement systems can provide more reliable, real-time data across various terrains, contributing to better hazard prediction. Abstract should summarize the contents of the paper in short terms, i.e. 150–250 words.

Keywords: Snow Depth · Avalanches · Electromagnetic Signals · Early Warning Systems

S. K. Ghosh et al. (Eds.): VEDA 2024, SPPHY 424, pp. 342–351, 2025.
https://doi.org/10.1007/978-981-96-5174-0_50

1 Introduction

Avalanches threaten life and infrastructure in mountainous regions, especially in countries like India, where the Himalayas and other high-altitude areas experience frequent snowstorms and unstable snowpacks. Recent avalanche events, such as Mount Kun (Oct 2023), Nathu La (Apr 2023), Gulmarg (Feb 2023), Draupadi ka Danda (Oct 2022), Siachen Glacier (Feb 2022), Lahaul-Spiti (Mar 2021), and Chamoli (Feb 2021), made headlines due to significant loss of life and property. These events underscore the urgent need for advanced monitoring and early warning systems to mitigate the risk of such disasters. Despite ongoing efforts, timely avalanche prediction remains challenging due to the complex interaction of weather, snowpack, and terrain conditions in avalanche-prone areas.

Traditional avalanche monitoring techniques, such as ground-based visual inspection and manual snowpack testing, often fall short of providing accurate, real-time data over vast and remote regions. Consequently, applying electromagnetic wave technologies, including radar, microwave radiometry, and satellite-based sensors, offers promising solutions to overcome these limitations. These technologies enable continuous, high-precision snowpack monitoring and, when integrated with machine learning models, can provide valuable early warnings for potential avalanche risks.

Recent advancements in Frequency-Modulated Continuous Wave (FMCW) radar, Synthetic Aperture Radar (SAR), and Ku-band radar have demonstrated their ability to detect snow depth and snowpack instability, which are the key factors contributing to avalanche formation. For example, FMCW radar operating at 60 GHz allows for high-resolution measurement of snowpack layers [1]. SAR, when combined with deep learning models, enhances snow depth prediction in challenging terrains like forested or highly inclined areas [2]. With machine learning, microwave radiometry provides large-scale snow classification and depth estimation, making it suitable for real-time avalanche risk monitoring across broad geographic regions [4].

In light of recent avalanche incidents in India, this review explores the potential application of these advanced electromagnetic wave technologies for snow depth measurement, aiming to improve avalanche monitoring and early warning systems. By integrating these technologies with real-time data processing and machine learning models, future avalanche prediction systems could significantly reduce casualties and enhance disaster preparedness in high-risk regions.

2 Principles of Snow Depth Measurement and Avalanche Prediction

2.1 Avalanche Formation Dynamics and Key Risk Factors

Avalanches are complex natural disasters resulting from snowpack instability, weather conditions, and terrain features. They occur when the forces acting on the snowpack, primarily gravity, exceed the snow's internal cohesion and bonding strength, causing a rapid downslope movement of snow. Avalanches can be classified into several types, with slab avalanches being the most dangerous, characterized by a cohesive layer of snow breaking loose from underlying weaker layers. Key factors that contribute to avalanche formation include:

Snowpack Structure. The snowpack is often composed of multiple layers of snow with varying densities, temperatures, and moisture levels. A weak, unstable layer within the snowpack can be a failure point when stressed by additional snow or external factors like wind or human activity [1].
Weather Conditions. Heavy snowfall, rapid temperature changes, and wind significantly contribute to avalanche risk. Fresh snow or rain can add weight to the snowpack, increasing stress, while temperature fluctuations can alter snowpack stability [6].
Terrain and Slope. Avalanches commonly occur on slopes with an incline of 30° to 45°. Steeper slopes lead to faster-moving avalanches, while gentler slopes may allow snow to accumulate without immediate risk of movement [2].

These risk factors make snow depth and snowpack structure critical for avalanche prediction, highlighting the importance of accurate snow monitoring technologies.

2.2 Role of Snow Depth Measurement in Predicting Avalanches

Snow depth is one of the most fundamental indicators of avalanche risk. It provides insights into snow accumulation, snowpack stratification, and the likelihood of snow instability. Measuring snow depth and understanding the layering within the snowpack allows for better assessment of:

Snow Accumulation Rates. A sudden increase in snow depth over a short period indicates rapid loading of the snowpack, often leading to instability and a higher risk of avalanches [1].
Weak Layer Identification. Combined with radar and remote sensing technologies, snow depth measurements can identify weak layers within the snowpack. These weak layers, often formed from buried surface hoar or faceted snow crystals, are critical triggers for slab avalanches [5].
Snowpack Stability. Snow depth data, when integrated with other snow properties (temperature, moisture), helps evaluate snowpack stability. Sudden temperature changes and moisture content variations can weaken snow bonds, making real-time snow depth monitoring crucial for avalanche forecasting [7].

2.3 Overview of Traditional and Emerging Technologies

Traditional methods of measuring snow depth involve manual probing and snow pits, where snow layers are physically examined. While accurate, these methods are labour-intensive, localized, and unable to provide continuous real-time data across large or remote regions. To address these limitations, electromagnetic wave-based technologies, as described in this section, are increasingly used to measure snow depth and monitor snowpack dynamics with high precision and efficiency [8].

Radar Technologies. Frequency-Modulated Continuous Wave (FMCW) radar and Synthetic Aperture Radar (SAR) are widely employed for snow depth measurements. These radar technologies utilize electromagnetic waves to penetrate the snow and measure depth and density variations. FMCW radar at 60 GHz has been used for high-resolution snowpack characterization, while C-band SAR is effective for larger-scale remote sensing of snow depth, even in forested regions [9].

Microwave Radiometry. Microwave radiometers measure snow depth by analyzing the natural emission of microwave radiation from the snowpack. Combined with machine learning algorithms, radiometry data provides large-scale snow depth and classification insights, helping identify unstable snow conditions [4].

Ultrawideband (UWB) Radar. UWB radar, particularly from airborne platforms, offers a solution for wide-area snow depth measurement in remote regions. Its ability to operate across a broad frequency range allows for precise measurement of snow depth and snowpack structure [10].

3 Electromagnetic Wave Technologies for Snow Depth Measurement

The emerging technologies discussed in the previous section have improved snow depth measurement's accuracy and scalability, enhancing the ability to predict avalanches in real-time. The measurement of snow depth and snowpack characteristics has significantly advanced with the adoption of electromagnetic wave technologies. These technologies leverage the ability of electromagnetic waves to penetrate snow, providing high-resolution data on snow depth, density, and layering, which are crucial for accurate avalanche prediction. This section reviews the critical electromagnetic wave-based technologies used for snow depth measurement, discussing their principles, applications, strengths, and limitations.

3.1 Frequency-Modulated Continuous Wave (FMCW) Radar

Frequency-Modulated Continuous Wave (FMCW) radar emits a continuous signal whose frequency varies over time. The reflected signals from the snowpack are analyzed to determine the distance to the snow surface and layers within the snowpack, allowing for precise depth measurement. FMCW radar at 60 GHz has shown remarkable accuracy in detecting snowpack layers due to its high resolution and sensitivity to snow-density changes [1].

Applications. FMCW radar is particularly effective for localized snowpack monitoring, providing detailed information on snow layer stratification and instability.
Strengths. High-resolution data can detect fine variations in snowpack structure.
Limitations. Limited penetration in very dense or wet snow, which can attenuate the radar signal.

3.2 Synthetic Aperture Radar (SAR)

Synthetic Aperture Radar (SAR) uses the motion of the radar antenna over a target area to create high-resolution images, enabling detailed snow depth measurement over vast areas. C-band SAR, operating between 4 and 8 GHz, is widely used for remote sensing applications, including snow depth estimation in complex terrains such as forests and mountainous regions [2].

Applications. Effective for large-scale snow monitoring, particularly in regions with challenging direct access.

Strengths. Ability to penetrate cloud cover and vegetation, continuously monitoring regardless of weather conditions.
Limitations. Requires advanced processing techniques and interpretation of data, which can be complex.

3.3 Ultra-Wideband (UWB) Radar

Ultra-Wideband (UWB) radar systems operate over a broad frequency range, providing a high-resolution snowpack profile. These systems are beneficial for airborne snow depth measurements, covering large and remote areas with precision. Airborne multi-channel UWB FMCW radar has been successfully deployed to measure snow depth in inaccessible terrains, such as high-altitude regions [3].

Applications. Large-scale snow depth measurement, ideal for remote or rugged terrains.
Strengths. Excellent coverage area, high accuracy, capable of detecting multiple snow layers.
Limitations. Performance can be affected by weather conditions; complex calibration is required for accurate results.

3.4 Microwave Radiometry

Microwave radiometry involves the measurement of natural microwave emissions from the snowpack. Passive microwave sensors capture the radiation emitted by the snow, which varies with snow depth and density. When combined with machine learning algorithms, microwave radiometry can effectively classify and measure snow depth over large areas [4].

Applications. Large-scale snow monitoring and classification, useful in diverse terrains and weather conditions.
Strengths. Capable of providing continuous, real-time data across wide regions.
Limitations. Susceptible to interference from other environmental factors; requires advanced data processing techniques.

3.5 Ku-Band Radar

Ku-band radar operates in the frequency range of 12–18 GHz and is used for high-precision snow parameter retrieval, including depth and density. Ku-band radars have been particularly effective in measuring snow on sea ice, providing critical data for polar research and monitoring [7].

Applications. Monitoring snow on sea ice, critical for polar regions and climate research.
Strengths. High-resolution data can capture fine details of the snowpack structure.
Limitations. Limited penetration in dense snow; may require high-power systems for effective operation.

3.6 Millimetre-Wave Technologies

Millimetre-wave technologies, particularly in the 30–300 GHz range, are emerging as a powerful tool for short-range snow depth measurement, especially for understanding the effects of snow on communication systems and atmospheric interactions [9].

Applications. Analyzing the impact of snow on millimetre-wave communications and short-range snow measurements.
Strengths. High-frequency range allows for detailed analysis of snowpack properties.
Limitations. High attenuation in dense snow; limited to short-range applications.

4 Integration of Electromagnetic Wave-Based Technologies into Avalanche Early Warning Systems

While electromagnetic wave-based technologies provide significant advancements in snow depth measurement, they also come with challenges such as signal attenuation, complexity in calibration, data processing and interpretation. Dense or wet snow can attenuate electromagnetic signals, particularly for higher frequency bands like millimeter waves and Ku-band. Accurate calibration of radar and radiometry systems is critical but can be complex, especially in varied terrains and environmental conditions. Advanced data processing algorithms and expertise in interpreting electromagnetic wave data are essential for accurate snow depth assessment.

Despite these challenges, integrating these technologies into comprehensive snow depth monitoring and early warning systems offers promising potential for improving avalanche prediction and response strategies.

Electromagnetic wave-based technologies can be integrated to provide real-time snowpack data to enhance the accuracy and effectiveness of avalanche early warning systems. This section outlines how these technologies, along with advanced communication solutions like LoRa, can be deployed to monitor and predict avalanche risks continuously.

4.1 Real-Time Data Acquisition

Electromagnetic wave-based technologies like FMCW radar, SAR, and UWB radar provide continuous, high-precision data on snow depth and snowpack structure. These systems enable real-time monitoring of snow conditions, which is essential for avalanche prediction in remote and inaccessible regions. FMCW radar, for instance, can detect snowpack instability and layering with high accuracy, providing early indicators of avalanche risk [1].

These technologies can be deployed on ground-based platforms or integrated into mobile units like drones or vehicles to provide continuous, real-time data on snow depth and snowpack stability. The data can be transmitted to centralized monitoring stations for further analysis and decision-making.

4.2 Machine Learning and Predictive Analytics

Integrating electromagnetic wave data with machine learning (ML) models significantly enhances the predictive capabilities of avalanche warning systems. Machine learning algorithms, such as deep neural networks (DNNs) and random forests, can analyze real-time data from electromagnetic sensors to detect patterns and potential avalanche triggers. For example, combining C-band SAR data with ML models has improved snow depth estimation and avalanche prediction, especially in challenging terrains like forests and mountainous regions [2].

ML models can continuously process incoming data from sensors, such as radar and radiometry, updating risk assessments in real-time. By recognizing changes in snow depth and snowpack instability, these systems can trigger early avalanche warnings, providing valuable time for authorities to take preventive measures.

4.3 Sensor Networks and Remote Sensing Integration

Deploying sensor networks that incorporate electromagnetic wave technologies, such as UWB radar and microwave radiometers, enables wide-scale snow depth monitoring in remote areas. These networks can provide continuous data from multiple locations, feeding real-time information to central monitoring systems. Airborne UWB radar systems, in particular, are well-suited for snow depth measurement in high-altitude and inaccessible regions, offering both coverage and precision [3].

Combined with remote sensing platforms like SAR satellites and airborne sensors, ground-based radar systems create a comprehensive monitoring system. These systems can continuously track snowpack changes over large areas and provide real-time data for avalanche risk assessment.

4.4 LoRa Communication and Data Fusion for Avalanche Early Warning

Effective communication is critical for transmitting real-time data from remote sensors to central monitoring stations. LoRa (Long Range) communication technology is ideal for this purpose, offering long-range, low-power data transmission capabilities, especially in remote, avalanche-prone areas. LoRa operates over distances of up to 10–15 km and consumes minimal power, making it suitable for regions with limited infrastructure. LoRa-enabled sensors, such as FMCW radar, UWB radar, and microwave radiometers, can continuously monitor snow conditions and transmit the data to central hubs via LoRa gateways [11].

Beyond communication, the integration of data fusion techniques enhances the accuracy of avalanche early warning systems. Data from multiple electromagnetic wave-based sensors (e.g., radar, radiometers, SAR) can be combined with meteorological data to understand snowpack dynamics comprehensively. By fusing data from different sources, these systems can generate more accurate predictions of avalanche risks and issue timely warnings to relevant authorities and the public [4].

LoRa communication can establish a network of sensors across large, remote areas, transmitting real-time snow depth and snowpack stability data to central monitoring systems. Once the data is received, data fusion algorithms can combine inputs from various sources to produce a detailed analysis of avalanche risks. This integration improves the reliability and scalability of early warning systems, enabling real-time decision-making and response.

5 Future Scope

Although significant progress has been made in integrating electromagnetic wave-based technologies into avalanche early warning systems, further research and development are essential to enhance their effectiveness and scalability. This section highlights key areas for future focus.

5.1 Advanced Data Fusion and Machine Learning

Current data fusion techniques, while effective, can be enhanced by developing more sophisticated algorithms. Future work should focus on integrating data from various sensors like FMCW radar, SAR, and microwave radiometers with machine learning models for more accurate avalanche prediction. Advanced AI models, such as deep learning, can improve the real-time processing of complex snowpack data, leading to more reliable risk assessments.

5.2 Scaling LoRa and IoT Networks

LoRa's long-range, low-power communication capabilities make it ideal for transmitting sensor data from remote areas. Future research should focus on expanding LoRaWAN networks and integrating IoT technology to support larger, more complex sensor networks in avalanche-prone regions. Enhancing network architecture and energy efficiency will allow for better scalability and continuous monitoring [11].

5.3 Improving Electromagnetic Sensors

While current sensors provide reliable snow depth measurements, there is room for improvement in terms of accuracy, range, and resilience in harsh conditions. Future sensor development should aim to increase resolution and signal penetration, particularly in dense or wet snow, and miniaturize devices for broader deployment in inaccessible areas [3].

5.4 Autonomous Drones and UAVs

Drones equipped with electromagnetic wave sensors can offer mobile, high-resolution monitoring in remote regions. Future research should explore improving the autonomy of these systems to operate in extreme environments without human intervention. Autonomous drones could play a critical role in conducting routine snowpack assessments in areas that are otherwise difficult to reach [3].

5.5 Real-Time Avalanche Risk Models

Advancing real-time avalanche risk models that integrate large-scale sensor data and weather forecasts will improve prediction accuracy. Future efforts should focus on developing faster, more precise models capable of processing vast datasets and issuing timely avalanche warnings [4].

6 Conclusion

Integrating electromagnetic wave-based technologies into snow depth measurement and avalanche prediction systems has significantly advanced the ability to monitor snowpack conditions and predict avalanche risks in real-time. Technologies such as FMCW radar, SAR, UWB radar, and microwave radiometers provide highly accurate, continuous data, enabling early detection of snow instability. Combined with machine learning models, these technologies can enhance the predictive power of avalanche early warning systems, making them more reliable and effective in preventing disasters.

LoRa communication and IoT networks have become valuable tools for transmitting data from remote, avalanche-prone areas to centralized monitoring stations. These scalable, low-power networks enable the deployment of extensive sensor systems in areas where traditional communication infrastructure is lacking, enhancing real-time monitoring capabilities.

As outlined in future research directions, challenges still need to be addressed. These include improving the precision of sensors, developing more sophisticated data fusion techniques, and advancing autonomous monitoring systems such as drones. Continued research and development in these areas will further enhance the capabilities of avalanche early warning systems, ultimately helping to reduce human casualties and protect infrastructure in vulnerable regions.

By advancing sensor technology, improving communication networks, and integrating advanced AI and autonomous systems, future avalanche monitoring systems will be more robust, providing real-time data and delivering early warnings with greater accuracy. These innovations will be critical in mitigating the impact of avalanches on both human populations and the environment.

Acknowledgement. The author is grateful to the CSIR-CBRI for granting the permission to publish this work. The financial support provided by CSIR-CBRI for supporting project No. NCP040304 is also gratefully acknowledged.

References

1. Wielandt, S., et al.: Characterizing snowpack with 60 GHz FMCW millimeter-wave radar sensors. In: 2023 57th Asilomar Conference on Signals, Systems, and Computers, Pacific Grove, CA, USA, pp. 1245–1250 (2023)
2. Yu, G., Gu, L., Li, X., Fan, X.: Forest snow depth estimation based on optimized features and DNN network using C-Band SAR data. IEEE Geosci. Remote Sens. Lett. **21**, 1–5 (2024)

3. Kolpuke, S., et al.: Airborne multichannel UWB FMCW radar for snow depth measurements. IEEE Trans. Geosci. Remote Sens. **62**, 1–18 (2024)
4. Bogerd, L., et al.: Classifying microwave radiometer observations over the Netherlands into dry, shallow, and nonshallow precipitation using a random forest model. J. Hydrometeorol. **25**(6), 881–898 (2024)
5. Merkouriadi, I., et al.: Solving challenges of assimilating microwave remote sensing signatures with a physical model to estimate snow water equivalent. Water Resour. Res. **57**(11) (2021)
6. Meloche, J., et al.: Altimetric Ku-band radar observations of snow on sea ice simulated with SMRT. In: EGUsphere 2024, pp 1–30 (2024)
7. Stefko, M., Hajnsek, I., Leinss, S., Floury, N.: Automated measurements of coherent backscatter enhancement at Ku-band for snow parameter retrieval. In: IGARSS 2024—2024 IEEE International Geoscience and Remote Sensing Symposium, Athens, Greece, pp. 1671–1675 (2024)
8. Hedrick, A., et al.: Data from: direct insertion of NASA airborne snow observatory-derived snow depth time-series into the iSnobal energy balance snow model (2020)
9. Abdulali, M., Salous, S.: Snow attenuation measurement for short-range mmWave fixed link. In: 2024 4th URSI Atlantic Radio Science Meeting (AT-RASC), Meloneras, Spain, pp. 1–4 (2024)
10. Jehangir, S.: Ultrawideband antenna solutions for weather radar applications (2023)
11. LoRa Alliance: The LoRaWAN specification. LoRa Alliance Technical Paper (2023). Available: https://lora-alliance.org, last accessed 2024/11/06

Pedagogical Review of Asymmetry of Dielectric Helix-Supports of a Travelling-Wave Tube

Amitavo Roy Choudhury[1(✉)], A. K. Sinha[1], and B. N. Basu[2]

[1] CSIR-Central Electronics Engineering Research Institute, Pilani 333031, India
arc@ceeri.res.in

[2] Supreme Knowledge Foundation Group of Institutions, Mankundu, West Bengal 712139, India

Abstract. Electromagnetic analysis of asymmetry of support rods of a helix slow-wave structure that causes the stop-band in the structure dispersion characteristics is reviewed. The asymmetry caused the lowering of the RF phase velocity and-cutoff frequency and the increasing of the stop-bandwidth. The results obtained by the electromagnetic analysis agreed to those predicted by the simulation code HFSS (high frequency structure simulation) within 1.5%. The present study helps in the design of a of a wideband helix-traveling-wave tube.

Keywords: Traveling-wave tube · slow-wave structure · asymmetry of helix-supports · dispersion characteristics · stop-band · HFSS

1 Introduction

Helix-travelling-wave tubes (TWTs) hold considerable promise for utilization in wide-band applications, particularly in the context of electronic warfare systems. However, the stop-band in the ω-β dispersion characteristics of the slow-wave structure (SWS) comprised of a helix supported by identical dielectric rods with wedge, circular, or rect-angular cross-sections puts a bar on the maximum device operational frequency. The sheath-helix [1] and tape-helix models [1, 3] are two prominent approaches to the field analysis of helical SWS. In various studies, the discrete dielectric rods have been simplified to continuous dielectric tubes, with effective dielectric constants calculated based on the volume fraction occupied by the support rods in the structure [4, 5]. These studies neglect the azimuthal harmonic effects produced by the support angular periodicity of the support rods. Loshakov [6] studied these effects taking the support rods to be radially unbounded, for the sake of simplicity of analysis. Basu et al. [7] considered these effects adding practical relevance to the analysis considering the structure to be radially bounded by an overall metal envelope. However, their study ignored the effect of angular asymmetry of the support rods which was subsequently taken into account by Joo et al. [8]. This effect of angular asymmetry of the support rods is revisited in this pedagogical review.

Section 2 presents the analysis in sheath-helix model, partitioning the helix to metal envelope region into n wedge-shaped regions, taking into account the asymmetry caused

S. K. Ghosh et al. (Eds.): VEDA 2024, SPPHY 424, pp. 352–358, 2025.
https://doi.org/10.1007/978-981-96-5174-0_51

by (i) angular offset due to misalignment of the dielectric helix-support rods and (ii) dielectric offset due to the dielectric constant variations from rod to rod. Results and discussion are presented in Sect. 3 presenting the dependence of the cut-off frequency and stop-band width on the angular offset angle and the validation of the analysis against HFSS simulation. This study with respect to angular offset is extended to the dielectric offset as well in this section. The present work is concluded in Sect. 4 highlighting the major findings of this review.

2 Analysis

A continuous dielectric tube that possesses a uniform effective relative permittivity $\varepsilon_{r,\text{eff}}$ can replace the discrete dielectric wedge-shaped helix-support rods. The effective relative permittivity is given by [1–4]:

$$\begin{aligned}\varepsilon_{r,\text{eff}} &= \frac{(\varepsilon_r)(A_s) + (1)(A - A_s)}{(A_s) + (A - A_s)} \\ &= 1 + (\varepsilon_r - 1)\frac{A_s}{A} = 1 + (\varepsilon_r - 1)\frac{\varphi N}{2\pi}\end{aligned} \tag{1}$$

where ε_r is the support dielectric constant. A_s is the area of cross-section of the dielectric support rods. A is the area of the cross-section of the helix to envelope region. φ and N stand for the support rod wedge angle and number of support rods, typically three, respectively, as shown in Fig. 1. The physical model of analysis is depicted in Fig. 2. Hence, the following dispersion relation is obtained in the sheath-helix model [1]:

$$\frac{k \cot \psi}{\gamma} = \left[\frac{I_0(\gamma a)\, K_0(\gamma a)}{I_1(\gamma a)\, K_1(\gamma a)}\right]^{1/2} D_{\text{lf}} \tag{2}$$

where

$$D_{\text{lf}} = \left[\left\{1 - \frac{I_0(\gamma a)\, K_0(\gamma b)}{K_0(\gamma a)\, I_0(\gamma b)}\right\}^{-1} \left\{1 - \frac{I_1(\gamma a)\, K_1(\gamma b)}{K_1(\gamma a)\, I_1(\gamma b)}\right\} \times \left\{1 + (\varepsilon_{r,\text{eff}} - 1)\, \gamma a\, I_0(\gamma a) K_1(\gamma a) \left(1 + \frac{I_1(\gamma a)\, K_0(\gamma b)}{K_1(\gamma a)\, I_0(\gamma b)}\right)\right\}\right]^{-1/2} \tag{3}$$

is the structure dielectric loading factor. Here, γ stands for the radial propagation. a and b are the radii of the sheath-helix and metal envelope, respectively.

The angular periodicity of the dielectric wedge supports of the helix give rise to the azimuthal harmonics. We can account for these harmonics by considering in the analysis the RF dependence as $\exp(jm(2\pi/\Theta)\theta) = \exp(jmN\theta)$, where m and $\Theta = 2\pi/N$ are the angular-harmonic number and the support azimuthal periodicity, respectively, N being the number of dielectric supports of the helix [1, 7, 8] (Fig. 1).

Taking into account the angular harmonics produced as a result of the angular periodicity of the dielectric rods, we can write the following electric field E and magnetic field H expressions:

$$E_{zs} = \sum_{m=-\infty}^{\infty} [A_{s,m} I_{mN}(\gamma r) + B_{s,m} K_{mN}(\gamma r)] \exp(jmN\theta) \tag{4}$$

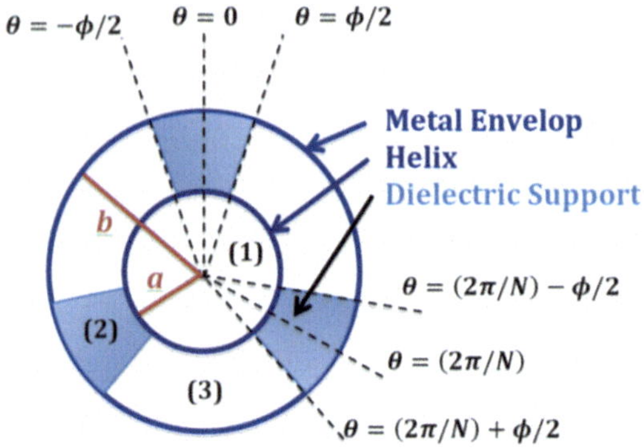

Fig. 1. The free-space region (1), inside the helix, the support rod region (2), and the free-space region (3), between two consecutive three support rods [region 1: $0 \leq r \leq a$; region 2: $a \leq r \leq b$; $-\varphi/2 \leq \theta \leq \varphi/2$; region3: $a \leq r \leq b$; $\varphi/2 \leq \theta \leq (2\pi/N) - \varphi/2$ with $N = 3$] [7, 8].

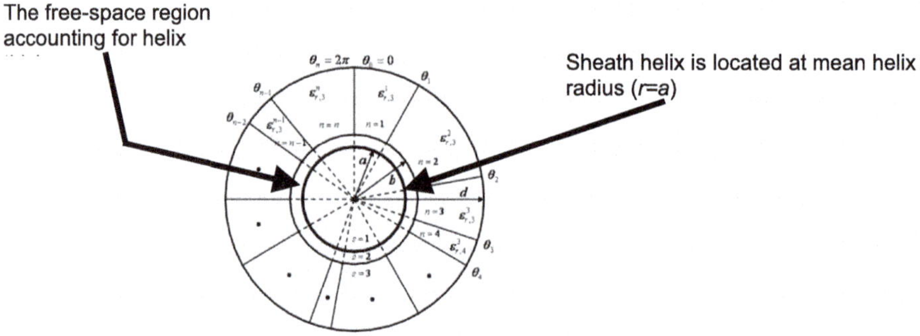

Fig. 2. General azimuthally asymmetric structure model showing the sheath helix and the metal envelope, located at $r = a$ and $r = b$, respectively. n denotes the wedge-shaped regions between the helix and the envelope. θ_n and ε_{rn}, represent the support wedge angle and dielectric constant (relative permittivity), both of the nth region, respectively, the free-space region devoid of a dielectric having a unity dielectric constant [8].

$$H_{zs} = \sum_{m=-\infty}^{\infty} [C_{s,m} I_{mN}(\gamma r) + D_{s,m} K_{mN}(\gamma r)] \exp(jmN\theta) \tag{5}$$

$$E_{\theta\, s} = \sum_{m=-\infty}^{\infty} \{(-j\omega\mu_0/\gamma)[C_{s,m} I'_{mN}(\gamma r) + D_{s,m} K'_{mN}(\gamma r) - (mN\beta/(\gamma^2 r)] \times [A_{s,m} I'_{mN}(\gamma r) + B_{s,m} K'_{mN}(\gamma r)]\} \exp(jmN\theta) \tag{6}$$

$$H_{\theta\, s} = \sum_{m=-\infty}^{\infty} \{(j\omega\varepsilon_s/\gamma)[A_{s,m} I'_{mN}(\gamma r) + B_{s,m} K'_{mN}(\gamma r) - (mN\beta/(\gamma^2 r)] \times [C_{s,m} I'_{mN}(\gamma r) + D_{s,m} K'_{mN}(\gamma r)]\} \exp(jmN\theta) \tag{7}$$

where the subscripts s = 1, 2, and 3 are for the regions 1, 2, and 3, respectively, as depicted in Fig. 1.

One can now write the boundary conditions at the interfaces between the regions 1, 2, and 3 as follows [1]:

$$E_{z1} - E_{z2} = 0 \ \ (r = a;\ -\varphi/2 \le \theta \le \varphi/2) \tag{8}$$

$$E_{z1} + E_{\theta 1} \cot \psi = 0 \ (r = a) \tag{9}$$

$$E_{z2(3)} + E_{z2(3)} \cot \psi = 0 \ (r = a) \tag{10}$$

$$H_{z1} \tan \psi + H_{\theta 1} = H_{z2} \tan \psi + H_{\theta 2} \ \ (r = a;\ -\varphi/2 \le \theta \le \varphi/2) \tag{11}$$

$$H_{z1} \tan \psi + H_{\theta 1} = H_{z3} \tan \psi + H_{\theta 3} \ (r = a;\ \varphi/2 \le \theta \le (2\pi/N) - \varphi/2) \tag{12}$$

Similarly, at $r = b$ (metal envelope), the boundary conditions to be used in the analysis are:

$$E_{z2(3)} = 0 \tag{13}$$

$$E_{\theta 2(3)} = 0\,. \tag{14}$$

The relevant field constants of the field expressions (4)–(7) can be used in terms of a single constant $A_{1,m}$ with the help of the boundary conditions (8)–(10). Hence, we can proceed in the analysis, following the steps—substitution of field expressions in terms of $A_{1,m}$ into the boundary condition (11); multiplication of the expressions by $\exp(-jqN\theta)a$; and integration between the limits $-\varphi/2$ and $\varphi/2$. Subsequently, we can perform the addition of this integration with the integration similarly acquired using Eq. (12) in place of Eq. (11), though this time taking the limits as $\varphi/2$ and $(2\pi/N) - \varphi/2$ (Fig. 1). The subsequent sum obtained is [7]:

$$\alpha_q A_{1,q} + \sum_{m(\neq q)=-\infty}^{m(\neq q)=\infty} \delta_{m,q} A_{1,m} \tag{15}$$

The expression (15) may be divided into one part referring to $m = q$ and another part referring to $m \neq q$, where q is an integer, α_q and $\delta_{m.q}$ being the functions of the structural parameters [7]. For the lower order integral values of q and m, the said expression (15) can be used to write:

$$\alpha_0 A_{1,0} + \delta_{1,0} A_{1,1} + \delta_{-1,0} A_{1,-1} = 0 \tag{16}$$

$$\delta_{0,1} A_{1,0} + \alpha_1 A_{1,1} + \delta_{-1,1} A_{1,-1} = 0 \tag{17}$$

$$\delta_{0,-1} A_{1,0} + \delta_{1,-1} A_{1,1} + \alpha_{-1} A_{1,-1} = 0. \tag{18}$$

The following expression is obtained if we put the condition for non-trivial solutions of (16)–(18) that the determinant with the elements as the field constant coefficients in their left becomes zero after algebraic simplification and using the relations $\delta_{1,0} = \delta_{-1,1}$, $\delta_{0,1} = \delta_{0,-1}$ and $\delta_{1,-1} = \delta_{-1,1}$ [7]:

$$\alpha_0\alpha_1\alpha_{-1} - \alpha_0\delta_{1,-1}^2 + \delta_{0,1}\delta_{1,0}(2\delta_{1,-1} - \alpha_1 - \alpha_{-1}) = 0 \quad (19)$$

Nevertheless, in practical instances, the first term of the left of (19) dominates over the remaining two, thereby enabling one to write [7]:

$$\alpha_0\alpha_1\alpha_{-1} = 0 \quad (20)$$

which gives three possibilities: $\alpha_0 = 0$, $\alpha_1 = 0$, and $\alpha_{-1} = 0$. However, the solution of $\alpha_0 = 0$ exhibits a significant variation across a broad spectrum of frequencies from the solution of $\alpha_1 = 0$ and that of $\alpha_{-1} = 0$, from the standpoint the applications in a TWT. From $\alpha_0 = 0$, we obtain the dispersion relation (2).

3 Results and Discussion

The frequency range between the lower branch and the upperbranch of the dispersion characteristics makes the stop-band (Figs. 3 and 4).The TWT operates as an amplifier without oscillations up to the cutoff frequency which is lowered by the structure angular asymmetry. The asymmetry also results in a higher propagation constant and, correspondingly, a lower RF phase velocity (Fig. 3). The results obtained by electromagnetic analysis agreed with those obtained by HFSS within 1.5% (Fig. 4).

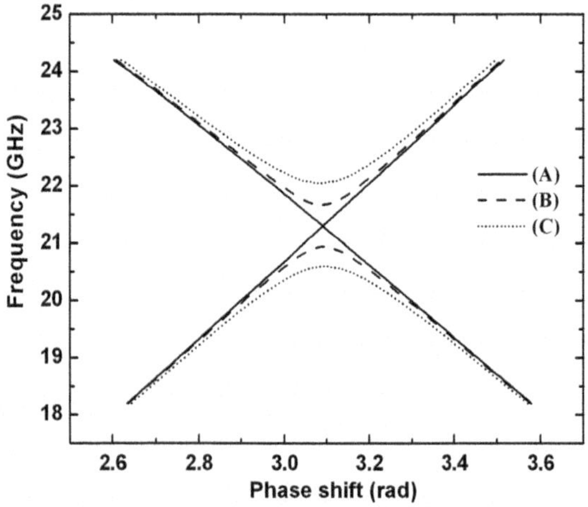

Fig. 3. Stop-band behavior of the structure with angular asymmetry for angular offset angles $\Delta\phi = 0°$, 10°, and 20° (curves A, B, and C, respectively), taking $a = 0.7$ mm, $b = 1.5$ mm, $\varepsilon_r = 6.5$, and $\theta = 60°$ [8].

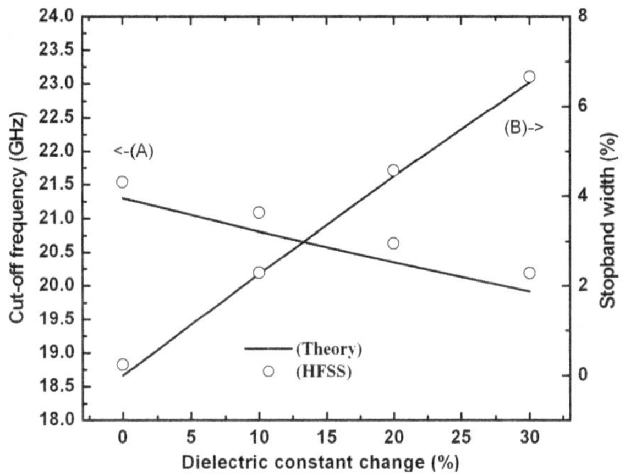

Fig. 4. Cutoff frequency (A) and stop bandwidth (B) versus dielectric constant change $\Delta\varepsilon_r$ characteristics showing the comparison between theory (based on electromagnetic analysis) (solid lines) and HFSS simulation (circle points) ($a = 0.7$mm, $b = 1.5$mm, $\varepsilon_r = 6.5$, and wedge angle $\theta = 60°$) [8].

4 Conclusions

The occurrence of the stop-band near the π-point in the dispersion characteristics of dielectric loaded helical SWS of a TWT is attributed to the asymmetry of dielectric helix-support rods. This asymmetry stems from the irregular angular positioning of the dielectric helix-support rods during the process of the assembly of the SWS as well as from the different dielectric constant values of the rods. The results derived from electromagnetic analysis were validated against those obtained with the help of HFSS simulation, within a difference of merely 1.5%. It is felt that the tape-helix model incorporating space-harmonic effects rather than the sheath-helix model used here to make the analysis simple might have given still closer matching between the analytical and simulated results.

Acknowledgment. The problem was taken up at the behest of Dr. Amarjit Singh, the erstwhile Director of CSIR-CEERI, Pilani, during a visit of one of the authors (BNB) to CSIR-CEERI, Pilani under the Distinguished Visiting Scheme of CSIR.

References

1. Basu, B.N.: Electromagnetic Theory and Applications in Beam-Wave Electronics. World Scientific, Singapore, New Jersey, London, Hong Kong (1996)
2. Basu, B.N.: Equivalent circuit analysis of a dielectric-supported helix in a metal shell. Int. J. Electron. **47**, 311–314 (1979)
3. Sensiper, S.: Electromagnetic wave propagation on helical structures. Proc. IRE **43**, 149–161 (1955)

4. Kapoor, S., Raju, R.S., Gupta, R.K., Joshi, S.N., Basu, B.N.: Analysis of an inhomogeneously-loaded helical slow-wave structure for broadband TWT's. IEEE Trans. Electron Devices **36**, 2000–2004 (1989)
5. Jain, P.K., Basu, B.N.: The inhomogeneous dielectric loading effects of practical helix supports on the interaction impedance of the slow-wave structure of a TWT. IEEE Trans. Electron Devices **39**, 727–733 (1996)
6. Loshakov, L.N., Ol'derogge, E.B.: Propagation of slow electromagnetic waves along a helix with dielectric supports. Radio Eng. Electron Phys. **13,** 45–51 (1968)
7. Basu, B.N., Jha, R.K., Sinha, A.K., Kishore, L.: Electromagnetic wave propagation through an azimuthally perturbed helix. J. Appl. Phys. **58**, 3625–3627 (1985)
8. Joo, Y.-D., Sinha, A.K., Park, G.-S.: Electromagnetic wave propagation through an azimuthally asymmetric helix slow wave structure. Jpn. J. Appl. Phys. **42**, 7585–7593 (2003)

Dr. Amarjit Singh—A Phenomenon

B. N. Basu(✉)

SKFGI, Mankundu, West Bengal 712139, India
bnbasu.india@gmail.com

Abstract. The contribution of Dr. Amarjit Singh in the progress of science and technology in India is presented. The President of India conferred the Padma Bhusan award on him. He served as the Director, CSIR-CEERI, Pilani for about quarter of a century, by far the longest period in the history of CEERI. His early education, higher education, and professional career are discussed.

Keywords: Amarjit Singh · Padma Bhusan · CSIR-NPL · CSIR-CEERI · Phagwara · Pilani · vacuum electronic Devices · Punjab University · Harvard University · Michigan University · Stanford University

1 Introduction

Padma Bhusan Dr. Amarjit Sing (Singh)—son of Sardar Jagdish Singh and Ishar (Kaur)—was born on 19th November 1924 in Ramdas near Amritsar, India. Singh breathed his last at Davenport, United States on 17th August, 2021. Singh significantly contributed to the research and development of vacuum electronic devices (VEDs) in India. The President of India in 1985 felicitated Singh by Padma Bhusan Award. The Vacuum Electronic Devices and Application Society (VEDA Society), India is celebrating the Centenary Birth Anniversary of Singh on 19th November 2024 at CEERI, Pilani. Singh served the Central Electronics Engineering Research Institute (CEERI), Pilani for about quarter of a century—first from 1959 to 1962 and then, after about a year gap, from 1963 to 1984 [1–3].

It dates back to the period 1954–1956 when Dr. NB Bhat joined as Planning Officer of CEERI. Immediately next, Dr. BH Wadia served CEERI during the period 1957–1959 as Assistant Director-cum-Officer-in-Charge. Immediately later during the period 1959–1962, Singh took over the same position from Dr. Wadia. Next, Dr. DL Subramaniam, during the period 1962–1963, took the responsibility as Director-cum-Officer-in-Charge of CEERI. Singh served CEERI for more than twenty years during the period 1963–1984—for the longest period—in its history. During the post-Singh period Dr. GN Acharya, Dr. WS Khokle, Dr. RN Biswas, Dr. S Ahmad, Dr. Chandra Sekhar served CEERI for the periods during 1984–1989, 1989–1993, 1993–1999, 1999–2003, and 2003–2015, respectively. After the retirement of Dr. Chandra Sekhar, it was Dr. RK Sinha, the Director of CSIR-CSIO, to take the Additional Charge as the Director of CEERI during 2015–2016. For brief periods, Professor S Chaudhury and Professor Raj Singh became the Director of CEERI and Acting Director, during the periods 2016–2018

S. K. Ghosh et al. (Eds.): VEDA 2024, SPPHY 424, pp. 359–363, 2025.
https://doi.org/10.1007/978-981-96-5174-0_52

and 2018–2019, respectively. Thereafter, during the period 2019–2020, Dr. RK Aswal, the Director of CSIR-NPL, took the Additional Charge as the Director of CEERI. From 2020 onward Dr. PC Panchariya is continuing as the Director of CEERI.

The author of this presentation carried out nonlinear Eulerian analysis of beam-plasma amplifier as part of his doctoral research. Singh had special interest in beam-plasma interaction. During 1962–1963, Professor JE Rowe at Michigan University assigned Singh the task of developing a high-frequency source by passing an electron beam between two closely-spaced electrodes, which essentially involves beam-plasma interaction [4]. The issue was identified as the device that involves the interaction between an electron beam and a plasma medium. This device resembles the Haeff tube or a double-beam amplifier, in which two electron beams with slightly different unperturbed velocities interact, while closely mixing with each other. The implementation of such a device would eliminate the necessity for the slow-wave structure required in a traveling-wave tube (TWT). The doctoral mentor of the author of this presentation, Professor NB Chakrabarti of IIT-Kharagpur, suggested the author of this presentation to carryout research at CEERI, Pilani, which was then being spearheaded by Singh, who served CEERI, Pilani as its Director for the longest time. In Sects. 2 through 4, early education, higher education, and professional career of Singh have been discussed. The presentation is concluded in Sect. 5.

2 School Education

Sardar Jagdish Singh (JS), the father of Singh, was a well-respected science teacher at a high school in the city of Phagwara in Punjab, India. JS motivated his son Singh to engage in scientific studies [4, 5]. JS was known for his demonstrations of science experiments, such as the electrostatic Wimshurst machine generator. He also taught Singh at home on how to experimentally generate oxygen and validate the principles of steam engine originally established by James Watt. Furthermore, Singh benefited from the guidance of his school teacher Sri Banwari Lal, who provided him with the access to books from his personal library. This additional reading enriched Singh's scientific knowledge beyond the conventional school curriculum. The experiment conducted by Hertz particularly inspired Singh igniting his passion for communication engineering [4, 5].

3 College Education

Singh studied BSc in Physics Honours and MSc in Physics at Sikh National College, Lahore under Punjab University, Lahore. Singh received MSc degree from Punjab University in 1945.C.V. Raman evaluated and highly praised the MSc thesis of Singh titled "A small shielded transmitter." For higher studies, Singh availed of the "Sargent Scheme" and moved first to Ohio State University (OSU) and then to Harvard University (HU). Singh studied Radio Engineering at OSU and Applied Physics at HU.

Singh received his Masters in Engineering Science and Doctor of Philosophy degrees in 1947 and 1949, respectively, both from Harvard University. Professor Roland WP King supervised the doctoral research of Singh on the topic: Extension of the tuning range of a magnetron [4, 5].

4 Professional Career

Singh joined the Physics Department of Delhi University (DU) as Lecturer in Radio Physics. He served DU during 1949–1953. He taught postgraduate and undergraduate students at DU. There one subject of his teaching was Electronics and another subject of his teaching was History of Science. He was not designated to teach the so-called 'pure' Physics class, as his experience in developing magnetrons at Harvard University was classified as his expertise in 'applied' Physics. Singh successfully operationalized one of the four SCR 584 (fire-control) radar systems that had been previously acquired from defense disposals and were stored in the Physics Department of DU. Singh also intended to reuse the magnetrons used therein for the purpose of developing a medical LINAC. Singh had then collaborative interactions with CSIR-NPL at New Delhi in the development of magnetrons. In view of his two-year contract with DU, Singh could not join NPL then as a regular Scientific Officer, in the post advertised by Dr. KS Krishnan, the first Director of NPL [1–5].

At CSIR-NPL, Singh discussed the magnetron development technology, which was rooted in his doctoral studies at Harvard University. The talk impressed Dr. KN Mathur—a close associate of the erstwhile NPL Director Dr. KS Krishnan—who was present in the audience of the lecture delivered by Singh. Due to the interventions of Dr. Mathur and Dr. Krishnan, the hurdle of the two-year notice period of DU could be overcome and Singh joined CSIR-NPL and served the laboratory for around six years from 1953–1959. Fortunately, the magnetrons of SCR 584 radar sets at DU continued to be available with Singh for his research at CSIR-NPL [4].

Singh developed the interdigital magnetron with a wide tuning range at CSIR-NPL [4, 5]. There he worked with Dr. Nagesh C. Vaidya. Later on Dr. Vaidya—as Singh had done earlier—shifted from CSIR-NPL, Delhi to CSIR-CEERI, Pilani. Subsequently, Dr. Vaidya shifted from CSIR-CEERI, Pilani to Electronics Engineering Department of Banaras Hindu University (BHU). The Centre of Research in Microwave Tubes (CRMT) was established there by Dr. Vaidya. On invitation from Professor Vaidya, the present author joined CRMT, initially taking leave on lien, and subsequently resigning, from the Regional Institute of Technology, Jamshedpur and worked at CRMT as Professor until the author's retirement from BHU [5].

During the stay of Singh at CSIR-NPL, many distinguished visitors, including the first Indonesian President Sukarno, visited the lab of Singh. Also, with the initiative of Dr. Krishnan, Pt. Jawaharlal Nehru visited Singh's laboratory, during Pt. Nehru's participation in a conference of CSIR directors [4] at CSIR-NPL. Later on, following a selection procedure, Singh was offered the position of Assistant Director at CSIR-CEERI at Pilani. At this stage, he faced the difficulty of taking a decision on his joining CEERI, Pilani, since at about the same time Professor M. Chodrow offered him a postdoctoral research position at Stanford University. Dr. Krishnan advocated for Singh's acceptance of the CSIR-CEERI position and accordingly attempted to convince Singh's wife Ms. Surinder Kaur the merit of Singh's joining CEERI. This advocating of Dr. Krishnan took place during one of Ms. Kaur's evening strolls when Dr. Krishnan met her by chance. Singh came to know then that that CSIR-CEERI had significantly better resources for vacuum tube development than CSIR-NPL. This is because CEERI received a significant grant from TCM (acronym for Technical Cooperation Mission) of USA. This motivated Singh to join CEERI in 1959 [4, 5].

Singh at CEERI, Pilani came across Dr. BH Wadia, who had then started constructing, in the CEERI campus, 52 residential quarters and the 110-ftCEERI tower. The tower at the Bell Laboratories prompted Dr. Wadia to build the CEERI tower. The tower could be used for long-distance microwave communication [4].

The equipment and instruments obtained from the grant of TCM were distributed at CEERI from the standpoint of their respective uses among four labs and one workshop of CEERI. The first lab was Physical Electronics lab led by Dr. BH Wadia. The second lab was Special Circuit lab led by Dr. AK Kamal. The third lab was Audio Engineering lab led by Dr. DL Subramaniam. The fourth lab was Vacuum Tubes lab first led by Singh himself, then by Dr. OP Gandhi and subsequently later by Dr. SSS Agarwala after Dr. OP Gandhi had left CEERI. The grant was also met available to the CEERI Workshop led by the Principal Foreman Mr. Jaswant Singh.

During the presence of Singh as the Director of CEERI, the construction work of residential quarters was expedited. Singh also created and/or expanded the other amenities in the CEERI campus such as the school approved by the school board, dispensary for medical treatments, auditorium, open-air theatre, canteen, guesthouse, hostels for trainees, bank, post-office, TV tower, market complex, and community centre with indoor and outdoor game facilities.

With the combined expertise of Singh and Dr. OP Gandhi, the Vacuum Tube lab of CEERI successfully designed and produced a batch of 75 S-band, 500-kW magnetrons. These magnetrons were delivered to the Indian Navy. Additionally, this lab developed S-band, 1.0-MW magnetrons for the Indian Air Force. These magnetrons ranged in six frequency variants from 2910 to 3100 MHz. The technology for manufacturing this category of magnetrons was later transferred to Central Electronics Limited in Sahibabad. This significant accomplishment by CSIR-CEERI, under Singh's leadership, received commendation from Jawaharlal Nehru, the former Prime Minister of India. Subsequent developments by CSIR-CEERI included the X band, 200-kW coaxial magnetron, as well as S-band tunable pulsed magnetrons with power ratings of 1.0, 2.0, 2.6, and 3.0 MW. Furthermore, this lab developed a 35-GHz mm-wave magnetron. Recently, CEERI made the technology transfer for a 2.6-MW, S-band magnetron to M/s Panacea Medical Pvt. Ltd. in Bengaluru for production [6]. Currently, CSIR-CEERI is investigating the development of low-power continuous wave magnetrons, spatial-harmonic magnetrons, and systems based on low-power continuous wave magnetrons. The narration of Singh's success in developing various vacuum electronic devices, in addition to magnetrons, is documented in [7] and [8].

Further, it was mentioned in section I that Singh, at the behest of Professor JE Rowe, visited Michigan University for studying beam-plasma interactions. Singh visited at Michigan the Bell Telephone Laboratories then. There Singh sensed how fast the development of solid state devices (SSDs) was taking place. This prompted Singh to strengthen the SSD division at CEERI.

After superannuating from CEERI, Singh continued his research in the area of VEDs—depressed collectors, in particular, for the improvement of the efficiency of gyrotrons—and he found it convenient to do this research at the University of Maryland.

Singh cheered the untiring effort of Dr. AK Sinha, who established the gyrotron lab at CEERI. In his effort, Dr. Sinha received a great support from Dr. Chandra Shekhar, the erstwhile Director of CEERI, and Dr. SN Joshi, the then National Coordinator of the DST-sponsored multi-institutional project on the development of a42-GHz, 20- kW, long-pulse gyrotron for Indian tokamak in fusion plasma heating application[5]. The author of this presentation was one of the members of the DST Steering Committee for monitoring the said project.

5 Conclusions

A brief review has been made to highlight the outstanding contributions of Singh to the development of science and technology in India. Versatile in all disciplines of electronics engineering, Singh had his own area of research in the area of VEDs. The team led by him had the credit of developing the first magnetron deliverable to Indian Navy and Indian Air force. The President of India felicitated him with the Padma Bhusan award for his contributions to the development of science and technology. It is hoped that an account of the life and work presented here will motivate the youngsters to devote themselves to the progress of India in science and technology.

Acknowledgment. The author expresses his thanks to Professor S Raghavan, Dr. Surendra Pal, and Mr. Raj Singh who motivated me to make this presentation. He extends his gratitude to Professor S.C. Dutta Roy, Dr. S.N. Joshi, Dr. Chandra Shekhar, Dr. Udit N. Pal, Dr. S. Mourya, and Dr. P.C. Panchariya for giving him their suggestions from time to time in the making of this document. The author's appreciation goes to Mr. Rohit Singh (Knowledge Resource Centre, CEERI, Pilani) and Ms. Sreelatha Menon (Universities Press, Hyderabad) for providing their editorial guidance to the author in preparing this presentation document.

References

1. Joshi, S.N.: Private communication
2. Shekhar, C.: Private communication
3. Pal, U.N.: Private communication
4. Singh, A.: Towards a shining India: recollection and reflections of a research scientist. Private communication
5. Basu, B.N.: Journey of Amarjit Singh from Phagwara to Pilani. IETE J. Educ. **64**, 71–78 (2023)
6. Mourya, S.: Past and present status of the magnetron development in the country and the efforts at CSIR-CEERI leading to the product development for the users. In: Proceedings, Webinar 2 on VED Thinkers Group platform (Ed. V. Kesari, Raj Singh, and B.N.Basu)
7. Available: www.ceeri.res.in
8. Available: www.bnbasu.com

Author Index

S. K. Ghosh et al. (Eds.): VEDA 2024, SPPHY 424, pp. 365–367, 2025.
https://doi.org/10.1007/978-981-96-5174-0

The manufacturer's authorised representative in the EU is Springer Nature Customer Service Centre GmbH, Europaplatz 3, 69115 Heidelberg, Germany. If you have any concerns regarding our products, please contact ProductSafety@springernature.com

Printed and bound by CPI Group (UK) Ltd, Croydon, CR0 4YY
01/10/2026
02225935-0010